Human Factors of
Information Technology
in the Office

WILEY SERIES IN INFORMATION PROCESSING

Consulting Editor
Mrs Steve Shirley OBE, *F. International Limited, UK*

Visual Display Terminals
A. Cakir, D. J. Hart, and T. F. M. Stewart

Managing Systems Development
J. S. Keen

Face to File Communication
Bruce Christie

APL—A Design handbook for Commercial Systems
Adrian Smith

Office Automation
Andrew Doswell

Health Hazards of VDTs?
B. G. Pearce

The New Revolution
Barrie Sherman

**Human Factors of Information Technology
in the Office**
Edited by Bruce Christie

Human Factors of Information Technology in the Office

Edited by

Bruce Christie

Human Factors Technology Centre,
ITT Europe Engineering Support Centre,
Harlow, Essex

JOHN WILEY & SONS

Chichester • New York • Brisbane • Toronto • Singapore

Library of Congress Cataloging in Publication Data:

Christie, Bruce.
  Human factors of information technology in the office.
  (Wiley series in information processing)
  Includes index.
  1. Office practice—Automation—Psychological aspects.
  2. Information storage and retrieval systems—Psychological
  aspects. 3. Telecommunication—Psychological aspects.
  I. Title. II. Series
HF5547.5.C487 1985 658.4'038 84-20903
ISBN 0 471 90631 X

British Library Cataloguing in Publication Data:

Christie, Bruce
  Human factors of information technology in the office.—
  (Wiley series in information processing)
  1. Office practice—Automation 2. Information storage
  and retrieval systems 3. Psychology, physiological
  I. Title
  658.4'0388'019 HF5548.2

  ISBN 0 471 90631 X

Printed in Great Britain by
St Edmundsbury Press, Bury St Edmunds, Suffolk

Contents

Preface   ix

PART ONE: INTRODUCTION

Chapter 1 Introduction   1
  Bruce Christie and John McEwan
  The Key Role of Communications in Human Life   2
  The Information Technology Market   3
  National and International Initiatives   6
  The Front Line: The User–system Interface   11
  The Nature of Systems Psychology   15
  Scope of the Book   18
  Complementary Books   23

Chapter 2 The Office: A Psychological Perspective   24
  Bruce Christie and M. Polly Kaiser
  Introduction   24
  The Organizational Context   26
  The User   39
  Conclusions   56

Chapter 3 The Office: A Historical Perspective   57
  Jack Field
  Introduction   57
  The First Offices   57
  Two Office Revolutions   59
  Conclusions   72

PART TWO: PRODUCT TRENDS

Chapter 4 Overview   75
  Bruce Christie
  Introduction   75
  Type A—Person to Person—Communication   76
  Type B—Person to 'Paper' to Person—
    Communication   80

Type C—Person to 'Intelligent' Machine—
   Communication   87
Integration at the User Interface   88
The 1983 Office System Scenario   89
The Future   95
Conclusions   96

Chapter 5 Electronic Meetings   97
            Bruce Christie and Marco de Alberdi
            Introduction   97
            Types of Meetings   102
            Effects of Media   105
            Users' Attitudes   120
            Conclusions   125

Chapter 6 Personal Information Systems   127
            Bruce Christie
            Introduction   127
            Some General Psychological Principles   128
            Some Characteristics of Paper-based Systems 133
            Advantages of Paper-based Systems   135
            Early Steps Towards More Effective Electronic
               Environments   137
            Hypotheses and conclusions   143

Chapter 7 Shared Information Systems   145
            Bruce Christie
            Introduction   145
            Identifying a Need for Information   146
            Formulating the Need   149
            Selecting a Source   153
            Using Online Systems   156
            The Emergence of Videotex   161
            Beyond Current Systems: Intelligence Applied
               to Relevance   163
            Conclusions   168

Chapter 8 Decision Systems   170
            Marco de Alberdi and James Harvey
            Introduction   170
            Decision Types and Processes   171
            Key Roles in Decision Making   174
            A Computer Based Model of Knowledge Rich
               Decision Making   181
            Conclusions   185

PART THREE: PRODUCT USABILITY

**Chapter 9 Assessing Product Usability: A Psychophysiological
Approach** 189
Anthony Gale
Introduction 190
Physiological Measures and Criteria for Their
Use in Applied Contexts 192
The Psychological Significance of Some Key
Measures 194
Studies of Work Stress 202
Psychophysiology and the VDU 205
Concepts of Information Processing 206
Psychophysiology, General Systems Theory, and
Situated Action Theory 208
Conclusions 211

Chapter 10 Dialogue Design Guidelines 212
Ian Cole, Mark Lansdale and Bruce Christie
Introduction 212
User Requirements 214
Selecting an Appropriate Type of Dialogue 217
Guidelines for Input to the System 223
Guidelines for Input to the User 228
Guidelines for Facilitating User–system
'Meshing' 234
Bibliography 239

**Chapter 11 Beyond Diaglogue Design Guidelines: The Role
of Mental Models** 242
Mark Lansdale
Introduction 242
Conceptual Representations 250
Implied Dialogue Models—Cognitive Factors in
Command Languages 259
Cognitive Models 266
Conclusions 270

PART FOUR: INTRODUCING SYSTEMS INTO ORGANIZATIONS

**Chapter 12 Introducing Office Systems: Guidelines and
Impacts** 273
Bruce Christie
Introduction 273
Guidelines 273

Impacts   286
Conclusions   296

Chapter 13 Identifying Future Office Systems: A Framework
            for a New Methodology   297
            Jack Field
            Introduction: Brakes on Progress   297
            A Framework for a New Methodology   300
            Conclusions   311

PART FIVE: CONCLUSIONS

Chapter 14 Conclusions   315
            Bruce Christie and John McEwan
            Introduction   315
            Product Trends   318
            Product Usability   322
            Introducing Systems into Organizations   324
            Conclusions   325

References   326

Author Index   341

Subject Index   347

Preface

This book has been produced by the Human Factors Technology Centre, part of the Research Centre at ITT Europe Engineering Support Centre (ESC), with contributions from various associates including: John McEwan (Manager of the ESC Research Centre), James Harvey (Manager of the Artificial Intelligence Technology Centre within the ESC Research Centre), Professor Anthony Gale (Professor of Psychology at Southampton University) and Dr. Jack Field (Head of Survey Research at the Open University). The Human Factors Technology Centre is part of a larger Human Factors capability within ITT and reflects the organization's long standing commitment to design products to the highest standards. In the rapidly emerging area of information technology and in particular office systems this increasingly includes addressing human factors.

It is a new area of research and the book does not provide many answers, but if it serves to raise awareness of some of the issues it will have served a useful purpose.

The authors appreciate the organizational support provided during the preparation of the book.

Remaining weaknesses and errors remain the responsibility of the authors.

Human Factors Technology Centre
ITT Europe ESC-Research Centre
Great Eastern House, Edinburgh Way
Harlow, Essex CM20 2BN, England

Part One
Introduction

Chapter 1

Introduction

BRUCE CHRISTIE AND JOHN McEWAN

The HAL computer in *2001: A Space Odyssey* could communicate with the humans in its environment in readily acceptable ways, including spoken natural language, and the humans could communicate with the computer in very natural ways, including spoken natural language—the computer could even accept drawings and read lip movements. And it was intelligent, perhaps even sentient. HAL was fictional but there are teams of scientists in different places around the world, especially in Europe, Japan and the USA, working on making something approaching HAL a reality within ten years. Whether they will be successful or not remains to be seen but it is certain that whatever they do succeed in achieving will affect all of us—at work, at home, in education, in health care, in all major areas of our lives.

This book is not specifically about HAL-like machines. It is about something more than that. HAL was confined to a spaceship. If it could communicate with people or machines beyond the confines of the ship, that was not portrayed as significant. This book, in contrast, is concerned with the development of a communications and information network that will span the entire globe and beyond, linking humans and intelligent machines in a human–machine system of distributed intelligence. The global system will be made up of the interconnection of smaller networks. The total system will combine both human and artificial intelligence in a system so complex that no single person and no single machine will fully comprehend its capabilities.

This book is about the very earliest steps. It is about the systems psychology of those earliest steps. What lies in the future is for the reader to speculate on.

THE KEY ROLE OF COMMUNICATIONS IN HUMAN LIFE

One of the characteristics that distinguishes us humans from the other animals with whom we share this planet—and it may be the most significant of such characteristics—is our communications capability. Other animals communicate with one another, of course, but they lack the flexibility and richness of human communication. A cat or dog, for example, can communicate that it is 'content' or 'excited', but it cannot explain why. Bees can inform one another of the location of flowers, but they are not generally believed capable of discussing the implications of a motorway on the local bee population, or the risks to bees of nuclear power stations or missile sites.

We humans are unique on this planet in the power of our language and in terms of our ability to communicate through space and time. On 13th June 1983 we communicated with Pioneer 10 as it left our solar system on its journey into outer space. It may still be capable of delivering our message to other life forms it may meet long after our planet has died or been killed.

Perhaps the most significant milestone in the development of scientific, technological, and cultural achievement was a communications milestone—the invention by Gutenberg in 1455 of the moveable-type printing press. This event led to a dramatic acceleration in the rate of change of all fields of human endeavour. Before that invention, scientists and others had no way of communicating the results of their work except by the slow processes of word of mouth, personal letters, and similar methods. After the invention, they could share their work more easily and rapidly, reducing unnecessary duplication of effort and building upon each other's work.

About 400 years later, in 1876, Alexander Graham Bell achieved his first electrical transmission of intelligible speech. The Age of the Telephone was born. One has only to compare the 400 years between Gutenberg's invention and the introduction of the telephone with the 100 years between the latter event and the landing on Mars to see that the rate of technological achievement has increased remarkably.

> Commenting on the effects of communications technology, Cherry (1978) has gone so far as to suggest that 'societies can develop and advance only as far and fast as they can acquire, use and maintain systems of communication: systems of acquiring, recording, assimilating and disseminating information' (p. 31).

Now we move from the Age of the Telephone to the Age of Information Technology (IT), and the rate of change has increased again. In recognition of this move into a new phase of development of our communications capability, the UK Government declared 1982 as Information Technology Year, with the slogan 'There is no future without IT', and in 1983 the Commission of the European Communities launched the pilot studies for its European Strategic Research Programme in Information Technology (ESPRIT), the main programme beginning in 1984 (more or less in parallel with the UK Government's Alvey Programme, also directed at stimulating the IT industry).

The new emphasis on IT promises to make more powerful technology more readily accessible to more people than ever before. Ordinary people will be able to use sophisticated technology not just to communicate with other people, as in the past (e.g. the telephone system), but with machines as well.

In the early sixties communicating with computers was a specialist job and the only means were through a set of buttons and lights on a console and punched paper tape or cards. A typical application of computers in a service industry—the Stock Exchange—was the production of tables of information in a short period of time, such as the calculation of financial indices for a daily newspaper. Here the system analyst/programmer, as soon as the Stock Exchange closed, had to get the hand written data, punch it onto paper tape, visually check the accuracy of the punching and input the punched tape into the computer. After an hour a punched tape would be produced and this was used to drive a flexiwriter to print up a master. Example calculations had to be done to ensure that the computer had performed the calculation correctly. If the form in which the information was handled changed, the program in the computer had to change. The same programming course was given to the computer operators as to the system analysts/programmers. Without understanding the workings of the computer, it was as impossible to program it as it was to operate it.

The future in which Information Technology will have an impact on all of us will be characterized by:

— much easier human–machine communication
— integration of computers into communications networks in a much more fundamental way than ever before
— emergence of intelligent knowledge-based systems.

We are moving from a world in which we could communicate with one another easily through space and time, and with some difficulty to relatively unintelligent but useful machines, to a world in which we can also communicate easily with truly intelligent machines.

Only time will tell what the next ten years hold in store, but in this book we take a look at some of the key factors that will determine it.

THE INFORMATION TECHNOLOGY MARKET

Growth in information occupations

It has been clear for some time that we have been moving towards an information society, as the graphs in Figure 1.1 show. We have moved on from Stage I, a society in which agriculture predominated; we have moved through Stage II, in which industry predominated; and we have just entered Stage III, dominated by information and the 'service industries', e.g. banking, insurance, consultancy services of various sorts, and other services where the handling of information is the main activity. This shift in emphasis has created the conditions for the birth of Information Technology and we can expect to see many changes in our everyday lives, reflecting the impacts of the electronic systems which form the basis of our new society. The emergence of word processors, bank service tills, personal computers, video games, videotape recorders, videodiscs, Ceefax, Oracle, Prestel, 'talking cars'—these are just the very earliest and most obvious signs of a new technology that will change our lives in the coming years.

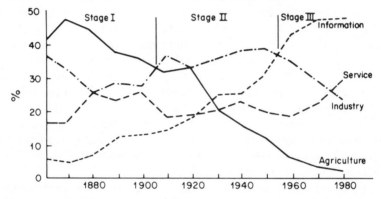

Figure 1.1: Four-section aggregations of US workforce 1860–1980, using median estimates of information workers (per cent) (from Parker, 1976). (From Edwin B. Parker, 'Social implications of computer/telecoms systems', Telecommunications Policy, 1, (1), 3–20. Reproduced by permission of Butterworth Scientific Ltd., Guildford, UK.)

Impacts

The new technology impacts all the major areas of our lives:

Impacts on business and industry. It promises

  — to improve access to customers, both through space (e.g. shopping from home) and time (e.g. 24-hour services)
  — to improve service to customers (e.g. electronic 'expert systems' and 'electronic manuals' to support the service engineer)
  — to improve access to suppliers (e.g. electronic directories of suppliers)
  — to improve response time (e.g. faster and more reliable checking of credit, invoicing, electronic funds transfer)
  — to improve the company's image (e.g. faster response to customers' queries, higher quality of information provided)

— to improve the well-being of staff (e.g. challenging but less stressful work).

Impacts on education. It promises

— to provide more personalized direct tuition
— to make educational possibilities available to more people for a greater part of their lives
— to provide an opportunity for more people to have access to machines that can handle the relatively routine, well-defined aspects of problem-solving, enabling them to develop higher-order problem-solving skills.

Impacts on health. It promises

— to improve the efficiency of health administration
— to improve access to people needing care (e.g. communications into the home, for emergency and other purposes)
— to improve the match between services offered and health needs (e.g. by improving knowledge, through surveys and other methods, of what the needs are)
— to improve individual treatment (e.g. use of expert systems to support medics and paramedics).

Impacts on the consumer. It promises

— to provide shopping, health, educational and other services into the home
— to provide a wider range of entertainment possibilities
— to aid in the day to day running of the home.

The key position of office systems

Information Technology is a major market. According to a study done by PACTEL for the UK National Enterprise Board (PACTEL, 1981), it approaches £105 billion in annual shipments, worldwide, with IBM and ITT being the two most important suppliers.

The structure of this market is changing, with some areas growing much more rapidly than others. Office systems represent a major growth area. The PACTEL forecasts suggest, for example, that word processing is growing at more than four times the rate of growth of public network equipment (see Figure 1.2).

The Commission of the European Communities has estimated that for the next twenty years office systems will represent the largest single market for information technology, amounting to more than £50 billion annually by 1990. To give an idea of what this figure means, it is about five times the manufacturing and process industry sectors put together.

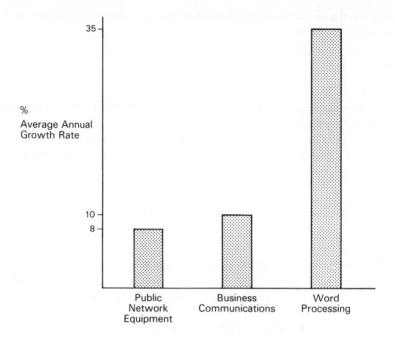

Figure 1.2: Comparative rates of growth of sectors within information technology, based on a study by PACTEL (1981)

NATIONAL AND INTERNATIONAL INITIATIVES

Recognizing the economic importance of capitalizing on an emerging market as large as Information Technology represents, both Japan and Europe have responded to the initial lead taken by the USA, and have launched their own initiatives.

The Japanese initiative

It is worth looking at the Japanese initiative in some detail as it forms much of the backcloth against which Europe is playing its part in Information Technology.

The Japanese had been observing trends in the USA for some time and had sent their own experts to key research centres throughout the USA to form a view of the future and develop their own programme, aimed at producing a Fifth Generation Computer. Before considering what that is, it is useful to review briefly the management priorities and general approach to research and development (R & D) that characterizes the Japanese approach to IT.

Japanese management priorities. While working in Japan during the late 1970's John Prentice of Japren (personal communication) visited a large number of Japanese companies. From interviews with the top management he was able to identify their list of priorities for their companies. They were in a different order from those of the Western industrial companies, and were as follows, in order of decreasing importance:

1. to increase world market share for their goods
2. to look after the welfare, education, and training of their engineers
3. to maintain the quality of the product and the company's good name
4. to improve profits for re-investment into research and development, and capital plant
5. to pay the equity holders
6. to have good market intelligence
7. to make use of the patent library to find new ideas.

This contrasts with American companies, which have made it their highest priority to pay the equity holders, which in times of recession has forced firms to adopt short term measures which reduce investment into R & D. In effect this has enabled Japan to improve its position during the 1974–75 recession and overtake the Americans by the mid-1980s.

In order to increase their market share, the Japanese first look at competitive products in detail and see how they can be improved, both in terms of quality and price, or if they can be replaced with new products with the same functions. Secondly, they endeavour to evolve the product so as to capitalize on their investment in the design and plant. Thirdly, they invest heavily in research and development so as to stay at the leading edge of technology.

Research and development. The Japanese see investment in research and development as being vital for future success. The Japanese companies are not chauvinistic when it comes to technology; they are quite willing to use patents from anywhere in the world and if need be to buy in new technology from outside of Japan.

R & D spending by companies is high. For example, the Japanese PTT is said to have been planning on investing US $391 million into R & D in 1983 (up by 6.9% compared with a 3.9% increase in income), the emphasis being on information processing ($50m), digital communication (US $34m), semiconductors (US $36m) and satellite communication (US $27m).

All R & D is target orientated to support single product technology teams. In this way the Japanese were able to capture the video recorder market, even though it took eight years to design and to solve the manufacturing problems.

Many of the technical advances made by the Japanese are driven by their need to solve the problems of isolation caused by their language and geographical position. The development of the Pentel pen, facsimile, dot matrix printer and the Fifth Generation Computer project (below) were all pushed by the need to cope with the Chinese characters of the Japanese language.

The Japanese companies work on the assumption that products have a single global market place. This allows them to seek new markets in the Near East, Middle East, and parts of Africa and Asia where non-Romanesque characters are used. Therefore once they have found the technology to solve the problems imposed by a non-Romanesque language, they will have an enormous market place. Further, they believe that technology which solves the non-Romanesque language problems will be easy to use for the Romanesque language.

For example, Matsushita Research Institute, Tokyo, have developed a new programming language AFL (A Fundamental Language) for 16 bit computers, to replace English based languages such as BASIC, PASCAL and COBOL. By developing this more natural programming language more akin to the syntax of Arabic, Korean, Thai, or Chinese, they will increase the market for the Japanese personal computers.

The Japanese are not averse to buying in technology from outside of Japan in order to help them meet their objectives. NTT, for example, has technical information exchange agreements with outside companies, e.g. IBM (optical disc material), and is looking for other partners from the USA and Europe.

The Japanese Fifth Generation Computer. The Japanese Fifth Generation Computer Programme was formally unveiled at a conference in Tokyo in October 1981. This programme has made it a Japanese national goal to become foremost in the IT industry by 1995. They aim to dominate the traditional computer industry and to establish a 'knowledge industry' in which knowledge itself will be a saleable commodity. The programme amounts to a minimum of about £244 million over the period 1982 to 1988, and could be much higher.

The objectives for the Fifth Generation Computer are:

— to increase productivity in low-productivity areas
— to save energy and resources
— to cope with an aged society.

The project is to set goals of leadership and creativity in R & D and promote research into:

— artificial intelligence
— automatic translation of multiple languages, and
— how to organize for R & D.

The fundamental goal is that by the 1990s computers will be able to solve problems using human-like reasoning. Such computers will need three main functions:

- a knowledge base function which can accumulate and retrieve knowledge
- an intelligent user–system interface which can understand problems input to it in forms similar to natural language
- an inference function which can make inferences based on accumulated knowledge.

The user–system interface crosses all the areas—and the study of the psychology of the user is paramount to this interface and to the use of artificial intelligence.

Currently the Japanese have set up three laboratories as part of the Institute of New Computer Technology (ICOT) in connection with the Fifth Generation programme.

The basic concept for Fifth Generation Computers came from work done in France and the UK. This work is held in high regard. However, a major item of the Fifth Generation Computer project is linguistics and here Germany, France and UK have a lead, although the Japanese will lead in the non-Romanesque langues if we are not careful. However the French are doing research into Arabic. Kumio Murakami, Chief of the First Research Laboratory at ICOT, gives a personal view of the work at ICOT (Murakami, 1983):

> Primarily, in the Japanese-style research and development approach, imitation and originality are not considered as two contradicting concepts; rather they are handled as a series of ideas; copying the idea will eventually lead to originality as the next step.

> We regard the on-going first three-year stage of the ten-year FGCS project as the discipline period which aims to find the possibilites for original research through examination of the various prospective concepts and mechanisms proposed to date. From the intermediate four-year stage, we will start building FGCS systems incorporating our own original concepts. Research which steadily pursues this methodology seems more suitable for the FGCS than that hastily aiming at nothing but originality.

> People enjoy seeing plum and peach trees struggling to bloom in the spring, but the late-blooming chrysanthemum waiting for autumn to come gives another pleasure to viewers. In this sense, the existence of research groups based on the late-blooming

approach can also be justified. We hope critics will be more patient to see the large chrysanthemum which we are going to breed. Then they will be able to judge our results as mere imitation or original work. I wonder how these critics will evaluate the FGCS project when we successfully hit the essential technology of new fifth and sixth generation computers with the first and second arrows.

The European response

Also observing the trends, and partly in response to the Japanese initiative, the Commission of the European Communities has launched the European Strategic Research programme in Information Technology (ESPRIT). This is a multi-million pound programme (about 7,200 person-years of effort over five years) jointly financed with European industry to accelerate the development and usage of information technology throughout Europe.

The ESPRIT programme is intended to complement national initiatives such as the Alvey Programme in the UK.

The Alvey Committee was set up in March 1982 by the Minister for Information Technology to advise on the scope for a collaborative research programme in IT and to make recommendations. It reported in October 1982, one year after the Tokyo conference.

The Alvey Committee believed that the Japanese had correctly identified the major advances in the technology which would be necessary to achieve their objectives. They also acknowledged the European ESPRIT initiative. After considering these initiatives and taking advice from many parties, including both industry and the universities, the Committee concluded with a recommendation that there should be a national programme for Advanced Information Technology (AIT), costing an estimated £350 million over five years (about £300 million pounds in industry and about £50 million in academic institutions).

ITT Europe was one of the first industrial organizations in Europe to recognize the importance of human factors, and in particular of systems psychology in setting up its Human Factors Technology Centre, predating the ESPRIT and Alvey programmes. Other companies are rapidly following suite partly in response to the ESPRIT and Alvey programmes.

The American scene

The US companies lead the research into human factors of office automation, although they have no equivalent of the Japanese approach to long term, highly targeted research and development. Current research initiated by Xerox at Palo Alto was with the objective of developing a prototype workstation for the creation of documents and in particular the layout of pages. The Xerox Star, released in 1981, was the first such commercial work-station to try to simulate current office activities with its 'desk manager'. Although not a commerical success for several reasons, including price, it has prompted the development in the USA (Apple, Apollo, Visi-On) and Japan (Toshiba) of the abandoning of the type and read mode to the use of the mouse and more visual graphical outputs. All of these systems provide superior ways in which to manipulate the presentation of information.

Wang Laboratories—currently a leader in word processing—set up its Advanced Systems Laboratories (ASL) with the mission to help Wang's largest customers to make the transition from the office of today to the 'office of the future'.

The ASL was established in 1980. Initially envisioned as a human factors research group, the ASL is further evidence of Wang's stated commitment to human factors as one of the key 'Six Technologies' of office automation (the other five being data, word, image, and audio processing, and networking).

Although the ASL still occasionally gets involved in basic human factors research, the organization's major thrust is in an area it calls *intellectual ergonomics*—using computer technology to increase the endurance and creativity of an organization's knowledge workers. ASL believe that a major stumbling block may well be the difficulty top management often has conceptualizing the opportunities offered by office automation. As long as office automation is viewed simply as an extension of traditional data processing, word processing, or communications, they argue, only incremental progress will be possible. Like an increasing number of others in this field, ASL see office automation as a qualitative change, with qualitatively new opportunities.

Apple Corporation have spent $50m on creating the Lisa personal computer. This $8k machine with 1m bytes of memory was launched in January 1983 in the USA and at the Hannover Fair. It followed the Xerox Star, but Lisa had an additional objective which was that a user would have to spend only one hour on finding out 'how to use' the machine. Other machines, notably the Apple Macintosh, have followed since.

The key role of the user–system interface

All of the initiatives considered above acknowledge a key need to improve the ease with which information technology can be used. The ESPRIT and Alvey initiatives in particular pick out research on communications across the user–system interface—communication between human and computer—as one of the key areas where research is needed.

THE FRONT LINE: THE USER–SYSTEM INTERFACE

Identification of the requirements of the user–system interface has been limited by lack of adequate theory in this area. Knowledge of the capabilities of the machine is well known. By using prediction of the costs of memory and processors (they halve every 2.5 years), it is possible to scale up the hardware to what will be available during the product life cycle. However it is only recently that the separation of the presentation of the output and to a lesser extent the input from the application has taken place (e.g. Apple Lisa and Xerox Star). This has meant that research into user behaviour at the user–system interface has been hampered by the overwhelming software effort needed to develop such interfaces. Now they are here, empirical research can speed up and if identified by psychology departments as an important area, a quantum leap should take place in our understanding of the psychological aspects of using computers. This will require, though, the running of more realistic and more relevant experiments. It often seems to be the case at present that a large part of the time is spent programming obsolete hardware to perform largely irrelevant experiments.

Approaches to the user–system interface

Different approaches to improving the user–system interface are possible, and this is reflected in the variety of work going on in key centres around the world. The work of any particular group typically combines more than one approach, but the groups differ in the mix of approaches they adopt. It is not the intention here to review the work of particular groups, but to discuss what is possible in terms of key dimensions that seem, admittedly subjectively, to emerge from considering the variety of work going on.

The concrete versus the abstract. New ideas can be presented in either concrete or abstract ways—as a physical model, perhaps, or a set of equations. For some purposes, a concrete example may be more effective. Certainly, a rule of thumb in management consultancy is to provide plenty of 'for instances', and some approaches to psychotherapy make extensive use of metaphor and analogy to facilitate communication. This may be effective because it helps, amongst other things, to make ideas more concrete.

The advantages of the concrete approach may reflect something very basic about human thinking. At the level of our species, it seems (from our point of view, at least) that the apes (the purported modern day equivalents of our ancestors) think in more concrete terms than we do. It also seems that children move through a stage of thinking in terms of 'concrete operations' earlier than they attain 'abstract operations'. An idea has often emerged in the mind of a genius as a concrete image first, only later to be translated into a set of equations or other abstract representation.

The Xerox Star, and its more recent distant cousins the Apple Lisa and Macintosh, may not represent the best approach to the user–system interface but by making abstract notions of 'ease of use', 'user friendliness', and so on, concrete, it has served to encourage more real interest in improving the interface. It is a concrete, real product, really on the market, and it shows in a very concrete way what can be done (though not by any means everything that can be done). In some ways, this may be worth more than many erudite publications in professional journals that deal with the same issues in a more abstract way.

The Star, Lisa and Macintosh are also concrete in another way. They do not require the user to interact with information in the abstract. Instead, the user is provided with concrete images of familiar objects, such as folders, pieces of paper, in-tray and out-tray, and so on.

The Star and its cousins are discussed further in Chapter 4. The important point here is that they emphasize a concrete approach to the design of the user–system interface. This does not necessarily mean that approach is best. The use of concrete images may help the user to whom electronic office systems are a very new field, but just as the child outgrows the stage of 'concrete operations' as his or her competence develops so the user may outgrow the use of concrete operations at the user–system interface. This is one of many areas where more research is needed.

Demonstrations versus tests. The value of demonstrations is illustrated well by work at the Massachussetts Institute of Technology (MIT) where new approaches to user–system communication have been developed, simulated and demonstrated in a specially designed environment called the Media Room. Although the work is sometimes criticized by those who would like to see more controlled experiments, the demonstrations caught the imagination of many people and stimulated developments elsewhere. Many of the ideas in the Xerox Star, Apple Lisa, Imperial College's 'Panorama Project' (Spence and Apperley, 1981), work at Queen Mary College, London (e.g. Lamming, 1979), and work elsewhere can be seen in the earlier work at MIT.

The stimulation effect of interesting demonstrations can be of value in helping to get things moving—it may even be necessary. It is closely related to the 'imagination games' that people play in other areas and that form such an important part of human psychology. It is a type of phantasy that can stimulate the imagination, just as science fiction can. Concepts such as teleportation, anti-matter, black holes, white holes, hyperspace, hyperdrive, and so on, are familiar to many people today, especially younger people. They have caught the imagination and become a kind of reality, part of the mythology of the Twentieth Century—so much so that many people would find difficulty in sorting the concepts just listed (white holes, black holes, etc.) into those that are 'pure science fiction' and those that are 'scientific'.

The main role of demonstrations is to capture the imagination, and are not often associated with rigorous tests. But imagination can only get us so far; we do need to test it against reality. In Freudian terms, we cannot afford to rely entirely on our id; the ego has an important role in human psychology in maintaining a degree of reality-orientation.

Pragmatic versus academic. Some minimal consideration of the user has always been necessary just to satisfy the pragmatics of putting a system together that actually does work and is sufficiently useable to be saleable. Whilst computer systems were novel this approach worked. Users had little choice; if they wanted access to computing power they had to accept the clumsiness and mystique associated with it. Now computers are not so novel, now the competition is fiercer, now users are not so accepting, this approach is no longer sufficient.

The early years of the 1980s have seen a marked increase in public exposure to computers. Most secretaries now know what a word processor is, and many use them. Anyone walking into a high street shop for a camera, television set, hi-fi equipment, or even magazines is likely to have been exposed to demonstrations and explanations of home computers. Most children know what a computer game is. If a manager's young son or daughter finds inexpensive computer-based games fun, why should an expensive office system meant to do serious work be difficult and aggravating? The users of the mid-1980s and beyond will be very demanding in what they expect of the user–system interface, and increasingly intolerant of awkwardness at the interface.

The Star and its cousins are examples of systems that have already emerged in response to this trend, and themselves contribute to raising users' expectations. The quality of the user–system interface is likely to provide the competitive edge in the mid-1980s.

Whilst the manufacturers are beginning to look for ways of improving the user–system interface, much of the longer-term research that is badly needed could be done more naturally in the academic institutions, where research is less influenced by short-term pragmatic needs and where there is greater opportunity to do the fundamental research necessary for developing the science on which future developments need to be based. Unfortunately, it seems that by the close of 1983 the academic world was only just beginning to wake up to the fact that the Information Technology Age was dawning, and they seemed to have been having difficulty in prioritizing their research to take account of the very rapid developments in technology that were taking place in industry.

The right kinds of links between industry with its largely pragmatic perspective and the academic world with its more fundamental scientific perspective could be of great value in accelerating the development of science and technology in this area.

Psychological versus physical. The emergence of systems psychology. The immediately visible, 'concrete', 'physical' aspects of a system are important. It is important that the user does not suffer skin complaints, headache, backache or undue stress in other respects as a result of poor seating, poor lighting, poor physical layout of the workstation, or similar factors. It is important that productivity is improved rather than made worse by the keyboard layout, the design of the speech input device, the type of screen used, the type of voice synthesizer, and so on. Perhaps because these are relatively 'concrete' and obvious needs, this is where classical ergonomics has focused.

These physical aspects of systems design are important but it is in the psychological aspects of the user–system interface that great strides forward remain to be made, and will be made by the close of the 1980s—either by Europe or by the USA or by Japan.

It is inconceivable to think of a human to whom his or her psychology is not central. A person without a human psyche is a 'vegetable' or 'animal', and difficult for other humans to relate to or to work with effectively. Neither can we work very well, on business matters or office work, with cats and dogs even though these biological systems are far more sophisticated than any robots built so far. 'Sophistication', either of biological or of electronic systems, is not sufficient. We need to develop electronic systems that have the right sort of psychology.

The electronic psyche should be compatible with and should complement the human psyche, so the two kinds of systems—electronic and human—can communicate with one another effectively and together form a team that is more capable and more productive than either electronic or human systems alone. The competition for developing the best user–system interface will be won or lost in the field of systems psychology.

THE NATURE OF SYSTEMS PSYCHOLOGY

Definition

Systems Psychology is a field of scientific psychology. It is concerned with applying the methods and models of psychology to the design of electronic systems for use by humans.

The use of the term 'systems' stems not just from the concern with electronic systems. It also reflects a basic axiom of the field, that:

> The human and the machine are two different types of systems (biological and electronic) that 'mesh in' with one another to form a third, hybrid kind of system, the human–machine system. The biological and electronic systems are different in kind but in a well-designed human–machine system they are complementary and support one another, to the benefit of the human.

Systems psychology is concerned with defining the functions of electronic systems that will indeed be of benefit to the human, and with specifying those aspects of the human–machine interface that facilitate the 'meshing in' of the two systems.

Focus on office systems

In this book, we focus on the systems psychology of office systems. The work is concerned with:

— defining the functions of business systems to benefit the humans who use such systems, in terms of their organizational roles; and
— specifying the human–machine ('user–system') interface to facilitate the 'meshing in' process.

This means applying the methods and models of psychology to aid in the design of business systems that will help their users do what they need to do in order to fulfil their roles within their organizations and satisfy whatever other needs are compatible with this.

Starting assumptions

Systems Psychology takes as given a particular role or class of roles within organizations as its starting point. If the starting assumption is that the role to be supported is that of the secretary, then it becomes the concern of systems psychology to develop a systems psychology requirements specification for a system to support the secretary. The development of a system that meets such a specification is the responsibility of others, although typically the systems psychologist works in a multidisciplinary environment in which there are good communications between the various specializations involved in a project; the systems psychology requirements specification will typically take account of what is known to be technically feasible for the time period concerned.

The development of such a secretarial support system may or may not have an impact on other roles within the organization (e.g. that of the clerical worker). If the manager is taken as the role to support, the design of a suitable system may or may not have an impact on other roles (e.g. that of the secretary). If the very highest roles within an organization are taken as the starting point, the design of a truly supportive business system in principle could have very far-reaching impacts throughout the organization. It is not primarily the systems psychologist's role to identify the roles for which products should be developed, that is primarily a marketing question—but, again, the systems psychologist typically works in a multidisciplinary environment where many different specialisms contribute to the organizational decisions that are made.

The extent to which organizations will accept electronic systems at various organizational levels and providing various degrees of support for different roles depends only partly on the systems psychology contribution. It also depends on the technical reliability of the product, its aesthetics and other design aspects, the kind of marketing done, general norms governing organizational attitudes about what sorts of systems organizations ought to be using, after-sales service offered, and many other factors. A high standard of systems psychological engineering of the user–system interface is increasingly becoming a necessary condition for a product's commercial success, but it can never by any means be a sufficient condition.

Key issues

The key objective of systems psychology is to define, for any given class of users, what an electronic system needs to do in order to support the user best, and how it should do it in terms of the interactions between the user and the system.

Steps towards this objective help to redress the bias towards a technology-driven approach that has characterized much product development in the past.

Technology has reached a stage where many different kinds of products are possible, bringing different technologies together in different combinations (e.g. a choice of speech input, 'mouse' input, touch screen input, voice output, large screen, small screen, colour, monochrome, and so on). Given the range of possibilities, it is possible to make a choice in terms of what is best for the user, not just what is technically feasible.

Key pressures on the user that need to be addressed include:

— Increasing pressure on organizations for greater productivity and greater efficiency, means users are required to achieve more in unit time.
— The above combined with a shift towards information occupations means users have to deal with increasing amounts of information and communication.
— The above is being exacerbated by the facility with which photocopiers, word processors, and electronic information systems allow papers to be created, copied and distributed, and by increasingly sophisticated communications networks that facilitate communications of all sorts.

Key issues in systems design from the user's point of view include:

— What does the user need to achieve from the organization's point of view?
— What else does the user need to achieve?
— What is the optimal allocation of work between the user and the machine?
— What is the optimal allocation of control between the user and the machine?
— What is the optimal allocation of information and knowledge between the user and the machine?

Design questions that arise from these issues, include amongst others:

— What services should the machine provide for the user, i.e. what should the machine do for the user? (See Chapters 2, 3, 4, 5, 6, 7, 8, and 13).
— What role can artificial intelligence and knowledge bases play in relation to these services? (See Chapter 8).
— What is the optimal way of representing the inputs the user makes to the machine? (See Chapter 11)
— What is the optimal way of representing the inputs the machine makes to the user? (See Chapters 11 and 12)

These questions cannot be answered adequately without considering the characteristics of the particular oganizational role that the user plays and the machine is designed to support (See Chapter 2), and characteristics of the individual user, including amount of experience with computer-based systems (see Chapters 9, 10 and 11).

More specific issues that follow from the questions and considerations above are discussed in subsequent chapters.

SCOPE OF THE BOOK

The developments with which we are concerned in this book affect everyone. There are four groups, however, for whom the book will be of special interest:

— manufacturers of information technology
— managers and trades unionists
— policy-makers, e.g. in government
— psychologists, ergonomists and other human factors specialists.

Interest to manufacturers

The market for information technology is very large, and office systems are the largest part of it. The industry is a rapidly growing, fast moving, dynamic high technology sector. The rate of change is so fast that more than half of the products on the market now did not exist even three years ago.

The rate of change is likely to continue to be high for some time. The cost of memory continues to fall, and performance to improve, as new technologies enable faster circuit speeds and greater chip densities. What products are capable of doing, in technical terms at least, is increasing all the time.

This has already led to a situation where the products offered by different manufacturers are highly competitive in terms of their cost, the capabilities they offer (e.g. communications, power typing, filing, and so on), and their physical parameters (e.g. amount of memory). There is little reason to suppose that this trend will not continue and the competition become even fiercer.

It is also evident, however, that with few exceptions the products that have emerged so far—whilst technically excellent in many respects—by and large have failed to adapt adequately to the new types of users they are intended to serve. The users of office systems typically are not computer specialists. They are interested in using electronic systems only as tools to do their office work more easily and effectively; they are not usually interested in investing much time and effort to become skilled computer specialists. The traditional kinds of user–system dialogues used in conventional computer systems can be powerful demotivators in the office environment, causing users to reject new systems.

There is an urgent need to develop methods of user–system communication that are better suited to the needs of the new types of user. Any product to make a significant step forward in this regard will inevitably improve its competitive position considerably as long as costs and technical capabilities also remain competitive.

Until very recently, the form of user–system interface used in office products still reflected too much the pre-history of information technology in data processing systems. Some new ideas have begun to emerge, but many more are needed. The new users of information technology—in the home and elsewhere, as well as in the office—do not wish to become computer specialists just in order to be able to use the new products. As well as requiring low cost, high capability, excellent reliability, and so on, they are increasingly also demanding high quality in the design of the user–system interface.

It is in the design of user-oriented interfaces more than in any other area that the competition will be won or lost.

Impacts on existing markets. It is not simply that information technology opens up new market possibilities. Important as that is, it is also the case that the new products will have secondary impacts on existing markets. Users' expectations about any electronic system will inevitably be raised by seeing what can be achieved with new products. The manager who sees a child playing with high resolution colour graphics displays as a game will be reluctant to accept that only poor quality monochrome can be offered on an office product meant to support real business. The possibilities for new types of services that the new products open up will also have impacts on existing products. Just as the convenience photocopier had a very marked effect on the use of carbon paper, so electronic copying and distribution of 'papers' can be expected to have an impact on the role of the convenience photocopier. Just as the voice telephone had an impact on the use of telegrams so the demand for new types of services—stimulated by the emergence of new kinds of office products—can be expected to have an impact on existing communications systems. These are just two examples; the total range of impacts is likely to be very broad.

Interest to managers and trades unionists

Whilst significant advances have been made in providing modern technology at the shop floor—especially in those countries which have subsequently taken larger and larger shares of world markets—the office until recently has remained the 'Cinderella' of the organization. That need no longer be the case. Successful managers in the 1980s and beyond need to be aware of what modern technology can provide for the office, and how best to capitalize on it.

The importance of this kind of understanding is becoming ever more important as the technology becomes more widely used, more sophisticated, and more encompassing. It is difficult to see how managers will be able to perform their role successfully without being competent in both the management and use of modern office systems, the new 'tools of their trade'.

Managers need to understand how to select the best equipment for their purposes, to ensure it meets all relevant criteria—human as well as technical—to understand the impacts of the new technology, the factors governing acceptance, and the relationships to user behaviour, understandings and feelings. As harsh experience has already taught many organizations, these user aspects will determine the success or otherwise of electronic information systems in particular organizations, and whether any given organization manages successfully to cross the threshold into our electronic future or whether it falls by the wayside, trampled by the competition.

Trades unionists also need to be aware of what the new technology can offer and how it can be used to most benefit. They also have a responsibility to encourage a user-oriented approach to the design of products with which their members may work. One important aspect of this might be to encourage adequate research to be done—in the universities as well as by industry itself—to help ensure that whatever guidelines or standards are developed for the design of the user interface they are based on the best possible research—that they are soundly based, in other words, and not based on speculation or intelligent guesses. Anything less would be counter productive, as requiring products to conform to the wrong standards in this area is worse than allowing flexibility. Unfortunately, the signs at the moment are that not enough research is being done and the pressure to develop standards may well result in the premature development of standards that are not always correct.

Interest to policy-makers

Understanding the psychological aspects of electronic systems is important to policy-makers. This applies not just to office systems, but to education, health care, industry, employment, and all other areas of life where government or other bodies have a responsibility to monitor new developments, understand and evaluate their potential impacts, and influence our choice of futures. High-rise buildings, automobiles, Three-Mile Island, and many other examples show where greater understanding and more careful analysis might have helped to achieve something better. It would be prudent of us to learn from our experiences and make sure we base our plans in the information area on a firm foundation of adequate knowledge and understanding.

There are already some moves in this direction in Europe as a whole—through the ESPRIT programme—and in the UK in particular (through the Alvey initiative). It is recognized that there may be some considerable advantage in bringing universities, research institutes, manufacturers, software houses, and others together to carry out pre-competitive research and development activities. Such activities are seen as forming an important part of an industrial strategy aimed at strengthening the manufacture and use of information technology, promoting the creation of a technology base from which can be developed products, processes and services facilitating successful competition on the world market to acquire important market shares.

Interest to psychologists, ergonomists and other human factors specialists

Psychology is the science of human behaviour, including what people do, think and feel; and is therefore the appropriate discipline for an understanding of user behaviour, 'users' simply being humans in a particular context. Behaviour, however, never occurs in the abstract; it always occurs in an environment of some sort. Psychology recognizes that human behaviour is a function both of the person concerned and of the environment. A person who reports the colour of a disk held against a dark background as being 'white' may report it as being 'grey' when it is held against a white background. A person who can deal effectively and bravely with a street mugger may need help when faced with a spider in the bath. A person who was effective in the office environment of yesterday may or may not adapt well to the electronic environment of tomorrow, and the person who had to struggle hard to keep up in yesterday's school may show great talent in the electronic environment of today and tomorrow.

Psychologists need to give proper attention to developments in electronic systems because these systems represent the most significant change in the human environment since the industrial revolution. Any psychologist with the slightest concern for 'ecological validity' will see the need to take account of the new 'electronic environment' in theory development and experimentation. It is all the more important to do so because the changes have a direct impact on that uniquely human quality to which we referred earlier—our ability to communicate. Applied psychologists will need to understand the psychological aspects of electronic systems in order to derive implications for their own specialisms in education, clinical settings, or organizations.

The design of user-oriented interfaces to electronic systems raises many questions which psychologists are already in a position to answer on the basis of existing psychological knowledge, and opens up many new avenues for further psychological research.

Ergonomists and other human factors specialists who have traditionally focussed on the more physical aspects of what was known by that now rather antique sounding term 'the man–machine interface' also need to turn their attention to the new kinds of 'machines' with which an increasingly large proportion of the working population work. Some evidence of moves in this direction in the UK is already apparent in efforts to develop an active community of interested researchers in this area. A meeting held by the Ergonomics Society in London in June 1983 marked the first significant step towards this goal.

Psychologists, ergonomists and other human factors specialists have a responsibility to ensure that necessary research is done and that the results of research are made known to manufacturers and others concerned with the development of new products. It is especially important that national and international standards bodies are given access to research findings and given every encouragement to use these findings when developing standards and guidelines. Standards or guidelines based on inadequate empirical research would be most likely to prove damaging rather than helpful.

Topics addressed

The next two chapters provide a psychological and historical perspective on office systems. It is essential to consider what it is that electronic systems are intended to do, or else they will be 'solutions looking for a problem'. What the systems of interest in this book are intended to do is to support user behaviour in the context of 'office work'. The two chapters together provide a context in which particular issues are discussed in more depth in later chapters.

The remaining chapters are organized into three main sections dealing with key application areas, the concept of usability, and introducing systems into organizations. The chapters on key application areas do not attempt to cover every important area in depth but four rather different areas are chosen to illustrate the principles and approach of Systems Psychology; these are: electronic meetings, personal information systems, shared information systems; and decision systems. Whatever the particular application, it is essential that the user–system interface is as usable as possible. What is meant by 'usability', how it can be assessed, and some of the things that can be done to achieve it, are covered in the chapters on usability. The chapters on introducing systems into organizations discuss the need to identify appropriate systems, and consider guidelines for introducing the systems and impacts they can be expected to have.

A concluding section reviews the main points and considers the direction in which electronic systems are moving and what the future may hold.

COMPLEMENTARY BOOKS

There are a number of complementary books which the interested reader may wish to consult, and more are appearing almost weekly. References are given later at appropriate points in the text, but the following are good examples.

Card, S. K., Moran, T. P., & Newell, A. (1983) *The Psychology of Human–Computer Interaction.* London/New Jersey: Lawrence Erlbaum Associates. The scope of the book is narrower than the title might suggest, concentrating on perceptual-motor and cognitive aspects (especially in regard to text editing) but it is a useful book and provides some interesting insights into some of the research that has been going on at Xerox Palo Alto Research Center.

Doswell, A. (1983) *Office Automation.* Chichester/New York: John Wiley & Sons. Part of the Wiley Series in Information Processing, this provides a broad overview of the office automation field, covering human factors and other aspects, organized into four main sections: 'concepts', 'office systems', 'applications', and 'implications'.

Feigenbaum, E. A. & McCorduck, P. (1983) *The Fifth Generation: Artificial Intelligence and Japan's Computer Challenge to the World.* London/California: Addison-Wesley. This is a very readable book giving an overview of what the Japanese Fifth Generation Computer initiative is about. It contains some useful appendices giving information on key research centres, operational expert systems, and related information.

Jarrett, D. (1982) *The Electronic Office: A Management Guide to the Office of the Future.* Aldershot: Gower Publishing Company Ltd. This provides a readable overview of electronic office systems.

Sime, M. E. & Coombs, M. J. (1983) *Designing for Human–Computer Communication.* London/New York: Academic Press. This is a useful book of readings by key researchers from the USA, Canada, Sweden, Scotland, and—mostly—England. The book is organized into two main parts, dealing with 'the user interface' (issues such as natural language, 'user growth', database query) and 'the task interface' (medical consultation, air traffic control, and other application areas).

The following two books in preparation at the time of completing the present volume may also be of interest.

Christie, B. (Ed.) *Report on ESPRIT Preparatory Study on Human Factors of the User–System Interface.* Amsterdam: North-Holland.

Gale, A. & Christie, B. (Eds.) *Psychophysiology and the Electronic Workplace.* Chichester: John Wiley and Sons.

Chapter 2

The Office:
A Psychological Perspective

BRUCE CHRISTIE AND M. POLLY KAISER

INTRODUCTION

Go into any organization and you will see people. Organizations are, in fact, primarily about people working together. They may or may not use various kinds of electronic and other machines to help them, but primarily they are people working together. It is important to consider what is involved in this in order to understand what it is that electronic systems need to be designed to support. Without such an understanding, such systems are solutions in search of a problem.

If you go into the offices of any organization you will see immediate evidence of the way that office work differs from, say, work on the shop floor. The office worker is much more involved in communicating with other people and with handling information in the form of 'papers' and other forms than is the worker on the shop floor (see Figure 2.1).

In this chapter we consider the nature of what office workers do, and the opportunities and needs for electronic systems to help them do it.

Since the advent of the microchip and associated developments in technology, computer-based systems have become increasingly accessible to greater numbers of people. They now encompass many aspects of our lives: games, bank service-tills, teletext, home computers, are just a few examples. And they are appearing more and more in offices—as word processors, information retrieval systems, personal computers, and other systems to be used

Figure 2.1: The office is primarily about people communicating and processing information. (Reproduced by permission of ITT Industries Ltd.)

as tools by office workers rather than by computer specialists. With an increasingly wide range of non-specialist users, a user-oriented approach to the design of the user–system interface has become even more important than in the past.

The electronic systems with which we are concerned are potentially much more powerful than traditional, paper-based office systems. But to convert this potential into actual benefits it is necessary that the systems be designed to serve actual organizational and user needs, and be designed to be useable by actual users (e.g. managers, secretaries) and not by some hypothetical type of 'user' (e.g. who naturally understands and knows how to deal with 'syntax error' or a screen that suddenly goes blank for no apparent reason). The user–system interface must be designed to serve users the way they actually are, not how some systems designers in the past might have felt they ought to be.

Proper understanding of the user and what he or she is trying to do can help to guide the design of the user–system interface such that the systems concerned can be easier to use and more geared to real user needs, so the suspicion which some users may otherwise feel about the systems can be reduced, and the benefits of the systems can be seen more clearly.

In this chapter we shall consider user behaviour at two levels: first, the organizational context in which the user works as a member of an organizational team; secondly, the individual user, with particular needs, limitations and capabilities.

THE ORGANIZATIONAL CONTEXT

Three key principles

The Systems Psychology view of the user of electronic office systems recognizes that the user operates within an organizational context. This view acknowledges the following important principles, which set a framework for understanding user behaviour and the role of electronic systems.

The need to consider organizational objectives, functions and roles. Organizational objectives, functions and roles help to explain what an organization aims to achieve and how it goes about it. The work an individual within an organization does is done within a framework defined by his or her organizational role. Some behaviour is appropriate to the role and other behaviour is not. The nature of the role determines the nature of the communications between that person and others within the organization—for example, a receptionist typically spends a lot of time greeting people, and perhaps responding to telephone calls, but would not normally be expected to interrogate technical databases or give presentations, which are more characteristic of some other organizational roles. The organizational roles defined for users are intended to enable organizational functions to be carried out, e.g. visitors to be made welcome, customers to be billed. The carrying out of these functions helps the organization to attain its objectives, e.g. to survive and if appropriate to make a profit. It is for use within this context that office systems need to be designed. If designed and implemented well, they can help users to perform their organizational role well, so that organizational functions can be carried out and organizational objectives achieved. Conversely, if designed badly or implemented badly, such systems can have deleterious effects.

The key role of communications. The relationship between the individual and the organization cannot be maintained without adequate communications. Communication is absolutely essential to all organizations. Without communication between its various parts, an organization ceases to be an organization—it becomes a mere concatenation of parts. Communication between the organization and its environment is also vital. Without that, the organization cannot respond to a customer's order, purchase materials, implement a change in tax laws, discover new business opportunities. Communication is essential, and the standard of an organization's communications is a key to its success or failure. Electronic systems can help to maintain and improve communications if they are designed well and implemented well, but can damage communications if they are not.

The principle of symbiosis. The individual remains a part of the organization just as long as a symbiotic relationship is maintained. In such a relationship the organization helps the individual to satisfy some of his or her needs—for example, to earn money, to play a worthwhile role in society, to make friends. In return, the individual helps the organization to meet its needs. Electronic systems can help to maintain and improve these symbiotic relationships if they are designed well and implemented well, but they can damage those relationships if they are designed badly or implemented badly. In the former case, such systems are of direct or indirect benefit to both the organization and the individual; in the latter case, they directly or indirectly have a negative effect on both.

We shall look at these three principles in a little more detail in the remainder of this chapter in order to understand the nature of 'user behaviour'—the behaviour of people using electronic office systems.

Organizational objectives

Organizations vary in terms of their overall objectives. For example, many organizations share a common objective of making a profit—but some organizations are non-profit making. Many organizations have organizational growth as an objective, but for some (e.g. Civil Service) this is not always appropriate. Most organizations have survival as an objective, but even this is not always the case—some organizations (e.g. businesses set up to serve a specific training purpose, and some organizations—e.g. War Department— set up during special circumstances), are intended to exist for only a limited time.

The corporate objectives of an organization are usually fairly broad or philosophical. The following is an example from a manufacturer of business forms (from O'Connor, 1976):

A. To maintain a position of leadership in the business forms field and to attain a position of leadership in the fields of promotional printing, source data collection, and other related peripheral equipment and data services.

B. To maintain itself in a sound financial condition and to obtain for its shareholders a fair and proper return on their investment, commensurate with the risks inherent in the nature of its business.

C. To provide its employees with good working conditions; to pay wages and salaries in line with those prevailing in its industry and the local industries in communities in which our plants are located, for similar work requiring like responsibilities, experience and skills; and to provide employment as secure and as steady as is practical, commensurate with the risks in its industry.

D. To be a good neighbor in the communities in which it is located and to foster and promote safety and health and other civic activities directed toward the fundamental improvement of those communities.

E. To provide a continuity of management to perpetuate the successful operation of the company.

F. To strengthen its position in the worldwide markets through the extension of 'know-how' agreements and by other means, if sound business judgement and conditions so warrant.

Organizational objectives vary between organizations and perhaps between cultures. Prentice (personal communication, 1983) has suggested that the following represent the key objectives of Japanese manufacturing industry as a whole, in order of priority (and see Chapter 1):

— expand market share
— welfare of staff
— good name of the company (quality)
— profit, in order to invest in research and development
— pay equity holders.

Whatever the overall objectives of an organization are, they are served by setting up appropriate organizational functions.

Organizational functions

A typical set of organizational functions includes: accounting; purchasing; legal function; production; marketing; sales; research and development; personnel; and others.

The particular set of organizational functions defined depends upon the type of organization, and how they are organized into departments and sections depends upon the particular organization.

Electronic office systems must serve the organization's functions if they are to be successful and so it is of interest to learn how many different functions there are that need to be served, and what the different functions involve—what the commonalities are, and what the important differences are. This is one of the research issues addressed by ESPRIT referred to briefly in Chapter 1.

Organizational roles

The various functions of an organization are carried out by people. These people often use machines of various sorts—increasingly electronic—but even so it is people who run the organization, not machines or anything else. It is possible in principle, of course, to imagine an organization run entirely by machines with no human presence or involvement at all, but notwithstanding the increasing use of robotics and other intelligent systems such a 'Cylon Empire' (as depicted in the feature film and television series *Battlestar Galactica)* is still a long way off and many would consider that to be an appropriate situation. So long as we always design systems to serve humans rather than to be served by humans we can, as Asimov recognized, reap enormous benefits from the new technology without fear of creating an electronic 'Frankenstein'.

In Systems Psychology, a 'better' system is a system that serves the human better. This means moving away from the situation typical of traditional data processing environments where the user had to adapt to the machine to be able to get any benefit from it at all, to a situation where the machine is designed to fit in with what is most natural to the user.

The user's role, therefore, is to use the machines available to carry out the functions of an organization; and the machine's role is to help the user to do that.

Organizational roles can be differentiated further. The role of a manager is different from the role of a clerical worker. The role of a marketing manager is different from the role of a personnel manager. In general, organizational roles can be distinguished in terms of two main factors:

— level in the organizational hierarchy
— organizational function.

Roles are not synonymous with people, of course. In a small business (and in some large organizations) one person may play many different roles. People also exchange roles over time. It is said to be a characteristic of many Japanese companies that as a matter of policy people are moved from one position to another every few years, and this is said to improve mutual understanding, respect and communication.

Impacts of electronic systems on users' roles. Electronic office systems impact users' roles in a number of different ways, and the impacts can be positive or negative, as follows.

Role perceptions. There may be differences in the way different people perceive the same role. For example, the head of a research department may encourage a scientist to do more research (sent role), and the scientist may perceive the message as 'publish as much as you can' (received role).

Differences in role perceptions can produce problems both for the people concerned and for the organization as a whole. The only way of avoiding or overcoming these problems is through better communications. Electronic systems need to be designed to improve communications, not to make them more difficult. This seems obvious enough, but one implication is that, for example, one needs to be cautious about setting up electronic systems that discourage informal, quasi-social communications—the drink in the bar, the chance contact in the corridor, the informal chat following a meeting can all be helpful in clarifying roles as well as in other ways.

Role conflicts. Various kinds of conflicts can arise connected with the roles that people play within organizations. For example, Payne (1981) distinguishes between: inter-sender conflict, where the expectations of two or more role senders are incompatible—e.g. a manager's superior might press her to put more effort into finding funds for the department whereas her subordinates might expect her to be a creative researcher; inter-role conflict, where two or more of the roles we occupy are in conflict—e.g. a manager and a friend, a member of the organization and a member of a division within the organization; and intra-sender conflict, where the same role sender has conflicting expectations—e.g. to increase output and to improve quality.

Bertrand (1981), in a study of female managers, found—as one might expect—that role conflict was associated with reduced job satisfaction. Brauer (1980) also obtained evidence for such an association in a study of middle management in an educational organization; Lamble (1980) obtained similar findings in a study of agricultural extension agents; and Shrivastava and Parmar (1977) obtained similar findings in a study of supervisors in textile mills. Taken as a whole, these and other studies also suggest role conflict can have deleterious effects on job performance.

One would hope that improvements in communications technology could be used constructively within organizations to improve communication and reduce unnecessary role conflicts.

Role overload. A person is said to experience role overload when (s)he is unable to meet the legitimate expectations made about that role (Payne, 1981). In an office environment this is closely related to what we might call 'information overload'.

For example, if a manager is given responsibility for more people, (s)he will need to communicate with those people, directly or through intermediaries. This will tend to mean extra meetings, telephone calls, letters, reports, and so on.

Now, it is obviously only possible in an eight hour day to spend eight hours processing information (or doing anything else). Unless the manager concerned takes steps to reduce the amount of time spent on meetings, telephone calls, etc. per person, whilst still processing relevant information sufficiently to carry out his or her responsibilities effectively, (s)he will go into information overload, discussed further below.

There is some evidence (Bateman, 1980) that individuals who experience feelings of role overload are more likely to exhibit coronary-prone behaviour patterns, poor job performance and low job satisfaction. A study by Smeltzer

(1982) also suggests that both role overload and role conflict are important contributors to work stress.

Information overload. This arises when the level of legitimate requests, directives, instructions, appeals, orders, and information needing to be processed—and/or the amount of processing that needs to be done, meetings to be held, reports to write, and so on—exceeds the capacity of the user to deal with it. Interestingly, there is some evidence (O'Reilly, 1980) that people may not appreciate they are in an overload situation in its early stages. At least for a while, the high level of information may be associated with increased feelings of satisfaction, even though decision-making performance has declined. This may reflect users' (erroneous) feelings that the extra information means they are less likely to make mistakes, even when the reverse may actually be the case.

Office systems need to be designed by manufacturers and implemented by user organizations in such a way as to reduce the possibility of information overload. This can be done in principle through the use of electronic intelligence to sift, sort, repackage and prioritize information inputs to the user, through the use of electronic processing power to help the user process information more easily, through the use of systems such as word processors and more advanced systems to help the user produce the required outputs more easily, through the use of teleconferencing and other systems to reduce the amount of time spent on meetings, and in other ways.

Poor design or poor implementation of office systems could have the reverse effect.

The organization as information processor

The view of organizations to be considered here emphasizes information processing. This is in contrast to the energy model of organizations that was popular during the 1970s (e.g. Katz and Kahn, 1978). The energy model was based on the idea that organizations are open systems that survive only as long as they are capable of maintaining negative entropy, that is, 'import in all forms greater amounts of energy than they return to the environment as product'. Katz and Kahn drew an analogy with various physical devices such as electric motors and transformers. Such devices all extract a price, usually in the form of heat, in changing energy from one form to another—electrical to mechanical, alternating current to direct current, and so on.

Katz and Kahn suggested it is appropriate to ask of organizations, as of electrical transformers, how much of the energy that is input into the system is output as product. The ideal would be one hundred per cent, but this is never attained in practice. There is always an 'energy price' to be paid. In the case of physical devices such as transformers this is often heat. In organizations, too, energy is 'wasted' or 'lost' at various stages between input and output.

Energy inputs to organizations were said to include: people, as sources of energy in carrying loads, operating machinery, and in other ways; materials which have already extracted an energy price in their manufacture and distribution; and energy proper in the form of steam, electricity, and so forth.

The concept of efficiency played a central role in the energy model. Improved efficiency meant, amongst other things, the creation of a surplus. The surplus could be used in various ways, e.g. to increase profit, to facilitate organizational growth, to promote survival, and so on. It would depend upon the organization's objectives.

Katz and Kahn adopted a very broad definition of the term 'energy' in developing their model, and it could be taken to include information if one wished to do so. However, there is some value in proposing an 'information model' of organizations as an alternative to the 'energy model'. It seems more appropriate as we enter our 'information society'. It helps to focus attention on the key role of information and communications in modern organizations.

The energy model was based on the idea that organizations need energy in various forms in order to survive. The information model asserts that organizations do not even exist, let alone continue over time, without information. That information is the 'glue' that holds the parts of the organization together, and is manifest in the communications that occur between the various parts of the organization.

We have noted above that organizations are not random, accidental concatenations of components. They are organized. They 'contain information', as a document or a picture may be said to 'contain information' by comparison with a random array of words or picture elements. But whereas (conventional) documents and pictures are static, an organization is dynamic.

There are constant, ever changing communications between its various parts, and with the environment.

The energy model was suited to the industrial society, in which manufacturing organizations were a much more important part of the economy than they are now. It is not surprising that Katz and Kahn's examples were drawn from such organizations. They compared two hypothetical companies concerned with manufacturing baseball bats; they discussed automobile manufacture; they compared hypothetical companies manufacturing television sets (see Katz and Kahn, 1978, pp. 330 ff.). These sorts of organizations lend themselves to the energy model, as they are primarily concerned with using energy to make gross changes in the physical form of materials.

Now we are entering our information society where the predominant type of organization (e.g. banks, insurance companies, consultancies, government departments, research establishments, publishers, software houses, educational institutions) deals in information, not energy and materials.

The information model seems better suited to this type of organization.

Figure 2.2 shows a few of the many communications links that might define an insurance company and its links to its environment. Whilst this is a gross simplification of any real insurance company it helps to give a flavour of the variety of types of information requirements (billing, receipts, policy documents, etc.) that form the substance of organizational communications.

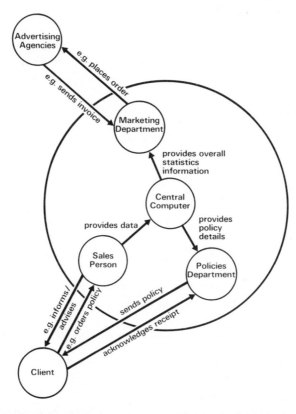

Figure 2.2: A few of the many communications links that might define an insurance company and its links to the external environment. (Reproduced by permission of ITT Industries Ltd.)

Even in manufacturing organizations, the information model applies to some extent. In the manufacture of television sets, for example, there are many information-related aspects of organizational functioning, such as: the design of the sets, the ordering of materials, stock control, marketing activities, the personnel function of the company concerned, and many others.

The two models are closely related, the differences being more a matter of emphasis and precise definition than fundamental viewpoint; if defined sufficiently broadly, 'energy' could be interpreted as including information as one particular form, albeit the most important form in our post-industrial society. The relationship between the two models is shown in Figure 2.3 (from Christie, 1981).

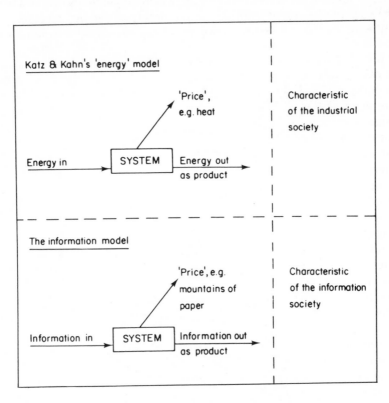

Figure 2.3: The energy model and information model of organizations compared (from Christie, 1981)

Impacts of electronic systems on organizational information processing. Three main classes of impacts of electronic systems on organizations have been proposed in terms of the information model (Christie, 1981):

Information input. All other things being equal, and assuming constant output from the organization, electronic systems will tend to reduce the amount of information input by selecting, pre-processing and packaging the information input according to users' needs within the organization. This requires appropriate application of artificial intelligence and satisfactory user–system communication. If the systems are not designed well from this point of view, their effect would be much more likely to increase the amount of information input (because of the ease of creating, distributing and accessing electronic information), creating information overload on the user and reducing organizational performance.

Information processing. Just as in Katz and Kahn's model the organization extracted a 'price' for the conversion between inputs and outputs, so the information model acknowledges that a price is paid. In Katz and Kahn's model, the price was an energy price; in the information model, it is an information price—mountains of paper, in traditional offices. Some idea of the size of that price is given by the results of a survey by Engel *et al.* (1979) on the corporate headquarters of a US multinational. They found that of the documents in office in-trays only 25 per cent originated from outside the company, and of the documents in office out-trays only 19 per cent were destined for outside the company. Another indication that the price is significant is an estimate (Jarrett, 1982) that the average office worker controls 20,000 pieces of paper—and, what is worse, adds 5,000 items a year whilst disposing of only 3,000. All other things being equal, electronic systems will tend to reduce the 'price' of processing information by reducing the need to create multiple copies of 'papers' to be stored in personal files in multiple locations; users will be able to access common files as and when needed, according to their security clearances. This assumes, amongst other things, a high standard of communications between users and electronic files, including appropriate information retrieval dialogues. If the systems are not designed well, the ease of creating electronic copies of information would be much more likely to increase the number of copies made and so increase the price of organizational information processing by adding to the complexity of the total system.

Information output. All other things being equal, including constant information input to the organization, electronic systems will tend to increase the amount of information output from organizations by making it easier to handle information. For some years now, word processing has made it easier to merge information from different documents, to edit text, delete columns and rows from tables of numbers, and so on. Emerging systems using many different media and applying artificial intelligence take this trend even further, enabling significant improvements in this aspect of organizational productivity.

Communication structures

The information processing done by organizations is done by users in communication with other users and machines in the organization. The information pathways defined by the communication links between users and electronic or other systems within the organization are not random. They reflect a mixture of organizational planning (the formal organization) and relationships that evolve unofficially according to particular circumstances (the informal organization). Electronic systems need to be designed and implemented in such a way as to facilitate the development and maintenance of optimal formal and informal communication structures.

Figure 2.4: Type A and Type B communication. (Reproduced by permission of ITT Industries Ltd.)

The communication structures within an organization can be regarded as being built up from two main types of unit: the 'source/sink–channel–sink/source' unit, and the 'source–store–sink' unit. These are shown in Figure 2.4, where the former type of unit forms the basis for Type A communication, and the latter for Type B.

The 'source/sink–channel–sink/source' or 'Type A' Unit. A two-party telephone conversation or face to face conversation is an example of a Type A unit. At any given point in time, one person (source) is acting as a source of information for the other person (sink). Communication is through a channel (the telephone line, or direct face to face communication). In the case of a telephone conversation, the information is predominantly verbal but there is also some non-verbal information communicated (e.g. intonation, pauses, hesitancies). Simultaneously, the 'sink' or 'receiver' is providing feedback to the 'source' so both parties play the role of 'source' and 'sink' simultaneously. In a telephone call, the feedback is restricted (e.g. indication the 'sink' is listening, indication the 'sink' is trying to interrupt), but is richer in face to face communication. The communication is highly interactive, not just because of the feedback but also because of the ability of the two parties to switch roles very rapidly and frequently. In pure Type A communication, there is no record of the communication other than in the memories of the parties concerned.

The 'source–store–sink' or 'Type B' Unit. A letter or telex is an example of Type B communication. A person acting as source writes the letter, which then acts as a store or record of the communication, and a second person acting as sink reads the letter. Simultaneous feedback is not possible, and rapid role reversal is not possible. Unlike Type A communication, there is a record of the communication. This is often copied and so the information may be stored in many different places—often unknown by the source, and often retained for much longer periods than the source might have guessed. Unlike Type A communication, there is significant uncertainty about whether or when the sink will receive the information and who else apart from the intended sink will access the information from store, and over what period of time this will remain a possibility. Memos, reports, notes, books, and similar items are all vehicles for Type B communication.

Type A communication was probably the earliest form of communication used by humans, is the type used by babies, and is the type preferred by adults for many situations (e.g. many people refuse to talk to an answering machine, and not many people would consider an exchange of letters to be an adequate substitute for a recruitment or counselling interview).

Type B communication became very popular with the advent of the postal service and the printing press. Since the latter event five centuries ago, Type B communication has increased in a positively accelerated manner. The advent of electronic systems for the office promises or threatens to push the rate of increase still higher.

Networks. Type A and Type B units form the building blocks for complex networks made up of sources, stores, sinks, and communications channels, in which roles may change frequently. Psychological research has focussed traditionally on rather simple networks—relatively small, relatively homogeneous, and where roles remain relatively stable over time. They have often been based on written communication but typically have failed to explore the broad organizational implications over time of having a record of the information communicated, i.e. they have often used Type B communication but looked at it in an artificial situation and from a Type A viewpoint. Some of the networks which have received special attention are illustrated in Figure 2.5.

The experiments done have shown that the type of network set up can affect both user attitudes and task performance. The complexity of the task is an important factor in which type of network is appropriate; if allowed to, networks will often evolve towards a particular type, depending upon the nature of the task. For a review of this research, see Baird (1977).

There is some evidence (e.g. Kano, 1977) that which type of network is the most effective depends partly on the amount of information it has to handle. If this is so, it could have important implications for the use of electronic systems. Such systems may be expected to increase the amount of information flow considerably, and so an organization's conventional network structures may no longer be appropriate. It is also possible to draw the implication that electronic networks should be designed to allow the functional networks to be redefined as and when necessary to accommodate the possibility of significant changes in the overall volume of information at different stages in the life of the organization and its electronic network.

There is some evidence (Rice and Case, 1982) that computer-based messaging systems can increase users' communication networks.

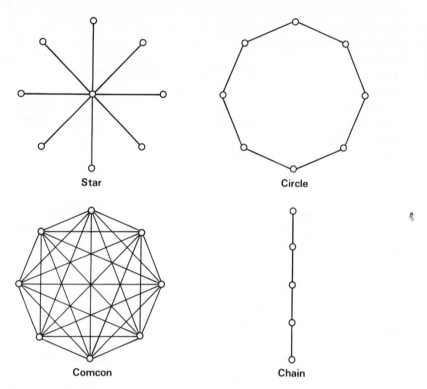

Figure 2.5: Some networks which have received special attention from researchers

Communications media. Real networks within organizations typically involve many different sorts of channels, using different media (e.g. face to face, video, voice, paper). The media available, like the pathways available, also affect user attitudes and task performance; and, if given a choice, users will often express a preference for using one medium rather than another. For example, research reviewed in Chapter 5 has shown that for Type A (interactive person–person) communication users will often prefer to meet face to face than use a telecommunications system; if they are required to use a telecommunications system, they will often express more positive attitudes toward video systems than audio systems, but in terms of task outcome audio systems are often just as satisfactory as video systems.

Implications for systems psychology: functional and physical networks. The technology available in the past tended to keep some networks separate from others—to define special purpose networks. The telephone network was used for voice conversations, letters were sent by mail, telexes used the telex network, data communication was done by mail or using special telecommunications networks, meetings depended upon the road, rail and air networks, and so on. This required users to segment their work artificially. They could not easily access a remote database, talk to the people there, and exchange documents with them simultaneously; these things would have to be done separately, and work was organized to take account of this. This situation is no longer the case. Modern switching equipment and telecommunications networks are emerging which allow voice, data, pictures, text, and so on, to be handled simultaneously and to be switched to 'end-user terminals' which allow the users concerned to interact with the network directly rather than through an intermediary. Users can send a voice message or other message from their home or car directly to the people with whom they wish to communicate; meetings can be held remotely using telecommunications, and the users concerned can access databases relevant to the meeting that may be spread all around the globe; in these and other ways, users now have access to much more powerful networks than ever before.

The design of modern telecommunications networks requires designing both the physical networks and the arrangements for the functional networks that are overlaid. The physical network is made up of the physical components in the network and the physical connections between them. The functional networks overlaid on the physical network are what the users see. To give a simple example to illustrate the distinction: the physical network might be based on a star or circle (see Figure 2.5); one set of users using the network might be constrained for some purposes to communicate in a chain—they see the functional network as a chain; others might communicate in a comcon; and so on, all on the same physical network. Different users will also see the network differently according to the range of services (things they can do using the network) and media they have available, depending on their particular interfaces.

Imagine an organization selling complicated video arcade games machines. The sales people out in the field who send orders back to the network from their remote locations, the maintenance engineers who use the network to use expert systems for trouble shooting and to refer to video 'manuals' for guidance on how to repair the machines, the manager who receives electronic mail through the network, the word processing operator who uses the network to prepare documents, the network manager who needs to allocate the resources available on the network appropriately—these people and others will have rather different views of the network, and their functional networks need to be designed with as much care and attention as the physical network.

THE USER

Having provided a view of the organizational context in which the user works as a member of an organizational team, we turn our attention in the remainder of this chapter to a consideration of the user as an individual 'office worker'.

Users' roles

We have seen above that the functioning of the organization depends upon the people within the organization fulfilling their organizational roles. Understanding the nature of users' roles is a key to understanding what office systems need to support.

A technique for defining user roles is to ask users to complete the following:

> If I were to fail to do any of the following I would be failing in my role as (job title).

The following is an example of a hypothetical manager's role defined in this way:

  a. to ensure sufficient funding for the department
  b. to ensure the department meets its work commitments
  c. to maintain departmental morale
  d. to identify new opportunities for the department
  e. to encourage the professional development of department personnel
  f. to promote a favourable image for the department within and outside the organization.

The following is an example of a hypothetical secretary's role:

  a. to ensure that documents are prepared properly and in time
  b. to maintain an adequate filing system
  c. to do necessary administrative work
  d. to deal with enquiries when the manager is engaged or absent
  e. to screen communications to the manager.

This way of defining users' roles provides a basis for systematically comparing roles at different levels within the organization, and between different organizational functions. Commonalities can then be examined, requirements for electronic systems identified.

No systematic survey and analysis of this sort, based on a representative sample of offices, has yet been published.

User functions

Users fulfil their organizational roles by setting up and performing key functions. These can be defined by asking users to complete the following statement for each role element:

In order to achieve (a) I ...

Taking the first role elements, (a), for the manager and secretary above as examples, role functions might be:

manager: 'In order to ensure sufficient funding for the department I

— keep in regular contact with previous clients
— organize presentations to potential new clients
— promote awareness of the department within and outside the organization through seminars, colloquia, conferences, and other means'
— bid for funds through the formal channels and mechanisms defined by the organization.'

secretary: 'In order to ensure that documents are prepared properly and on time I

— allocate work to our word processing pool
— do some typing myself
— send some typing out to bureaux during very busy periods.'

The user performs these functions by interfacing with the rest of the organization and with the world beyond the organization.

The user interface

The interface between the user and the rest of the world (including the rest of the user's organization) is shown in abstract form in Figure 2.6, and in more concrete form in Figure 2.7.

All communications between the user and the rest of the world by definition occur only through the user interface.

The structure of the user interface is usually complex, but the interface is structured. It usually provides the following main 'channels' or 'areas'. A typical example of that part of the interface that is physically associated with the user's office or desk area would include:

— a door through which people can and must enter in order to communicate with the user directly (Type A communication)
— one or perhaps two telephones through which people can and must communicate if they wish to communicate with the user interactively by telephone (Type A communication)
— an 'in' and 'out' tray which provide the main 'Type B' channel to and from the user
— an area of the interface which provides an interface to the user's 'personal information space'. This is usually highly personalized and includes: filing cabinets, shelves, the desk top and other physical areas.

People other than the user typically find great difficulty in using it if they need to try to do so, and if anyone 'tidies it up' for the user then the user subsequently finds it very difficult to use. This part of the interface is discussed in more detail in Chapter 6.

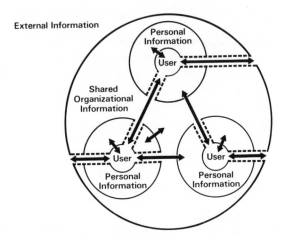

Figure 2.6: The user interface. (Reproduced by permission of ITT Industries Ltd.)

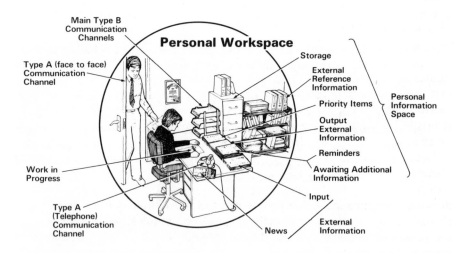

Figure 2.7: A typical user interface in a conventional office. (Reproduced by permission of ITT Industries Ltd.)

These various areas of the conventional interface have qualities that need not simply be copied in advanced electronic interfaces. The telephone can actively attract the user's attention but the personal information space does not—it is passive; in the advanced electronic interface, the personal information space may—for example—continuously monitor key items of work on which the user is engaged and actively seek to attract the user's attention if it contains something of relevance the user might wish to use.

User activities

By interacting with the user interface, the user can execute activities related to the functions that need to be performed in order to achieve the user's role objectives. For example, the hypothetical secretary above indicated that an important function was to 'allocate work to our word processing pool'. This would involve a number of specific activities, such as: requesting a deadline or assigning a priority, noting the author, assigning a job number, defining the format required, conveying the manuscript or other materials to the pool, and so on.

Various studies have been done of what activities occur in offices and how much time is allocated to them. Figure 2.8 (from Christie, 1981) presents the findings for managers and other office principals from a study by Engel *et al.* (1979). The study was done in the headquarters of a multinational corporation. The results give a flavour of the nature of the data that have been collected in such surveys, but it is difficult to draw any precise general conclusions about how time is allocated. Different studies give different results. For example, Jarrett (1982) shows how three different studies of how managers use their time give three different estimates of the amount of time devoted to meetings: 46 per cent, 35–45 per cent, and 22 per cent—the highest estimate being more than twice the lowest.

A general, qualitative finding that does seem to emerge from the various studies done, and it fits in with what less formal observation suggests, is that the more senior people in organizations tend to spend more time on meetings. In a study done of the Commission of the European Communities (Christie, 1981), the most senior grade of official studied spent almost four times as much time at meetings as did the most junior grade.

Detailed comparisons of the results of six of the key studies done are presented by Doswell (1983). These are of interest in suggesting possible differences between different kinds of organizations, but one needs to avoid drawing any firm conclusions because the studies were typically done on only a small sample of organizations and it is not clear to what extent the findings are generalizable. Also, whilst Doswell feels that the findings from studies done at different times have shown some consistency, the fact remains that they were done in the 1960s and 1970s and it is not clear to what extent they apply to organizations of the 1980s (even apart from changes arising directly as a result of introducing electronic systems).

No survey of user activities and functions based on adequate methodology and a representative sample of adequate size has yet been published, and this remains an important area for systematic research. Not only have the studies published so far not been based on representative samples of interestingly defined populations, they have not analysed user behaviour in a very satisfactory way, either; for example, Figure 2.8 contains no explicit reference to decision making, even though a Booz, Allen and Hamilton study found that over 77 per cent of the US managers they surveyed saw enhanced decision making as being a major benefit of office automation (Jarrett, 1982). (Decision systems are discussed in Chapter 8.)

Activities	Average percent of time			
	Level 1	Level 2	Level 3	All
Type A communication	38.2	26.8	19.5	26.5
Telephone	13.8	12.3	11.3	12.3
Conferring wih secretary	2.9	2.1	1.0	1.8
Scheduled meetings	13.1	6.7	3.8	7.0
Unscheduled meetings	8.5	5.7	3.4	5.4
Type B communication	38.3	47.4	44.2	44.2
Writing	9.8	17.2	17.8	15.6
Proofreading	1.8	2.5	2.4	2.3
Searching	3.0	6.4	6.4	5.6
Reading	8.7	7.4	6.3	7.3
Filing	1.1	2.0	2.5	2.0
Retrieving filed information	1.8	3.7	4.3	3.6
Dictating to secretary	4.9	1.7	0.4	1.9
Dictating to a machine	1.0	0.9	0.0	0.6
Copying	.1	.6	1.4	.9
Mail handling	6.1	5.0	2.7	4.4
Other	23.3	25.9	36.0	29.4
Calculating	2.3	5.8	9.6	6.6
Planning or scheduling	4.7	5.5	2.9	4.3
Travelling outside HQ	13.1	6.6	2.2	6.4
Using equipment	.1	1.3	9.9	4.4
Other	3.1	6.7	11.4	7.7
	100%	100%	100%	100%
Total number of principals	76	123	130	329

Level 1 = upper management
Level 2 = other management
Level 3 = non management

Figure 2.8: Allocation of time by office principals (from Christie, 1981, adapted from Engel *et al.*, 1979, by permission from *IBM Systems Journal*)

In general terms, one can expect office systems to have four main impacts
on office activities :

— Some activities, e.g. making a telephone call, will be done more efficiently. For example, it will be easier and less time consuming to set up
  a call, it will be possible to leave a voice message, and it will be possible
  to show one another 'papers' and make changes to them whilst on the
  'phone'.
— Some activities, e.g. photocopying, will tend to disappear as it becomes
  easier to use information in electronic form and to make electronic
  copies.
— Some activities will evolve into new types of activity. For example, users
  will be less restricted to the medium of paper for producing reports.
  It will be possible to 'write' reports that incorporate graphics, text,
  voice, motion video, small programs, and other forms of information
  in a more interesting and more useful package than can be done with
  paper.

Some 'new' activities will emerge. For example, just as 'calculating' appears explicitly in Figure 2.8, so with more sophisticated decision support
systems appearing, 'decision making' will appear as a more explicit and more
clearly bounded type of activity.

User behaviour through time: the behaviour tree

The activities in which the user engages in order to fulfil the various functions that need to be done to achieve role objectives do not occur all at the
same time or in a meaningless sequence. They occur as a stream through
time, and in an order that has some meaning.

At any given point in the stream of behaviour, the user can choose to
behave in different ways according to the options that apply. There are five
main classes of options:

— to wait (e.g. for a telephone call that is expected in connection with a
  proposed purchase)
— to act (e.g. sign a purchase order)
— to generate information (e.g. prepare a specification of the sort of
  product needed)
— to seek information (e.g. about relevant products—either from one's
  personal information space or from elsewhere)
— to opt out of the situation (e.g. by taking a coffee break, going into a
  daydream, or chatting with a colleague).

The choices made can be thought of as mapping out a route through a
'behaviour tree', the route being dependent upon information received and
the way it is processed. A generalized behaviour tree is shown in Figure 2.9.

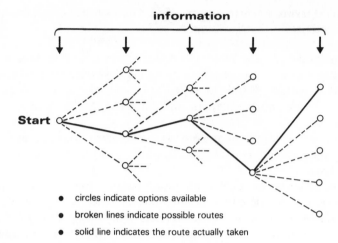

- circles indicate options available
- broken lines indicate possible routes
- solid line indicates the route actually taken

Figure 2.9: A generalized behaviour tree

It is possible to construct a 'high level' behaviour tree to describe a person's course through life and compare it with what (hypothetically) might have been. The choices made can be very significant. The choice of going to university or not, of which subjects to study, of which if any person to marry, generally affect the subsequent parts of the behaviour tree very noticeably. Opportunities once passed by are lost forever.

More detailed trees can be constructed for behaviour at the user interface. In designing an electronic interface, these can be made explicit or not, as the designer chooses. In videotex (e.g. Prestel), the choices available at any given part of the behaviour tree are made explicit by providing the user with a visible menu. This makes it easy for the user to know which options are available at any given point in time. On the other hand, a conventional videotex system such as Prestel typically does not provide the user with a broader view of the tree as a whole, or the part of the tree in which the user is located. The user can only guess what will happen if a particular option from the menu is chosen. This is a weakness in the system and often results in the user having to 'backtrack' through the database for missing information which is wanted and which does actually exist in the database.

It is worth noting a subtle point, that whilst the user may backtrack through a Prestel database (s)he cannot backtrack through the behaviour tree. Time moves in one direction only. If a user returns to the same Prestel page as seen previously, (s)he has not moved back through the behaviour tree—the situation is different, because the information available to the user (about what lies beyond that page in the database) is different, and so the nature of the choices available is different.

In some difficult situations, the user is not even helped by being reminded of what options are currently available. This is typically the case with an enhanced telephone system. The user may get a busy tone when trying to call a number. The system may in fact provide an option of instructing the system to keep calling the number automatically until it answers (the instruction being keyed in or delivered in some other way). The user will never choose this option if the only option (s)he remembers as being available is putting the 'phone down and dialling again later. Clearly, such an interface could be improved by having the system remind the user of what options are available.

Another question the designer of behaviour trees for user–system interaction needs to consider is how many levels to build into the tree, and how many options at each level. Figure 2.10 illustrates this point by reference to a hypothetical (and unrealistically simple) enhanced telephone system. The optimal design is not always clear without doing user tests to find out. Many videotex databases provide examples of trees that have so many levels that users often complain about the tedium of having to go through them all to find the information of interest. Some work by Snowberry, Parkinson and Sisson (1983) suggests that fewer levels and more options per level may be preferable, at least for up to about eight options per level, but the study was done using lists of options presented visually (menus) and it is not clear whether similar results would be obtained in cases where the user has to remember what options are available at each stage.

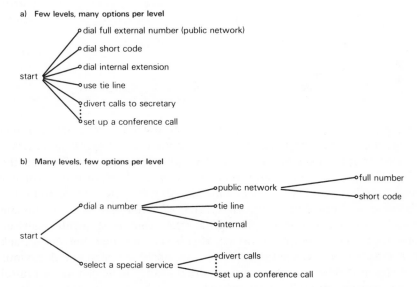

Figure 2.10: Two of a number of possible trees for a simple enhanced telephone system. (Reproduced by permission of ITT Industries Ltd.)

The user as information processor

The Systems Psychology view of organizations is that organizations are systems of inter-communicating users and machines, within a physical environment. The focus is on machines concerned with the creation, storage, transmission and processing of information.

The user of these machines is regarded as an information processor, and so is the machine. The point is made diagramatically in Figure 2.11. Both the user and the machine accept information inputs, process them and generate information outputs. Their joint information processing activities make up the information processing that is done by the organization.

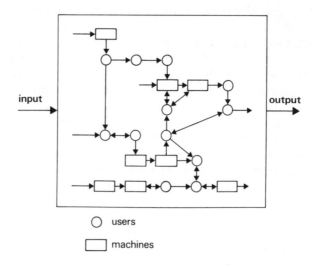

Figure 2.11: The organization, the user and the machine as information processors. (Reproduced by permission of ITT Industries Ltd.)

Despite the fact that the organization, the user and the machine can all be regarded as information processors, our understanding of the nature of the information processing is not the same for these three (see Figure 2.12).

In the case of conventional machines like typewriters, or even word processors, we have a very detailed understanding of how they operate. This is because we built them. With the advent of intelligent, knowledge-based systems, this is beginning to change a little. Our understanding of such machines as have been built already is still very detailed, but not quite as detailed as before because the machines are capable of changing themselves, e.g. generating new rules that were not programmed in when they were built.

Organizations are fairly well understood because to some extent they are, like machines, designed by us according to some deliberate plan; subsequent changes to the organization are also made according to a deliberate plan. The 'specification' for the organization is therefore known. But this is only a partial view. The organization rapidly develops an informal, partly invisible structure that is not completely known by any one person, and this affects the organization's functioning in important ways.

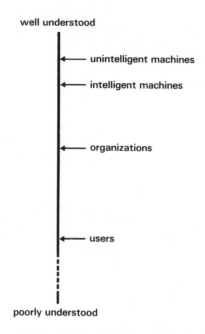

Figure 2.12: Differences in understanding
of machines, organizations and users

When it comes to the user, our understanding is much poorer. This is because we did not build the user. We have not been given the specification or the engineering diagrams. The best we can do is to infer how the user operates by observing user behaviour under a range of conditions we can choose. This is the area addressed by scientific psychology, and it is more difficult (because of the greater complexity of the user) than trying to produce an engineering diagram of, say, a microcomputer by observing what the microcomputer does (without lifting the lid off).

The view of the user as an information processor is the one adopted by modern cognitive psychology and some of the implications for systems design that can be drawn from research in cognitive psychology are discussed in later chapters.

The user as a human being: limitations of the information processing model

The information processing model of the user forms a basis for much current thinking in the design of user–system interfaces, but by itself it is not fully adequate for designing a system whose capabilities are compatible with the requirements of its intended users. The reason for this is that the simple information processing view of the user emphasizes just one aspect of the user, whereas in practice the system needs to be compatible with the user as a

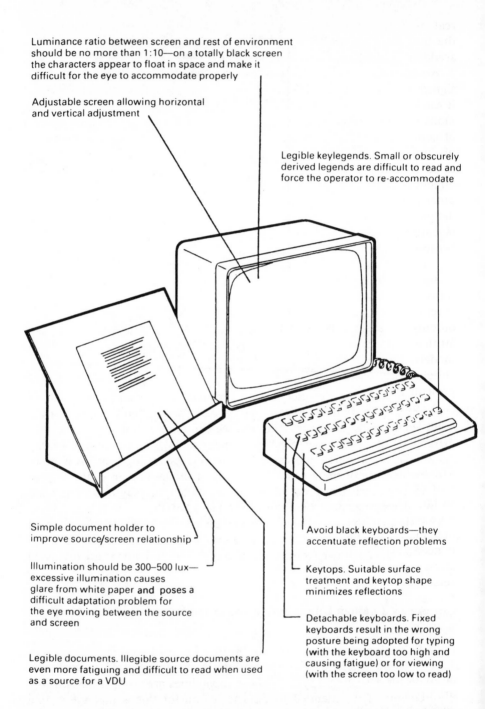

Luminance ratio between screen and rest of environment should be no more than 1:10—on a totally black screen the characters appear to float in space and make it difficult for the eye to accommodate properly

Adjustable screen allowing horizontal and vertical adjustment

Legible keylegends. Small or obscurely derived legends are difficult to read and force the operator to re-accommodate

Simple document holder to improve source/screen relationship

Illumination should be 300–500 lux— excessive illumination causes glare from white paper and poses a difficult adaptation problem for the eye moving between the source and screen

Legible documents. Illegible source documents are even more fatiguing and difficult to read when used as a source for a VDU

Avoid black keyboards—they accentuate reflection problems

Keytops. Suitable surface treatment and keytop shape minimizes reflections

Detachable keyboards. Fixed keyboards result in the wrong posture being adopted for typing (with the keyboard too high and causing fatigue) or for viewing (with the screen too low to read)

Figure 2.13: Some good keyboard and screen characteristics (from Doswell, 1983). (Reproduced by permission of APEX.)

real, multifaceted living system—a human being. Any approach to designing the user interface which ignores this fundamental observation will be of academic interest only and will fail to deal with the practical reality.

An example of the practical reality might be a system with a poorly designed interface which nevertheless is used a great deal because of the things it can do, the support provided to the user by other people, and the high motivation of the user. This was a fairly common situation in the early days of computers, but is much less common in the office environment. Manufacturers increasingly need to provide excellent user interfaces in order to be competitive with other products, and office workers are motivated in directions other than learning to grapple with cumbersome, awkward computer procedures. The reverse situation is now far more likely. A fairly well designed system might not be used because its intended users are not sufficiently motivated to use it, or because inadequate support is provided, or because to use it proves to be unduly stressful.

It may well be that cognitive psychology's information processing model of the user in principle can handle such issues, and no doubt some cognitive psychologists would argue that alternative models are unnecessary. It remains the case, however, that in practice cognitive psychologists focus on only some aspects of user behaviour. This no doubt partly reflects their attitudes and interests, and may partly reflect the appropriateness of the models—although one can use a screwdriver to chisel a piece of wood, a chisel might do an easier and better job, even though essentially the same principles are involved.

A complete approach to the design of the user–system interface will acknowledge and deal with the many different aspects of human behaviour. The following are among the more important.

Ergonomic considerations. The mechanical and physiological aspects of the human must be taken account of by the system. The human can only reach so far, can only move so fast, can only resolve a certain amount of detail on a screen. Failure to take due account of these human constraints when designing a system is likely to mean that performance will be poorer than it need have been, that the users will be put under unnecessary physical stress with all that that implies (e.g. backache, headache), and there will be reluctance to use the system especially when better, competitive products are readily available. Figures 2.13 and 2.14 (from Doswell, 1983, pp. 160 and 161) illustrate some of the more general design guidelines that have been developed on the basis of ergonomic considerations, although much more detailed guidelines are also available.

Arousal. People do not generally perform their work very well when they are tired or sleepy. They generally do better when they are alert and fully awake. But if they are too 'alert' and too 'awake'—'over-excited'—their performance tends to deteriorate again. People vary in terms of their general

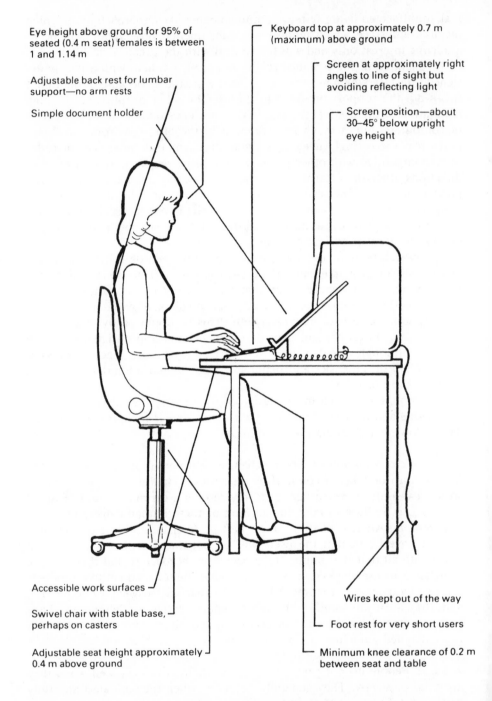

Figure 2.14: Some good workstation principles (from Doswell, 1983). (Reproduced by permission of APEX.)

'arousal level', and this affects their performance. The general form of the relationship postulated is shown in Figure 2.15. It has also been postulated (e.g. Eysenck, 1970) that different personality types vary in their habitual levels of arousal—specifically, that under the same conditions of task and environment, introverts tend to be more aroused than extroverts, and that within each group the 'emotionally labile' tend to be more aroused than the 'emotionally stable'. As the arousal potential of the task/environment situation is increased (e.g. by making the task more difficult, or the environment noisy), the labile introvert will pass through the optimum level of arousal before the stable extrovert. Starting at the 'low arousal' end of the continuum, this theory predicts that to promote maximum performance from the user the electronic system needs to stimulate the stable extrovert more than the labile introvert; this could be done, for example, by providing speech output in parallel with a visual display, or colour graphics instead of monochrome. This is one example of the way in which the needs of individual users differ, in ways that electronic systems do not.

Arousal and other psychophysiological concepts are discussed in more detail in Chapter 10.

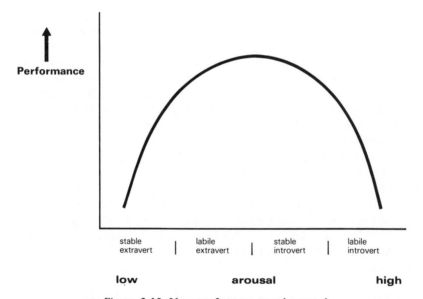

Figure 2.15: User performance and arousal

Motivation. The concept of 'motivation' refers to why people do the things they do. Some reasons are superficial (e.g. I made a telephone call in order to find the time of a train. I wanted to find the train time in order to make plans for a meeting. I wanted to go to the meeting ... because it was part of my job.). Some are deeper (e.g. I want to do my job well in order to ...). Two well-known theories of motivation are Maslow's and Herzberg's, discussed for example by Payne (1981). The Maslow theory proposes a hierarchy of basic needs which motivate human behaviour, which from bottom to top are: physical (e.g. hunger); need for security; social needs; ego needs or need for esteeem; need for autonomy; need for self-actualization. Herzberg, focussing more directly on job satisfaction, identified a different set of factors. One subset he called the hygiene factors. These helped to avoid dissatisfaction whilst not contributing positively to satisfaction; they were: salary or pay; relationship with peers; job security; status; company policy; working conditions; relationship with boss. The other subset he called the motivators; these did contribute positively to satisfaction, and included: interesting work; feelings of significant achievement; feelings of personal growth; and being responsible for worthwhile activities; along with the promotion and recognition that went with these.

Both the Maslow and the Herzberg theories have been criticized for their lack of empirical support (e.g. Campbell and Pritchard, 1976; Kline, 1983), but they remain popular. A more tentative but more empirical scheme is that of Cattell and his colleagues, and is discussed by Kline (1983). The Cattell scheme includes the following main needs, as well as other less important ones: food seeking; mating; love of comfort and warmth; gregariousness; parental pity; exploration; fear; self-assertion; pugnacity; acquisitiveness; career; self-sentiment; parental-family; superego; wife or sweetheart; and religion.

It is important to realize that: a) whatever the 'true' picture is exactly, it is clear that human behaviour is motivated in complex ways; and b) these factors always operate—all behaviour takes place within this motivational framework, 'user behaviour' being no exception. This is another practical reality that designers of the user–system interface should take into account. The human is not simply a computer that, once switched on, runs according to programme unless physically disrupted. What the user does is done within a total context of which the task and the electronic system form only one part. Electronic systems will be rejected by their intended users if they run counter to the user's motivational structure—for example, by reducing the user's self esteem, or raising fears, or inhibiting the user's gregariousness.

The need to consider 'user behaviour' as just one element in the user's life, influencing and being influenced by other elements, is discussed further in Chapter 10.

Stress. Stress is different from 'challenge'. People need a certain amount of challenge in their work or it becomes too uninteresting, too unenjoyable, and may not be seen as worthwhile. Stress, on the other hand, is generally to be avoided. It represents an overload condition.

In view of the complex and potentially conflictful motivational context in which the user operates, it is perhaps not surprising that under some circumstances the user's behaviour can start to deteriorate—both in terms of role performance and in broader terms—and the user may even show

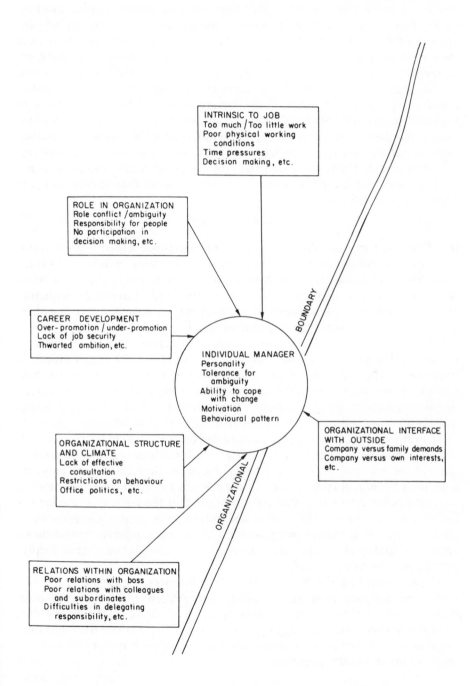

Figure 2.16: Sources of managerial stress (from Cooper and Marshall, 1978, p. 83)

physical signs of stress, such as disabling ulcers or coronary heart disease. As Blythe (1973) has said, 'certain stress-diseases are lethal, and can kill just as successfully as a well-aimed bullet'. Some estimates have suggested that the cost to the nation of stress-related diseases may be about 1 to 3 per cent of the Gross National Product in the USA. (Cooper, 1980, p. 49). Stress affects people at all levels of the organization, and—contrary to popular belief—may affect those lower in the hierarchy more than those higher up. Figure 2.16 (from Cooper and Marshall, 1978) illustrates the complexity of what is involved by showing the main factors influencing the level of stress experienced by managers. As we move into the Information Technology Age, it is essential that we design the new electronic systems to reduce the stress on the people they are designed to serve, rather than to increase it.

CONCLUSIONS

It is important to understand what it is that electronic office systems need to support; otherwise, there is a real danger of creating 'solutions looking for a problem'. What they need to support is user behaviour. This does not take place in a vacuum, and nor does the user's interaction with the electronic system. It takes place within an organizational context, and within the context of the user's life as a human being.

The organizational context can be described first in terms of broad organizational objectives to which everything else about the organization is subservient. These determine the general style of the organization, what is important and what is not. The organizational objectives are served by setting up the necessary organizational functions (e.g. marketing, accounting), and these in turn determine the roles that users need to fill.

Within the context thus defined, information is processed by users to fill their organizational roles, so that the organizational functions can be carried out and the organization can achieve its objectives. Both the organization as a whole, and individual users can be regarded as information processors.

User–system interaction also takes place within the context of the user's life as a human being. The user is not just an information processor in a simple sense but a human being with physical, physiological, motivational and other needs—needs that mean the user cannot sensibly be treated simply as another type of computer.

The successful design of the user–system interface requires that the interface be designed to serve the needs of the intended user in both the organizational and life context. Anything less is avoiding reality.

In the next chapter, we consider how the office has evolved from humble beginnings to the modern environment where electronic systems of various sorts vie for the user's attention.

Chapter 3

The Office: A Historical Perspective

JACK FIELD

INTRODUCTION

The future can be read in the events of the past. To the extent that this is true, it is helpful to consider what has been happening to the office in the past in order to gather clues about what the future will bring. This chapter presents one person's view of where offices have been and where they are going.

The historical perspective will not focus on the one office feature that is most ripe for immediate development. It will however give a sense of proportion to the umpteenth discovery of familiar territory and to this morning's new product announcement. Equally, a sense of direction will be gained from tracing the connection between technology and the offices of the past.

THE FIRST OFFICES

Adopting the historical perspective, we start as nearly as possible at the beginning. Figure 3.1 is a reproduction of the first generation of paperless offices, from the days of Moses and the Pharoahs.

In this office of the past (the left of the two chambers shown in Figure 3.1) scribes are entering on papyrus, using calculating frames, the accounts of grain being stored in two of the three bins in the right-hand chamber which they can glimpse through a screen.

One might say that clerical staff are scrolling their pages of input as they key in data from their screen. More important than glib analogies is how this primitive office epitomizes the differentiations and boundaries that define an office.

Office Separated from:	Integrated within Office:
The productive activities of storage, assembly, distribution, exchange, maintenance, service, etc.	Aggregating the representations of productive activities; and the interpretation and negotiation of their value.

Essentially an office both separates itself from productive or service activities and is linked to them; and within its four walls contains and merges the two distinct processes of aggregation and negotiation. The arrangement gives enough distance to oversee, abstract, record, cumulate and aggregate the production and service activities over space and time—so much of this, so much of that in each period. These representations of activities are what is being entered on the scrolls in Figure 3.1.

Figure 3.1: A very early office: Egyptian funery model from the tomb of Meket Re. (Reproduced by permission of the Metropolitan Museum of Art, New York.)

The distinction between the two processes within the office is derived from Goffman's analysis of situational encounters (1976). He distinguishes between processes used to solve practical problems on the one hand (the 'rational' or 'functional' processes) and processes by which meaning and value are arrived at by interpretation and negotiation on the other (the 'expressive' or 'ceremonial' processes). Goffman quotes from a book of etiquette to mark the distinction: '... indicate by your language that the performance (of a function) is a favour and by your tone that it is a matter of course.' The accurate recording of the productive activities is the functional process; but not everyone in 'the office' in Figure 3.1 is engaged in making a record. Is it too fanciful to suppose that the others would be—with due ceremony—comparing their observations and discussing the indications of good quality and high yield? Perhaps they would even be bargaining among themselves, feigning interest, expressing more confidence than they felt and adopting the customary strategems to test how firm the price might be, and what values would ultimately attach themselves to the figures being recorded.

We shall now go on to show briefly how the critical distance between office and production and the nature of the integration of the two informational processes—the functional one of aggregating facts and the ceremonial one of negotiating values—will respond to changes in underlying technologies.

TWO OFFICE REVOLUTIONS

The office revolution we are living through now would have seemed less unprecedented had we been amongst those who endured the changes in the office that followed on the industrial revolution, because there were:

Two office revolutions
1.   1840–1900
      Reaction to the Industrial Revolution

2.   1960 +
      Reaction to the Post-Industrial Revolution

Up to nearly the middle of the nineteenth century the prevalent office mode was pen and ink and face-to-face. Face-to-face in two senses:

— Domestic. The office was situated within the family in the domestic buildings of the farm or the front room of the residence.
— Staff. Clerks were at least as skilled in the management style of their day (as legal clerks are nowadays) as their employers and they worked alongside them, either in adjacent offices, or facing them on the other side of what are now Dickensian antiques—partners' desks.

The first revolution

Two kinds of new technology disrupted the cosy office of the past. First the technology of production, then, in response to it, but lagging behind

(Ogburn, 1964), the technology of information. At the turn of the seventeenth century, the new production technology of the industrial revolution transferred the basis of production from field and cottage to towns and factories, sucking in raw materials and grinding out finished consumer and capital goods for home and abroad.

Having solved the major production problems of industrialization and international trade, further expansion and efficiency stood poised for the nineteenth century's own explosion in information technology. Figure 3.2 shows, in its first column, how quickly the ability to convey the materials of information to and fro increased. In the second column, the ability to communicate—creating and distributing information outward—is shown increasing equally rapidly. At the end of this explosion in technology an office in 1880 looked like the picture in Figure 3.3.

	Conveying	*Communicating*
1840s	Penny Post	*Daily News* founded
	Morse Code
Iron & steel bridges	
1850s	6,635 miles of
railways in UK	Advertising Tax repealed
		D. Telegraph founded
Civil Service Commission
established
1860s	London Metro	Paper Tax repealed
	Cable to USA	Press Association
founded
	15,310 miles of
railways in UK	
1870s	Duplex Telegraph	Remington launched the
first typewriter (and
	London Telephone Exchange	eighty years later the
first US computer,
	Parcel Post	UNIVAC, for the Census
Office).
	Pullman Carriages	

Figure 3.2: Growth in ability to convey and communicate
information

It will be noted in Figure 3.3 that—as is often the case in the use of new machine technology—only one of two typewriters is in operation (the other is in the extreme left foreground). The single roll-top desk is occupied by the solitary man. One has the impression that everyone is too pressured to systematize who needs what, when and how: to regulate the

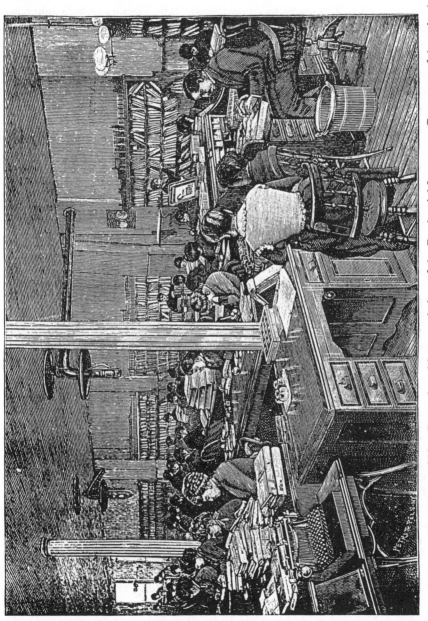

Figure 3.3: An office of the 1880s. (Reproduced by permission of the Prudential Insurance Company of America.)

traffic of files, which litter desks and line the walls; to decide whether working independently without aids or links leads to duplication and unchecked errors, despite surveillance from the front. In fact, the solution to pressure of work in the 1880s was to engage more staff. Starting from a low base, such an expansion—a doubling of office staff as a percentage of the total workforce by the 1890s—has not been seen in any other sector in the USA, unless in mobilization for war. While Figure 3.4 shows the investment in labour, to that must be added the investment in buildings. New staff were layered in towers and skyscrapers that were celebrated on letterheads, logos and labels.

	19th Century		20th Century	
	%	millions	%	millions
'60s	6	0.5	42.0	28.5
'70s	5	0.6	46.4	37.2
'80s	5	1.1	(46.6	44.6)
'90s	12	2.8	(46.8	44.8)

Figure 3.4: The percentage and number of office workers (USA)

The data in Figure 3.4 are from the USA Bureau of Labour Statistics (projections in brackets). The UK lacks equivalent data (Doswell, 1983, pp. 15–16); however, a British market research survey (IMR, 1982) found that only one in three of the British working population (7.5m out of 23.5m) work in offices (a smaller proportion than in the USA in the 1960s).

With the beginning of the twentieth century the office revolution paused to consolidate. Filing now tended to be separated-out from the office. Individuals or each work-group had their own desk, and offices shared a rudimentary organization: work was channelled from points of input through various hands to a checkout for quality and then output. The most dramatic innovation was the use of the telephone as a key means of input.

Use of the telephone for input reveals the discontinuity in processing between the electromagnetic system of the telephone and the social system of the office: between the functional and ceremonial processes. What seems at first a relatively simple substitution of media—a telephone for a personal visit or letter—carries implications both for practice and meaning. Firstly, a telephone receiver can pass only one call at a time, while several pieces of paper can be batched and rebatched and passed on simultaneously or consecutively as quickly as the staff feel is appropriate in the circumstances— the calls however must queue. The incoming call has to be identified (is it a wrong number, a hoax, a debtor?) and captured (logged and written down) before it becomes an order ready to be executed by those who could not overhear the call which has been processed for them. Further, the response to the telephone input can only be matched to the record of the incoming message, rather than to the message itself. As the call is processed it loses meaning. While the recorded or coded part of the message may contain the information needed to supply the order, background information (e.g. how the customer came to place the order, how satisfactorily previous orders had been completed) important for planning and development, has to be discarded. The time for personal contact during the processing both within the office and between individuals in the office and the customer has to be minimized (e.g. the link between information and cost is tighter than in the case of a visit or a letter; also the telephone must be free for incoming calls). The communication has become one-dimensional. The exploration of now unexpressed customer needs has to be hived off to marketing research where, perhaps even by telephone, enquiries now with an excess of ceremony are made (e.g. '... may we trouble you, if you would be kind enough to help ... how did you come to place an order? Could you please say how satisfied were you the last time?'...).

With greater task specialization two or more people split a job that previously had been one person's. Equipment manufacturers moved in to fill cracks opening up between tasks. Figure 3.5 from promotional literature (Osborn and Ramsey, 1918) shows first how the telephone can be used to do something beyond short-circuiting a sequentially ordered task, causing the stop–go queuing down the line as just described. The telephone here, in a small way, knits together media, time and space (voice contact with a note written previously; separate offices). Secondly, added office equipment creates external specialisms: the specialism of maintenance and servicing. For example the emphasis on the trivial sub-task of pencil sharpening implies adjustments and repairs to duplicators and typewriters which foreshadows visiting maintenance staff, engineers and consultants without which modern organizations grind to a halt. Next, office equipment is hinted to be not merely an instrument but a source of pride and, in the last item with an unconvincing reference to efficiency, a status symbol. Equipment serves the ceremony as well as the functioning of the office.

By the twenties, after sixty years of rapid development, the office had caught up with the factory as a centre of activity. As in the factory, the ideas of Frederick Taylor were applied. The National Association of Office Managers, enshrining his ideas of scientific management, was founded in 1919. By the twenties 'going to the office' had become the middle-class way of life.

While equipment manufacturers produced product improvements— posture-adjustable chairs, glarefree desktops, free-running filedrawers—and time-and-motion men produced data and formulae, what the first-generation office revolution accomplished was not merely an increase in individual output but a change in the character of that output. Replacing pen and ink with the information technology of type, file and phone allowed office work to be broken up into sets of routines for distribution among specialist staff separated by location and status. As office piled on office specialist managers had to be brought in to liaise and motivate office staff. The

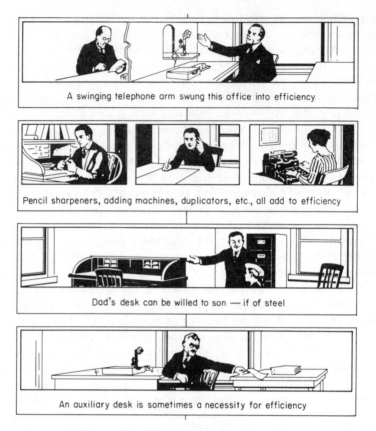

A swinging telephone arm swung this office into efficiency

Pencil sharpeners, adding machines, duplicators, etc., all add to efficiency

Dad's desk can be willed to son — if of steel

An auxiliary desk is sometimes a necessity for efficiency

Figure 3.5: Office equipment of the 1920s

distance from the factory floor, counter, checkout, or client—which originally allowed the office the necessary perspective—was now stretched by industrial and commercial demands and expanded by new procedures and equipment. The distancing of the office which once had provided it with a useful perspective now made it remote and unresponsive—unfinished business for the second office revolution to complete.

The second revolution

Let us for a moment go back to the earlier example of office telephone usage in the twenties. One could nowadays provide a terminal that allowed fuller details of the telephone order to be input without slowing down the order by providing the operator with a textpad or functionalized keyboard to output the order, sales entry and invoice. Simultaneously matching to lists of customer and product characteristics could take place (e.g. first, modified or repeat order; state of account, state of readiness and location of product ranges). Optional prompts to the operator to deal with the outcome of the matching could be provided (on the screen to avoid it interfering with the telephone conversation with the customer).

It is not so much that the input can be processed faster, but that the information can be conveyed, retained, analysed and distributed to all relevant staff at their location, in a format to suit both them and the nature of the total task. Functionally, the handling of the order is expanded into the handling of the product: the parts of the information are separated and brought together at the points where and when they are needed. Similarly, time is created for interpersonal or ceremonial procedures: staff can relate to each other so that they can develop a feeling of responsibility for what they are doing; self-management replaces unproductive supervision, exhortation or 'hustle'; managers are freed to attend to the boundaries of the system (e.g. how, at a justifiable cost, the number of incoming calls can be increased, how item availability can be extended, etc.). The boundaries of the office become less distinct, the technologies of automation, robotics and information begin to coalesce.

Today's practice falls a little short of all that: updating the matching list often fails to keep in step with requirements; instead of coming to feel more responsible, staff become apathetic and managers bored; customers feel uncertain about the information that is held about them — which often omits the one item they regard as vital. These may be transitional difficulties if we can get a better understanding of the nature of the office of the future.

The present

While the first office revolution was driven by a technology that concentrated production in factories and towns, the drive behind the second office revolution that is with us now, has been the technology of multinationalism. Multinational companies grew by absorbing the uncertainties of independent markets and reconstituting them as internally managed hierarchies, the 'Visible Hand' as Chandler (1977) puts it in contradistinction to Adam Smith's idea of the hand of individual self-interest invisibly steering the economy towards prosperity, published 201 years before (*The Wealth of Nations*, 1776). Similarly governments, through big-block organizations like the EEC and the World Bank, seek to restrain the market forces of price, supply, demand, employment, and inflation.

Multi- or supra-national organization of production technologies poses difficult management problems in measuring effectiveness, allocating resources and devolving the making of decisions to the appropriate level. While external competitive forces push organizations towards integration, internal accountability pulls management towards interior market systems (e.g. in private business, transfer-pricing and cost charge-outs for overheads; in public authorities, economic pricing, work schemes and income maintenance instead of subsidized welfare services and free goods).

Having come to depend on uniquely complex mechanisms to manage economic resources and relations, modern organizations are waiting for an information technology to provide the leap in administrative and decision-making capacity that will underwrite their survival.

Office dimension. Fewer markets mean more meetings. The Quickborner team of German office designers derived their ideas of office landscaping from analysing office information flows in the fifties; they doubled and trebled their meeting-space standards in the sixties and seventies. The parameters that until then had confined office design to tall towers, high ceilings and big windows—the controlling factors of light, noise and draught-free air—fell away with the technologies of air conditioning and illumination.

Similarly, telephone, copying and data-processing have become almost free goods in the office, so that private use or use for personal study or leisure (e.g. computer games) scarcely causes comment. Work spills over into home, travel and free time and vice versa. It is less necessary for office workers to be next to each other. Indeed it can be organizationally advantageous to be elsewhere: on the shopfloor, in the field, with the customer or client. Also the degrees of permanence of office components—the 'four s's' of shell, services, scenery (furnishings) and sets (movables) with their different depreciation or leasing rates (Eosys, 1982)—makes one want to consider exactly what are the essential physical accoutrements of an office. It is necessary only to have the means of fulfilling the necessary roles and integrating them. Figure 3.6 sketches the dimensions of the developing offices of the future, running between the extremes of the largely ceremonial and the mainly functional. As information technology has progressed, the office has been able increasingly to detach itself from production, and has developed according to its own laws. The location of offices is now to quite an extent a marketing operation, designed to impress customers and stock holders. Meanwhile functional office tasks—purchasing, sales, accounts, reports—are sketched out at home, perhaps using a datalink for at least some days a week. Where along the dimension developments will peak, overlap or collapse will depend on how prevailing information technologies will couple office functions to production and the required functional–ceremonial balance.

Production dimension. Production technologies vary along a dimension from prototype through batch and mass to process production (Woodward, 1965). In Figure 3.7 we have contrasted the prototype and process technologies (analogous to the contrast between a football match and a relay race). Where along the proto-process dimension a particular organization balances, is the outcome of the push-and-pull of external and internal forces and the drift-and-drag of conflicts both in the environment (e.g. within and between economic and regulatory forces) and inside the organization (e.g. within and between unions, sectional management interests and major shareholders).

In the abstract, the advantages of the two types of production technology

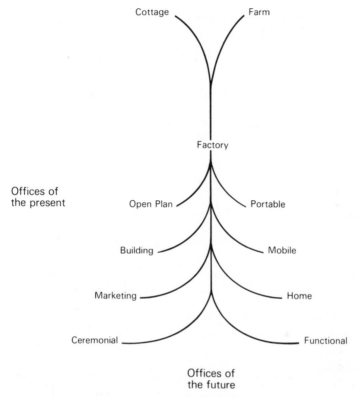

Figure 3.6: **Offices of the future**

are evenly matched. Any particular organization however is swayed by the push–pull drift–drag forces that always make some other position on the proto-process dimension more attractive than its current position. Organizations are set to scale all possible markets or in the case of non-profit services to cater to all-encompassing conditions of need. Production technology locks management into some niche along the prototype–process dimension. There day-to-day management restrains forward-looking management from venturing too far. Office technology however multiplies the ability of management to exercise a simultaneous presence over widely dispersed areas of responsibility and provides room for intervention and manoeuvre. Prototype technology through improved planning can move closer to giving suppliers advance warning of orders if not actually long term contracts of the kind that process technology can provide. Process technology can begin to tolerate greater input variation by suppliers and move controls towards later office-managed stages of the production process, as detection and

corrective feedback become more reliable. By more precise product specifications in prototype production, some single purpose components can
be modified for wider markets, and in process production extra finishes can
make some products answer certain purposes more exactly at a higher profit
margin. Figure 3.7 gives an overview of the impact of office technology on
production technology towards the two ends of the prototype–process dimension.

Prototype Technology (PTT)   **Process Technology (PT)**

One-off, customised product or service, each
created to meet the unique needs of a particular
market situation or configuration of known
characteristics.

Standardised flow of a uniform product or service
used in a wide range of unspecifiable situations.

Examples:

Examples:

New towns, Concorde, remedial treatments, policing,
farming, book publishing, brand advertising.

Cars, refining, brewing, food processing, magazine
publishing, classified advertising.

Because PTT tends towards the one-off, it provides
the opportunity to shop around for its supply. PTT is
supplier independent. **Office technology** creates the
opportunity for management to monitor and bank
the performance of competitive suppliers and
ancillary services, (e.g. time and cost sharing
schemes) temporary surpluses and slack and to build
buffers and margins into schedules to compensate
for supplier shortages and shortcomings and to
minimise stocks and facilitate forward ordering.

Because PT tends towards continuous production, it
demands long-term continuity in all the
characteristics of supply: PT is supplier dependent.
Office technology offers the opportunity for
management to reduce dependence by allowing
backward integration into the supplier domain (e.g.
pea harvesting by frozen food processors). Also, in
continuous processing deviations from standards are
often detected only at or near output or changes in
state, when it is too late to rescue the products of
that cycle. Off-line detection and control devices
more evenly distributed along the production line
would allow management greater discretion over
when to intervene and over how much weight to
attach to set-up and shut-down orders, etc. More
generally, the extra management capability can
stabilise the fluctuations arising from the interaction
of independently controlled, but dependent
subsystems (Singh, 1981).

PTT yields products and services that tend to serve
unique situations and that therefore (while adding
some monopoly value) are locked into markets too
narrow to generate the benefits of scale and
expertise. **Office technology** can break out of these
limitations by giving management the extra
capability to distribute work where there is capacity
and to search for and identify targets suitable for the
deployment of accumulated expertise and generate
some benefits of scale (e.g. in policing, staff are not
evenly deployed, but targeted on or shunted
between accident blackspots and high incident
areas, as demand-monitoring determines).

PT yields products and services that are relatively
standardised and therefore vulnerable to duplication
and price erosion. **Office technology** can help by
creating extra management capability for collating
specialised production lines to cater more completely
to client needs ("large companies will become like
federations of small enterprises"; Cadbury, 1981);
and by forward integration (e.g. salesmen
determining finishes packaging and delivery dates
instead of being bound by production devised
schedules; claimants preparing their own case at a
local terminal plugged into the Benefit Board's files
of regulations; Lynes, 1981).

Figure 3.7: The impact of office technology on the management of the two
production technologies

Office and production dimensions

Physically an office or member of staff may be distant from production
and service processes and from offices associated with an organization (e.g.
suppliers or clients). Yet via a local network, public transmission lines or

satellite, a manager in a distant office may be in more intimate contact with what is going on in his or her organization than when pacing the factory, shop or office floor. Wherein does distance reside? If one receives the right data at the right time about the right events and then fits these together properly, one can reconstitute a better picture of what is going on from a distance than from any single position within the organization. Judgements and values, however, still have to be applied to this picture. This can be problematic. What is right and proper resides in the interactions between people (whether they are face-to-face or miles apart).

FUNCTIONAL	CEREMONIAL
Conveys text and graphics in a range of formats.	Conveys confidence that text and graphics have been compiled by those most intimately connected with obtaining the data in collaboration with those who intend to use it.
Information immediately available or as part of a carefully worked out but rigid schedule.	Time-base and time-scale adapted to what the units of analysis and the decision-makers demand.
Adjustments and corrections for seasonal and other temporary or secular factors made automatically.	Whether to present raw and/or adjusted data and what adjustments to make arises from day-to-day deliberations about the nature of the data.
Adjusted and raw data presented simultaneously.	Opportunity to discuss reasons for adjustments between those who obtained and those who need the data.
Opportunity to interrogate and interact with files and output.	Why some kind of information is available and not another kind and whether the non-available data would be the more decisive is probed.
Format is easy to assimilate and may be based on decision rules that lead to unambiguous recommendations.	Information is presented with qualifications and alternative interpretations that make it difficult to assimilate and to act on because of implied knock-on, backwash- and side-effects which present **both** break-through and missed opportunities.
The sources and destinations of data are widely distributed.	Sources and destinations can be identified with known individuals who can illuminate the data. Text and graphics can be directed more precisely at specific audiences, avoiding eavesdroppers.
Information in identical formats can be sent quickly to many people to await their attention.	Information can be released slowly and changed on an individual basis to take account of the response obtained at every step (as in a conversation).
Information is held in files available on specification, which can then be searched merged, etc.	Information is held by people who are known specialists in certain information areas and who can refer to relevant, perhaps essential, information additional to what was specified.

Figure 3.8: Functional and ceremonial aspects of office information

In this social domain of the office the classical theory of information (Shannon and Weaver, 1949) is of limited use. It emphasizes transmission— the flow of information in a channel—which yields a content-neutral or sewage theory of communication: garbage in, garbage out. What it is that makes input and output worthwhile and significant is not part of the theory. In the nomenclature of this chapter, the new technology can be said to have been better at catering to the functional needs of the office than the ceremonial ones. During the expansion phase of information technology not standing on ceremony may be felt to be one of its advantages, if one also feels that ceremony is merely procrastination and procedural formality. These feelings may not however survive the current phase of initial expansion. Procrastination may even now be overtaken by information overload and procedural formalities replaced by rigid data-gathering and data-distribution.

Figure 3.8 contrasts the functional and ceremonial aspects of office information currently prevailing.

The mechanism underlying the ceremonial column in Figure 3.8 is articulated diagramatically in Figure 3.9 (based on Gratz and Salem, 1981).

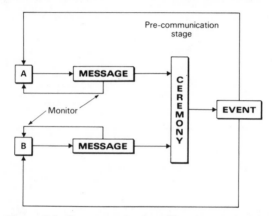

Figure 3.9: Ceremony in the office, based on Gratz and Salem (1981)

A and B are persons, offices or departments in communication with each other. What they are communicating about is an EVENT (e.g. anything from an occurrence to a sales statistic). The event itself is, potentially, infinitely complex: the simplest sales figure is at least a composite of many customer needs, credit facilities, pricing policies, promotional strategies, competitive activities, local, seasonal and long-term factors—and of how hard the salesperson tried and what made him or her try so hard (see Figure 3.10). For the message to be understood, A and B need some kind of shared understanding about their position *vis-à-vis* the event prior to communicating about the event itself. A stage for laying down some agreed assumptions about the nature, level and direction of the interaction is needed, of the kind usually signalled by gesture, stance, gaze, intonation and style of language and medium (e.g. memo, telephone, visit, etc.), before tackling aspects of the event itself. Otherwise initial misunderstandings about the nature of the communication will cloud the issues and delay or even prevent the ultimate understanding of the event. The speed, precision and attractiveness of the format may be less of an aid in communicating the essential message of a sales figure, than chatting to the sales manager as he or she takes one through the figures. A pre-communications stage is part of the ceremony of the office. That stage might have been laid down years ago, but without such foundations offices would have a shaky future. Figure 3.9 draws a distinction between the message sent and the feedback to the sender or sending group: within the boxes A and B a Russian doll arrangement must be imagined, because the A-B relationship is repeated for each relationship within the group. In other words, the message that is sent is the product of the relationships that exist within each box. Even in the limiting case where the box is occupied by one individual, such a person would be apt to say to him or herself '... shall I provide/ask for that information now or later; how much detail shall I provide; shall I specify its limitations or its generality; what interests will be served—my own (and which of them), my department's, some other perhaps antagonistic interest, short-run or long-term interests ...' and so on. It is not only written or speech communications that may be in continuous redraft, apparently in a drive for clarity. Often what

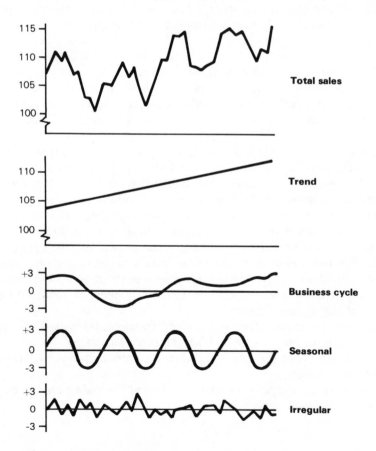

Figure 3.10: Decomposition of sales statistics into components (from House, 1981). (Reproduced by permission of Auerbach.)

is sought is a single message stuffed with multiple intentions or with the intention of pre-empting multiple interpretations.

CONCLUSIONS

'The office' is not a recent phenomenon that has suddenly emerged out of nowhere to make demands for new technology. It has a long history, dating back at least to the time of the Pharoahs, and to understand what it is that current developments in office systems need to address it is helpful to take cognizance of the historical context. Neither is it the case that it is only very recently that management has considered the potential for electronic systems. The new technology has been under the scrutiny of management literature for a quarter of a century. The recent upsurge of interest in 'office systems' needs to be understood against this backcloth.

From an historical perspective, the office can be seen as separate from production activities *per se*. Its role has emerged from a need to aggregate representations of production activities, and to interpret and negotiate their value.

Electronic office systems, then, need to support the processes of representation, interpretation, and negotiation. Two rather different processes need to be supported in fulfilling this role, and these are: on the one hand, the rational or functional; and on the other hand, the expressive or ceremonial. Electronic systems will fail to the extent that they disregard either one of these.

In performing its role within organizations, the office has gone through one 'revolution', as a response to the Industrial Revolution, and is going through a second 'revolution' now—one that started in the 1960s and can be seen as a response to the Post-Industrial Revolution.

Out of the second, current revolution, are emerging not a single 'office of the future' but many different types of 'offices of the future', running between the two extremes of the largely ceremonial and the mainly functional. The new, 'electronic' offices facilitate a better relationship between 'office' activities and 'production' activities—one that is less dependent on physical proximity. The new kind of relationship depends upon and reflects the possibility of better communications.

The new, electronic office systems will be successful only to the extent that they can handle adequately both the ceremonial as well as the purely functional aspects of the office to improve communications by facilitating shared understandings.

In the next few chapters we consider some of the main elements in these emerging 'electronic offices'.

Part Two
Product Trends

Chapter 4

Overview

BRUCE CHRISTIE

INTRODUCTION

'My God, it talks!' exclaimed the Emperor of Brazil—not in wonderment at a new supercomputer but, so we are told (Young, 1983, p. 2), in amazement at Alexander Graham Bell's demonstration of his primitive voice telephone at the Philadelphia Centenniel Exposition in 1876.

By the close of 1983, a wide range of office products in addition to the conventional voice telephone were widely available to meet two broad categories of user needs, for:

— Type A communication — person to person interactive communication in real time
— Type B communication — person to 'paper'/'paper' to person.

There was also a growing awareness of the possibility of a third kind of communication:

— Type C communication — person to 'intelligent' machine, including interaction with electronic 'knowledge bases'

but there was next to nothing available by way of Type C products to serve the general office worker. The personal computer, however, had arrived and provided a few minimal hints of what was to come; for the first time to any significant extent the user had local machine 'intelligence' available—of a

75

very primitive form—and this could be used in various ways, especially: to tailor the machine's communications to the needs of the individual user (e.g. converting tables of numbers quickly and easily into bar charts, or graphs of various sorts), and to provide primitive 'decision support' (e.g. 'what is' simulations). These primitive personal computers were a long way from the knowledge-based machines that were promised for the future, but they were a definite step on from the passive medium of paper which took no active part in the communications process but acted merely as an unintelligent vehicle.

TYPE A—PERSON TO PERSON—COMMUNICATION

Type A communication is interactive communication between people in real time, with no intervening functional store. If a tape recording or video record is made of a conversation, that record is an instance of Type B communication (conducted in parallel with the Type A communication). The difference between the two is clear if we consider a hypothetical example. During the conversation itself the people directly involved in the conversation can interrupt one another to ask questions or make requests—they can respond to one another. If one person asks another for a further explanation of some point or other, that person can respond to the request in some way. That is Type A communication. If a tape recording of the conversation is later played back, the relationship between the tape and the audience is different. The tape forms a functional store that intervenes between those involved in the recorded conversation and those listening to it. The possibility of real time interaction between the 'people' (the recordings) on the tape and the people in the audience does not exist. That is Type B communication.

There are a number of psychologically important differences between the two. For example:

— as indicated, a close 'meshing in' in real time of the behaviour of the various parties is characteristic of Type A communication but not of Type B;
— The only record of pure Type A communication is, by definition, in the biological memories of the humans involved and is subject to all the characteristics of human memory (e.g. selective forgetting, unconscious fabrication, qualitative and quantitative distortions), whereas, by definition, an external, literal record exists of Type B communication but this record contains only the objective, observable aspects of the communications and not the subjective, internal personal context that is available in human memory of a Type A episode;
— the people involved in Type A communication usually know at least in general terms who is involved in the communication episode (although

one can never be sure there are no 'eavesdroppers', e.g. on an extension line), but in Type B communication one can never be sure who will access the record, when (maybe twenty years later) or for what purpose;
— exchange of information and problem solving contingent upon that can be much faster with Type A communication than with Type B, and the style of the two types of communication differs in various ways (e.g. Type B communication tends to include many more complete, grammatically correct sentences than does Type A) (e.g. Chapanis *et al.*, 1972).

The distinction between the two can be made clearer if we consider the more subtle example of a radio 'phone in' programme. The communication between the radio station host and the person 'phoning in is Type A. Each can interrupt the other, and if the person 'phoning in says something obscene or which in some other way is embarrassing to the radio station it is received by the station and heard by the host who can respond to it. In constrast, the communication with the listener to the programme is Type B. The listener cannot interact in real time with the host (not without changing from being a listener to being a person 'phoning in), and the listener will not normally hear any embarrassing material—it will be edited out and the listener will hear some music or other 'filler'. The listener is not in fact 'eavesdropping' in Type A mode (although it usually seems that way), but is actually listening to a censored recording broadcast about five or ten seconds after it is made. In principle, by varying the delay, editing out of non-broadcastable material could be done invisibly so that listeners would not even be aware it had happened whilst probably perceiving it to be a complete view of a conversation taking place in real time, but this would create logistical problems in managing the broadcast which are not normally considered to be warranted. Such 'invisible editing' does, however, form a normal part of 'computer conferencing' considered later in this chapter.

So although Type A and Type B communication are often closely associated, they can be distinguished.

Type A communication is the earliest form of human communication, both phylogenetically (occurring before cave paintings, probably) and ontogenetically (the newborn generally communicating with its parents in real time rather than writing them notes). It is still the preferred type of communication for most adults in most situations, including in the office environment.

Type B communication really came into its own with the invention by Gutenberg of the moveable type printing press in 1455, and has been the main reason for the explosive growth of scientific and technical achievement. It is currently taking a further major step forward with the development of modern 'information technology'.

Probably the most significant step forward in terms of Type A communication occurred with the invention of the voice telephone.

The traditional voice telephone

England had its first public telephone exchange by the close of 1879, installed by the Telephone Company at 36 Coleman Street in the City of London. This signalled the dawn of The Age Of The Telephone.

The voice telephone network has by now spread right around our planet, enabling someone in London to hold a conversation with someone in Sydney, or someone in the wilds of Canada to talk with someone in rural India—someone in the United States to talk with someone in Moscow. The telephone system has also evolved, not just grown, over the last one hundred years. Three developments are especially noteworthy from the Systems Psychology point of view, as follows.

The video telephone

The addition of a visual channel to the voice telephone to give the people concerned an option of seeing one another would seem to be a natural extension of the voice telephone. It is not especially futuristic. The American 'Picturephone' was first demonstrated publicly over twenty years ago, and during the early 1970s market forecasts were projecting that by now the voice telephone would have virtually disappeared in favour of video communication. That has not happened, partly for technical and economic reasons, and partly for psychological reasons—although it is a 'good idea', it is not such a good idea from the user's point of view as to have warranted the kind of investment in new network equipment that would have been necessary to have made it a reality by now. The psychological benefits did not outweigh the economic costs, a point discussed in Chapter 5.

The situation is changing. Increasing interest in local area networks is a particularly important stimulus, because the equipment necessary for video telephony on the network comes with the network, at least for some 'broadband' networks. In these cases, video telephony is not the prime reason for having the equipment but is a useful additional benefit. Local area networks provide a means for efficient and flexible communications within a limited area, often a single building, and are considered again later in this chapter. Cable systems provide another opportunity for video telephony on a limited scale, and optical transmission systems promise more for the future. The key technical problem in all these cases is the need to be able to switch video signals with the same flexibility and efficiency as voice signals. Some attempts—for example, at the Massachusetts Institute of Technology (MIT)—have attempted to get around this problem by storing most of the visual information locally and only transmitting information about changes (e.g. about lip movements). This requires less bandwidth than full video and enables the transmission to be done over ordinary voice telephone lines; but although of some interest such attempts have not so far been very successful from the psychological point of view. Another approach is to send all the visual information that is needed, and to do so over the voice telephone system, but this means sending it relatively slowly (slow scan video) such that only still pictures can be transmitted (about one every minute or so at most), and so the visual aspect of the conversation is effectively non-interactive.

Teleconferencing

Interest in 'electronic meetings' or teleconferencing (both with and without a video channel) was high during the early and mid-1970s, starting in Britain and rapidly spreading to many other countries. The first paid call on the American Telephone and Telegraph Company's (AT&T's) 'Inter-city Visual Conferencing Service' took place on 25th July 1974.

Teleconferencing systems were typically studio-based, linking pairs or triads of studios seating up to five or six people in each, allowing meetings to be held between people in different cities using audio or audio and video channels. The limited number of studios used meant switching was not a problem.

Research, reviewed in Chapter 5, showed that the benefits to the user of teleconferencing were not as great as many had expected, and in particular that the marginal benefit of a video over an audio-only system was relatively modest. Teleconferencing never really took off and many of the more successful networks that were set up were audio-only. Possible psychological reasons for this are discussed in Chapter 5.

Despite this history, there is some evidence of a resurgence of interest in teleconferencing. Atlantic Richfield in the USA is a key example of a company which has recently invested significant resources in developing a nationwide network of video conferencing studios, and other companies have also shown interest in the concept. Local area networks may prove to be a further stimulus for video conferencing, especially where such networks may be linked by satellite communications (providing broadband channels over long distances).

Enhanced voice telephony

By far the most successful evolution of the voice telephone in terms of levels of penetration and use is what has become known as 'enhanced voice telephony' or simply 'enhanced telephone'.

This is the addition of a range of special services to the system which the user can access directly. These are possible because of the development of electronic exchanges and pushbutton dialling. The user can use the pushbuttons to send special codes to the electronic exchange which can then set up the service required. The services cover such things as:

— being able to have calls diverted to another extension
— having the telephone automatically keep trying a busy number until it is free and then ringing to let you know it has got through
— being able to use short codes to dial otherwise long numbers.

Some modern telephone systems are quite complex in terms of what they provide for the user (in some cases offering hundreds of 'features' to the user) and there is a significant task in training and guiding users in the use of the systems, and a significant challenge in devising new approaches to facilitating the interaction between the user and the system.

The voice telephone is no longer a simple communications device, even from the user's point of view, but a sophisticated system which forms an integral part of today's information technology.

TYPE B—PERSON TO 'PAPER' TO PERSON—COMMUNICATION

Since the late 1970s, there has been increasing evidence of a shift in the pattern of development of communications and information systems, and the emergence of a new family of office products—concerned with 'papers' and their electronic equivalents. The emergence of word processing as a major product area is a key example. Growing at about 35 per cent annually, shipment of word processors has been growing at more than four times the rate of growth of public network equipment (PACTEL, 1981).

Type B systems address the following functions in various mixes, according to the type of product:

— creation of 'papers' or similar items
— storage and retrieval
— accessing of public stores of information
— distribution/communication of 'papers' or similar items.

Emphasis on creation: word processors

Word processors are rapidly becoming a familiar feature of the modern office. They are effectively 'computerized typewriters', emphasizing 'power typing'—the ability to add, delete, change, move and reformat text quickly and easily. Although the earliest and most primitive 'word processors' had only paper output or perhaps a single line electronic display ('visual display' or 'screen'), a word processor is now normally expected to have a screen, often A4 in size and shape. Word processors come as:

— complete systems which share intelligence, storage and/or printers
— smaller but powerful, dedicated stand-alone units
— software packages that run on other machines, such as personal computers.

At the close of 1983, a good, dedicated stand-alone word processor could be purchased in the UK for around £6,000, and price relative to performance is falling. The world market for word processors is expected to be in excess of £4 billion per year by 1985.

Word processors in the past have emphasized the processing of text (hence the term 'word' or 'text' processor), but increasingly they are being given the capability of creating tables of numbers, graphics, and other forms of information as well as text. Some incorporate mathematical and other packages to assist in the preparation of reports and other items.

Emphasis on storage and retrieval

Word processors have to store information in order to carry out their 'creation' function effectively—especially in order to merge text and graphics from different sources to create new pages. Some incorporate a fairly sophisticated storage and retrieval capability that can be useful, for example, in relation to the mailing of papers to people selected from a mailing list according to specified criteria (e.g. income, age, location).

Other types of systems provide a much more powerful storage and retrieval capability.

Personal computers. Relatively inexpensive software packages can be bought for most personal computers, often based on a dialogue that follows the same general logic associated with using a card index system.

Microform systems. The use of hard disks ('Winchester disks') helps to expand the somewhat limited memory of a personal computer; or the computer can be used as an 'intelligent terminal' to a larger computer system. An alternative approach to storing large amounts of information is the use of microfiche, microfilm or similar media. This method makes it as easy to store graphics as numbers or text. The simplest systems are small stand-alone readers. More elaborate systems incorporate microprocessor technology to provide intelligence, enabling particular items (images) to be searched for under computer control. Some of the larger of such systems provide a means of finding an item in a database of up to four million or more pages in a matter of seconds. Items can be entered into the database using a camera, but it is also possible to create items by computer ('computer output microform' or 'COM'). Some systems allow the photographic image to be updated.

Videodiscs and optical discs. These can contain tens of thousands of pages of information (such as a 'film') that may or may not appear in 'page' form to the user. In the case of some 'films', the user can use 'slow motion', 'reverse', 'fast forward' and similar facilities; in accessing pages, the user can branch through the database in different ways (as in a programmed learning text or Prestel tree-structure) as well as accessing individual pages as required.

The question of what constitutes an appropriate dialogue for the interaction between a user and a personal information storage and retrieval system is addressed in Chapter 6.

Emphasis on accessing public or other shared databases

The kinds of storage and retrieval systems considered above are concerned with relatively limited, relatively well-defined databases mostly accessed by the people (or at least the organization) that put the items into the database. People also need to access public databases that are potentially very large indeed, and ill-defined.

Specialized databases. Online databases have been available for many years, traditionally accessed by teletype. They have been growing rapidly in number and the range of information covered, especially in the scientific and technical (STI) area. Two well-known and widely-used examples are Psychological Abstracts and Chemical Abstracts, available online through such systems as Lockheed Dialogue. Other sorts of information are also available— for example, in agriculture, energy saving and the legal area. The computer search techniques typically used are powerful but are not easy to use and they require specialized knowledge of the databases. Various networks of computers providing access to STI databases have been set up, the key example in Europe being the International Euronet Diane system set up by the Commission of the European Communities to provide access to a wide variety of online databases throughout the Member States.

Videotex. Videotex is a family of systems which allow users to use a simple numeric keypad or alphanumeric keyboard to send instructions over ordinary or high-speed telephone lines to a computer which then searches through a database to find the information required. The very first videotex system was viewdata, a British invention, and Prestel—British Telecom's public service based on viewdata—is currently the most successful videotex service.

The videotex user normally (but not necessarily) sends the instructions to the computer by selecting choices from a menu. The information in the computer is normally (but not necessarily) held in the form of 'pages' (screens of text and graphics), and is sent to the user over the same telephone lines for display on a television screen.

The level of detail possible on the screen varies with the particular system. At the moment, the European systems (Viewdata, Antiope, Bildschirmtext, and their derivatives) give the coarsest picture, and the Canadian, American and Japanese systems give the finest. 'Picture Prestel'—which has been demonstrated already as an experimental system—will provide for much more detailed pictures than provided by any of the current systems. (Picture Prestel will require significant local memory in the user's set, and high-speed lines if the build-up of the picture is to be rapid.)

Videotex was originally aimed at the home, capitalizing on the fact that it makes use of two very widely available items: the telephone and the television. The idea was to offer a national information service (Prestel in the UK) based on the public telephone system. For a variety of reasons, many related to the systems psychology involved, videotex has not yet taken off in this market and attention has now shifted towards the use of videotex in the office. With this new emphasis has come an increased interest in private systems, including the use of high-speed lines. Various videotex bureaux are also in operation, for example operated by GEC and SDL.

The main types of videotex systems include: viewdata (the British system, which forms the basis for the Prestel and Prestel International services operated by British Telecom); Bildschirmtext (FR Germany); Antiope (France); Telidon (Canada), and CAPTAIN (Japan). These all operate according to different standards, but the Commission of the European Communities is attempting to set up various initiatives to encourage interworking between the three main European standards, with the aim of stimulating the European videotex industry, and has defined a 'European Standard' as one element in this strategy.

Videotex systems do more than just storage and retrieval of information. By allowing users to send information to the computer in the form of 'response pages', they can provide for order-entry capability. Banking transactions are also possible, pioneered on the Bildschirmtext system in FR Germany, and a variety of other facilities are also available, especially the down-loading of computer programs ('telesoftware') from the main computers to local intelligence at the user's terminal. Videotex therefore represents one early approach to the functional integration of different services provided to the user, a topic addressed later in this chapter.

Shared databases within the organization. Accessing public databases is a special case of the more general problem of accessing shared information (e.g. archives within an organization). Some systems, e.g. videotex, can be used both for public databases and for shared information within an organization. The psychological aspects of shared information systems are considered in Chapter 7.

Emphasis on transmission

Copying and printing. The traditional means of distributing papers has relied on copying, printing, and the mail. Convenience photocopiers are now a familiar part of most offices. Intelligent copiers are now becoming available which add intelligence and communications capability. For example, they can send papers between one another over telecommunications lines, can be programmed remotely to produce multiple copies of complex documents, can collate documents in accordance with complex requirements (e.g. mixing different colours of papers, and mixing single-sided with double-sided copying according to a predetermined plan), and in other ways assist in the copying and transmission of documents.

Facsimile. Facsimile machines have been around much longer than intelligent copiers. One of their first applications was to the transmission of weather pictures. Papers are scanned electronically, and the electronic codes representing dark and light parts of the page are transmitted over telephone lines.

The costs of facsimile machines have fallen considerably, and their performance has improved. Modern machines can transmit an A4 page in around one minute, and produce a high quality copy. Because these machines are digital (unlike the earlier, analogue machines) they can be connected to other digital systems.

Teletex. Teletex is a recent upgrade of a much older system, telex. Telex is an international system available in around 200 countries, and with over a million subscribers. It offers a fast method for sending text from user to user, and is widely used. In the UK alone, the 85,000 subscribers made 91 million separate calls during 1980 within the country, and another 77 million internationally.

Telex uses telegraph circuits, but teletex uses the telephone network. It will eventually provide a means of sending information around the world on publicly available communication links. A further advantage is the wider set of characters that can be used. Using telex, one can only send messages in upper case. Teletex uses both upper and lower case characters, and provides a means for sending high quality letters or other pages of text.

Electronic mail. Facsimile is sometimes referred to as 'electronic mail', or forming a basis for it, but the term 'electronic mail' is normally reserved for the communication of information which has been coded by keyboarding, optical character recognition (OCR), or other means. Communicating word processors could form the basis for a simple 'electronic mail' system, but again the term is normally reserved for systems which incorporate some intelligence so that items can be sent from user to user according to priorities, 'put into' electronic 'in-trays', and so forth.

Speech store and forward. Products capable of storing and transmitting messages in speech form are now becoming available. Extra facilities and special applications are expected to follow. A speech store and forward system has been in service in San Francisco for some time, providing an 'intelligent' answering service. When someone calls to speak to a person who is out of the office, the call is redirected to a receptionist. The receptionist takes the call, and is provided with prompts on a screen. Voice messages can be stored and forwarded later on to the intended recipient. Callers using the answering service usually think they are talking to the secretary of the person they intended to contact.

Voice annotation. Some office communication systems now provide a capability for adding voice messages to text. In this way, voice messages can be added to a memo, for example. The visual information can be scanned by the user quickly to get the main points, and the user can press the 'voice' button to hear more about points of particular interest. Or a manager can annotate a letter or report that has been prepared by a secretary in order to indicate corrections or changes. These systems can also be used for dictation.

Computer conferencing. Electronic mail systems can be extended to form the basis for computer conferencing networks. These networks link users together in 'synchronous' or 'asynchronous' 'conferences' or 'meetings'. In a synchronous computer conference all the users are online at the same time, exchanging typed messages analogous to the exchange of spoken utterances during a face to face meeting. In an asynchronous conference users come online as and when convenient, and such conferences may extend over periods of months. In both cases, the computer can prepare a complete transcript of the conference. Other features are available as well—for example the capability for the computer to collect and report on the results of a vote or answers to a questionnaire. Private messages can be exchanged between users involved in the conference, and these messages are not included in the public transcript. Computer conferences— especially synchronous conferences—involve a degree of real time interaction between users which approaches the kinds of exchanges characteristic of face to face meetings and other Type A communication systems, but they also involve a functionally significant store (characteristic of Type B systems); they therefore form a bridge between these two types of systems.

Local area networks. Interest has been rising in recent years in the concept of local area networks—cable systems designed to connect various kinds of electronic office products together within a building or close group of buildings. The products can include word processors, intelligent copiers, computers, facsimile machines, and almost any other type of electronic office product. Different networks can be connected together to extend the communications capability still further.

Crossing the organizational boundary

A significant obstacle to the even more rapid uptake of Type B office products is essentially the same as one of the most significant factors limiting the growth of the video telephone in the 1970s. It is no use having a video telephone if the people with whom you wish to communicate do not have one. It is no use having an electronic mail system, or other communications-oriented Type B product, if the people with whom you wish to communicate do not have a compatible system. It is a 'chicken and egg' problem. No-one (to caricature the situation) wants to install such systems on a significant scale until they are in widespread use in offices generally, but they will not be in widespread use until many individual organizations start installing them. This is one area, amongst others, where Government could in principle play an important stimulating role through the provision of tax incentives, grants, or other incentives to organizations taking information technology on board in a serious way.

The problem facing an individual organization in this respect at the close of 1983 was three-fold:

— Both technical and human factors aspects of integrating different systems within an organization were only just beginning to be solved—it was often not straightforward, for example, to send items created on a word processing system through the organization's electronic mail system.
— Communication in the electronic medium with other organizations was often difficult or just not practicable even if they used electronic systems extensively themselves—because the systems were not capable of interworking.
— Communication in the electronic medium was not possible at all with most organizations, because most organizations used paper as their principal or sole information medium and there was no very good way of converting information on paper quickly, reliably and completely (e.g. including letterheads) into an electronic form that could be processed by an organization's information systems in the same way as if it had been created by a word processor.

We can expect to see significant strides forward in regard to all of these during the mid- and late-1980s. And even without such progress, the three factors combined still do not add up to a sound case for not introducing information technology. There is evidence (e.g. Engel *et al.*, 1979) that the large majority of information items to be found in an organization were created within the organization and never leave the organization—most items in fact never cross the organization boundary. If such items are necessary to such an extent, then at least the creation, transmission, and use of them can be made more efficient through the use of modern office technology.

TYPE C—PERSON TO 'INTELLIGENT' MACHINE–COMMUNICATION

One of the most exciting but also one of the newest elements in people's thinking that was beginning to emerge by the close of 1983 was the possibility of applying artificial intelligence to office systems, especially in the form of 'intelligent knowledge based systems' (IKBS).

Some hints as to what the future might hold—very minimal hints—were given by the rapid acceptance of personal computing. For the first time to any significant extent users had the power of a full computer at their fingertips. The cost of computer time was, for the first time, zero. Users could and often would spend hours at the computer, trying things out—and there would be no bill to pay at the end. What did they learn?

— Interacting with a computer need not be like interacting in batch mode with a mainframe computer from the 1960s. The Apple Lisa, for example, went to great lengths to 'look familiar', presenting a 'desktop', 'filing cabinet', and so forth—as discussed in a later chapter.
— Interacting with a computer does not necessarily require any significant knowledge of how computers work, or even of programming languages.
— It does not necessarily require much by way of keyboarding skills. (The mouse and single word recognition devices were available by the end of 1983.)
— The output from a computer does not have to be in a standard form but can be tailored to the needs and preferences of the individual user. (For example the Apple Lisa provided a means of presenting the same data to the user in various different ways—e.g. table of numbers, bar chart, graphs of various sorts.)
— Computers need not function in a simple, predictable fashion but can take information from different sources into account in deciding what to do (e.g. to which part of the program to branch) based on an 'if ... then' (if certain conditions are satisfied then do such and such) type of conditional logic.
— Computers are not infallible. They do give incorrect answers from time to time. This is (not often) due to malfunctions, which cause errors of the type $2 + 2 = 5$, and often due to inappropriate logic being applied or inaccurate or insufficient data being supplied.
— Computers are relatively unintelligent compared with humans. They are good at doing routine operations with a negligible risk of error, but in other ways they are 'stupid'. They are therefore to be regarded as useful tools, not oracles.

These learnings can be assumed to have led users to a different view of computer technology—a view that might be preparatory to the introduction of more 'intelligent' computers.

At the close of 1983, there seemed to be the promise—though not certainty—of a new direction in computing, based on the use of 'artificial intelligence' and 'knowledge bases'. The programs involved were thought by many to mimic the processes of human cognition to a significant degree, breaking away from traditional approaches to programming. Simplistic as this view was in psychological terms, it served as a significant motivating force for the development of better computers—more useful, more 'intelligent' tools.

Despite the promise, next to nothing actually existed by the close of 1983 that was of any significant practical value to the general office worker. Some interesting experimental systems (e.g. MYCIN, EMYCIN) could be demonstrated and a few commercially valuable systems existed in very specialized areas (e.g. the configuration of computer/telecommunications systems), but there was nothing of any real value to the general office worker. Just the promise.

Chapter 8 takes up this theme, specifically in regard to decision support.

INTEGRATION AT THE USER INTERFACE

One thing was becoming very clear by the end of 1983: The number of services being offered to the user was increasing markedly. The voice telephone itself was becoming a highly sophisticated system, with perhaps several hundred features available to the user. In addition, there was the whole range of Type B products—word processing, electronic mail, etc.—and the first hints of a new category of services based on artificial intelligence and intelligent knowledge based systems. How could the user be expected to cope with such a complex array of services *and* do his or her normal work as well?

The general consensus was that there was a need for integration. There were two equally important aspects to this:

— technical integration, so that various services could intercommunicate in a physical and informational sense
— human factors integration, so the user would see one system rather than a multitude of different systems (even if they could intercommunicate physically).

Further than this, a consensus was developing concerning the broad framework that should be adopted—in terms of the user's perception of the overall system. Diagrammed in terms of the essential elements, this is shown in Figure 4.1.

The emerging consensus, to which not all office systems conformed, was that the user interface should be hierarchical.

At the highest level, the user would switch on (log on to) the system to find a user entrance to the whole system. It would not be necessary explicitly to log on to the separate services. This topmost level would be called the 'desktop'.

The Desktop would be a visual representation of the conventional desktop. However, the most important functional aspect of it would be that it would be a list—a menu—of:

— services (e.g. word processing, electronic mail)
— items of information (e.g. memo, letter, report),

The user would enter the overall system by selecting a service or item of information. These would be indicated by a verbal label and an 'icon' or pictogram, and selection would normally be done by positioning a cursor using a mouse or a keyboard. In principle, the service would cover Type A, B and C services although in practice in 1983 they were usually (but not always) confined to Type B.

Having selected a service or item of information, the user would be led to the next level of the user interface hierarchy. If an item was selected the user would be asked what he or she wanted to do with it (e.g. edit it, file it, send it). If a service was selected, the user would be asked what use of the service (s)he wanted to make of it (e.g. create an item—what item?; file an item—what item?; and so on).

Each service would have its own particular dialogue. For example, (obviously) the dialogue for word processing would have to be different from the dialogue for filing and retrieval. But as far as possible terms would be used consistently in different dialogues, e.g. 'move' would mean the same thing no matter in which service the user happened to be.

THE 1983 OFFICE SYSTEM SCENARIO

Within the general framework described above, there was some consensus as to what an office system should look like to the user. The following description is abstracted from several different systems on the market at the end of 1983 and does not describe any particular system precisely.

Overall structure of the USI

The overall structure of the USI is shown in Figure 4.1. It consists of three main levels through which the user communicates with the underlying electronic system, as follows:

Level 1 allows the user to enter the system and gain a rapid overview of what is 'on offer'. It is at this level that the user tells the system which of the particular services (e.g. electronic mail, word processing) needs to be accessed.

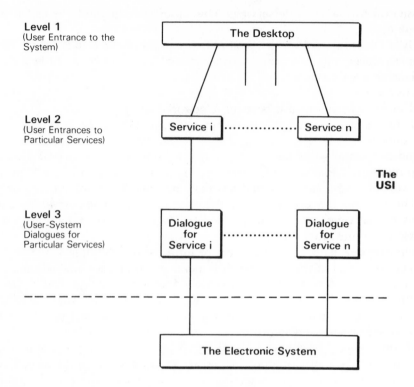

Figure 4.1: The 1983 office system scenario—the overall structure of the USI
(Reproduced by permission of ITT Industries Ltd.)

Level 2 provides access to the particular service the user has selected at
Level 1. The 'User Entrance' to the particular service might be
a menu, a form to fill in, a question to answer, a complex image to respond to in some way, or something else, depending on
the particular service. It allows the user to tell the system what
specifically the service needs to do (e.g. if the user selected word
processing at Level 1, then at Level 2 (s)he may tell the system
(s)he wants to create a new document—or edit an existing document, or set up a mailing list, or whatever).

Level 3 provides the user with a dialogue for achieving what has been
specified at Level 2—e.g. a dialogue which will allow the user to
create a new document.

The Desktop

During the very late 1970s and very early 1980s, the menu had come into
fashion as being a 'user friendly' means of allowing the user to select from
a number of different options. The emergence of videotex is probably the

classic example of this, and in videotex Level 1 presents an overall picture of the total system by means of a relatively simple menu.

By 1983 the Desktop had emerged as a pictorial development of the simple verbal menu. It was not adopted universally—the USI for the Wang Alliance, for example, was based on the more conventional type of menu—but its appearance in the marketplace (e.g. in the Apple Lisa) caused a great deal of interest as a possible way forward in improving the design of the USI. Instead of a list of verbal labels the user would be presented with a two-dimensional array of pictures ('icons'). These icons would be designed to indicate what lay behind them (e.g. an 'in tray' or a 'calculator') but to help minimize ambiguities they were often given verbal labels as well (e.g. the words 'in tray' under the icon representing an 'in tray').

The form of the Desktop varied from system to system. In some, the icons could appear anywhere in an invisible grid on the screen (e.g. the Apple Lisa) and icons could be placed 'on top of' each other; in others they would appear in specified columns and would be numbered, very much as in a conventional menu (e.g. the Sydis VSS).

In all 1983 versions of the Desktop, two kinds of icons appeared, representing:

- all the services (e.g. word processing, calculator) the user could access
- 'objects' (e.g. folders, documents) the user chose to leave 'on the Desktop'.

In addition, therefore, to providing a means for the user to select a particular service, the Desktop also provided a small personal, relatively unstructured information space, rather like the conventional office desk.

Because the Desktop effectively contained two menus—for services and for objects—on the screen simultaneously, the user could perform a few simple operations at the Desktop itself. For example, by selecting a particular document from the array (menu, set of icons) of objects, and selecting the 'filing cabinet' from the array of services, the document would be filed. By selecting the 'wastebasket' instead of the 'filing cabinet', the document would be deleted from the system (although exactly when depended on the particular system; sometimes it was as soon as a second document was put in the basket, sometimes it was when the system's invisible 'janitor' made its regular tour of the wastebaskets). If the user selected the 'out-tray' instead of the wastebasket, the document would be mailed. And so on. The selection of objects and services could be done by a mouse, keyboard, or other device depending on the particular system.

The services available

It is difficult to generalize about the relative importance of the many different services on offer in the various systems, as this depends partly on the particular type of user, but any system worth its salt offered a good selection from the following.

Word processing. Given that word processing had been at the forefront of office automation it is not surprising that a word processing service was considered essential in multi-service office systems. A distinction was often made between:

— word processing for support staff (i.e. word processing operators or secretaries), and
— word processing for the manager or professional.

It was often claimed that the latter category of users did not require and would not want the full word processing that a secretary or word processing operator would need. This was sometimes given as the reason for, amongst other things, providing the manager/professional with a half-page rather than a full page display. Such users, it was felt, would be concerned much more with final editing than with original typing.

The general logic of this argument, not usually articulated very clearly in any case, can be questioned, as follows.

First, a distinction should be drawn between 'managers' and other 'professionals'. Whilst the manager may not need to do a great deal of word processing, many other professional office workers do. Furthermore, many of them prefer to do it directly themselves than through an intermediary. This is possibly more true of some countries (e.g. FR Germany) than others (e.g. the UK) and may perhaps be truer of the younger rather than the older generations.

Secondly, even in those cases where the user wishes to have direct access to word processing only for editing purposes, a full page screen makes more sense than a half page screen. Only with a full page screen can the user see clearly exactly what the page will look like when it is printed out (and in 1983 most pages did end up being printed out).

Word processing, then, was seen to be an important service to offer but its implementation generally was not very well thought out.

Word processing aids. Some systems offered various aids in the preparation of documents, apart from word processing itself. These included, for example, a spelling dictionary to identify possible misspellings, and a readability index to identify parts of the document that might be difficult for people to comprehend.

Dictation. Consistent with the general logic that users would want to a) minimise their use of keyboard, and b) generally do their word processing through an intermediary, a dictation service was supplied in several systems. The user could dictate the material, this would be sent through the system to a secretary or other person who would transcribe it and send it back to the user for editing.

Voice annotation. One way of doing the editing was to attach little 'voice messages' to various parts of the text, explaining to the secretary exactly how to do the editing (e.g. 'make this the first paragraph', 'delete the third sentence down'). Voice annotation could also be used to create mixed media items; for example, a user might write a memo and add further voice comments to explain particular points in more depth. Or the recipient of an item such as a memo could return the memo to its originator along with a voice response attached to it. Voice annotation was done by the user dictating the message into the system and attaching the message to the appropriate part of the text by inserting a voice annotation symbol—typically an icon depicting a loudspeaker.

Voice messages. A simple extension of the principle of voice annotation was that the message need not have any text at all but could simply be a voice message.

Voice editing. For all these applications of voice, it was necessary to supply the user with an editing system. In the Wang Alliance and similar approaches to this problem, the user was presented with a visual display. Explained simply, this consisted of a bar which extended in length as the user kept speaking. When the user paused, a break would appear in the bar. The bar could also be marked in other ways. To edit the voice item, the user could move a cursor to any desired position on the bar and perform 'playback', 'insert', 'delete', and other editing functions.

Electronic mail. Implicit in all of the above services is the notion that users need to communicate with others. Electronic mail was provided to allow users to send any type of object (e.g. folder, voice message, voice annotated document) to one another. Electronic mail was represented at the Desktop by two icons representing an 'in tray' and 'out tray'. To send mail, users would put the objects into the out tray; to see what mail they had received they would 'point at' (e.g. move a cursor, or touch the appropriate key) the 'in tray'.

Telephone directory. An online telephone directory was often supplied from which the user could simply select a name in order for the system to dial the number automatically in order to establish a voice telephone conversation. If there was no answer from the called number the user could elect to send a short message through the electronic mail system. This message would be completed in part (e.g. the user's name, the date and time, the intended recipient) by the system automatically, and could be either text or voice.

Personal filing. The user needed to have somewhere to put mail received and copies of items made. There were two broad approaches to this—not mutually exclusive in principle but separated out here for clarity:

— Some systems (e.g. The Apple Lisa) provided a 'filing cabinet' in which users could deposit items by first putting the items into folders and then putting the folders into the filing cabinet (e.g. at the Desktop by moving the icon representing the folder over to the icon representing the filing cabinet). Folders could be put into folders, thus creating a simple hierarchical structure not unlike videotex (but with 'folders' instead of videotex pages). Users could look inside the cabinet by 'pointing at it', and inside individual folders in a similar way.

— Other systems put more emphasis on providing the user with a capability of finding items rapidly by combinations of keywords. The 'visual memory' of the Wang Alliance was an excellent example of this. The user did not have to worry about which folders to put in which, but simply told the system to 'file the item'. Later, when the item was needed again, the user would specify some particular combination of words (s)he could remember being in the item that would identify it (e.g. the intended recipient and the date, or the topic and part of an address), and the system would find it in a matter of seconds.

The simplest 'filing cabinets' were very primitive indeed and it is difficult to see how, using these, users could have coped with even moderate numbers of items on file.

Psychological aspects of personal filing and management of personal information space are considered in Chapter 6.

Access to shared databases. Systems often provided access to databases on a host computer in the organization or external databases.

Mathematics, graphics and other application packages. A variety of software packages were available to support, for example:

— calculations, often using a 'spreadsheet'
— preparation of various kinds of graphs quickly and easily from the same basic data
— preparation of charts and graphs for project management
— preparation of standard lists.

Calendar. The user could note important events, e.g. project deadlines, in an online calendar. This could also be used for finding mutually convenient times for meetings.

Personal computing. The user was often given a 'personal computing' service based on BASIC, PASCAL, C, or some other high level language. In some cases (e.g. Apple Lisa) a library of software routines was also available.

Input and output media

By the close of 1983, the keyboard and monochrome (40 column, 24 line) display were giving way to a wider range of possibilities for input and output, especially:

— mouse—a hand-held device connected to the main unit by a cable; the user could move a cursor on the screen by moving the mouse over the desk surface, and 'point to' or 'pick up' items on the screen (and perform other such simple operations) by pressing one of a small set of buttons (typically one or two) on the mouse
— touch screen—the user could touch an area of the screen as a means of input, e.g. to select an item from a menu by pointing to it and touching it
— joystick and other devices, mostly for moving a cursor around the screen but essentially for inputting x, y coordinates easily and quickly
— data tablets, light pens, and other devices—similar to the above in essence but more suited to drawing graphics
— voice—to be digitized for storage, transmission and retrieval; used, for example, in voice messages and voice annotation
— speech recognition—to be used to input instructions in a form on which the computer could act; usually (but not always) requiring training by the particular user, usually (but not always) requiring the individual words or phrases to be separated by silences, and typically operating with a vocabulary of around 100 or 200 words or phrases (but a wide range)
— high-resolution bit map display for graphics—typically approaching 512 by 512 pixels (but 1024 by 1024 pixel displays were already available)
— colour displays offering a palette of perhaps a million different colours with perhaps up to a dozen or so possible on the screen at any one time
— speech synthesis for producing output in spoken form
— special devices, e.g. tactile displays for blind users.

THE FUTURE

The future will reflect an interaction of a number of trends and pressures, especially:

— An improvement in the quality of applications software; the various services (e.g. personal information management) offered will take better account of the relevant areas of psychology. Various aspects of this are discussed further in later chapters, especially in regard to personal information systems (Chapter 6) and shared information systems (Chapter 7). There will, amongst other things, be a better understanding of how to use different sorts of input and output media appropriately in relation to different services. On the negative side, well-intentioned but inappropriate attempts at premature standardization may slow down real progress in some areas.

— Greater possibilities for incorporating artificial intelligence techniques into the services offered. This is discussed especially in regard to decision systems in Chapter 8.

— Better integration of the various services ('electronic meetings') services along with Type B ('papers'). Psychological aspects of electronic meetings are reviewed in Chapter 5.

CONCLUSIONS

As we move into the mid-1980s and beyond we can discern several important trends that will interact to shape office systems. These include especially

— an improvement in applications software to take better account of the psychology of the intended users
— greater possibilities for applying artificial intelligence
— better integration of services at the user–system interface.

In moving forward in the development of more user-oriented systems it will be important for system designers to be in a position to make use of the most recent research findings on the psychology of user–system interaction. They will need to have the freedom to incorporate these findings in their designs and to avoid well-intentioned but premature standardization. Whatever human factors standards are adopted need to be based on adequate research, not simple consensus.

In the next few chapters we look in more detail at some illustrative examples of the kinds of services office systems can provide and consider their psychological aspects. We begin with electronic meetings as this was the first area to receive serious attention from psychologists, in the 1970s.

Chapter 5

Electronic Meetings

BRUCE CHRISTIE AND MARCO DE ALBERDI

INTRODUCTION

We have seen in Chapter 2 that Type A communication—face to face meetings and their electronic alternatives (telephone, teleconferencing)—form an important part of what people in organizations do. Understanding this aspect of user behaviour, and the role that Type A systems can play in meeting user needs, is therefore essential for a complete understanding of the potential for electronic systems in organizations. Whilst the emphasis in systems development in the 1980s, and the emphasis of this book, is on 'papers' and their electronic equivalents (Type B systems), it is appropriate to begin a consideration of applications with what was historically the first area of concern: electronic meetings. The psychological research on electronic meetings formed an important phase in the development of systems psychology during the 1970s, and a consideration of the results of the work done is helpful in understanding the roots of systems psychological research.

Historical context

The first 'picture telephone' was demonstrated publicly in the USA just over twenty years ago. It was orginally seen as a replacement for the traditional voice telephone, and the early market forecasts were highly optimistic that this would be achieved by the late 1970s. For a variety of different reasons—including the capital investment required to upgrade the public telephone system to the necessary technical standards, combined with the results of psychological research indicating the benefits of adding a visual channel to be less than had been anticipated—for these and other reasons, the 'picture telephone' never really took off.

By the early 1970s, attention was shifting to a related but different concept—teleconferencing. The idea was that a lot of business travel could be avoided by using two-way television or similar systems to hold meetings.

An important stimulus for the development of research into teleconferencing came from the proposed dispersal of large numbers of civil servants out of London, and the associated concern with the concept of 'communications damage' (see Hardman, 1973). It was felt that an important disincentive against such dispersal could be the 'communications damage' done by 'stretching' the communications links between the offices that would relocate and those that would remain. The Civil Service Department decided to fund work at the Communications Studies Group, University College London, to investigate the extent to which 'communications damage' could be avoided by providing telecommunications links to replace the physical contacts lost as a result of relocation.

About the same time (1971–72), there was growing concern in the USA about the deteriorating quality of life in the inner cities—rising crime, violence, fire risks, poverty, disease, and other factors. This was seen as related to the high population densities characteristic of inner cities. Peter Goldmark (e.g. Goldmark, 1972a, 1972b) saw the New Rural Society as the answer to the problem, and telecommunications as the means to that end. He argued that:

> The stage is set for a communications revolution that will allow growing numbers of people to communicate rather than commute to their jobs. This will allow millions more people to live in the country rather than having to move to already overcrowded cities... it is our belief that all the necessary inventions have already been made and broadband communications systems now can be imaginatively applied to the needs of business, government, education, health care, and cultural pursuits to stimulate the development of the new rural society. The task is gigantic: it will present an urgent challenge to our youth, and all of us must direct at least part of our efforts to it. I believe the magnitude of this task will make going to the moon seem like a ferryboat ride.
> (Goldmark, 1972a)

Goldmark—who had been Director of CBS Research Laboratories until he retired, and had a string of important patents to his credit—was in a strong position to attempt the task. He saw a need for 100 million Americans to move out of the cities over a twenty year period, bringing new life to thousands of economically dying communities scattered right across the USA. And the US Department of Housing and Urban Development and the National Science Foundation were prepared to back him.

In the event, Goldmark was killed in a car crash only a few years after he embarked on this ambitious endeavour and the New Rural Society he envisaged never materialized. Even if he had not been killed it is doubtful whether he would have been able to succeed. The climate of opinion was changing and 'urban renewal' was becoming a more popular concept than population dispersal. Even so, the project did succeed in providing a further stimulus to research on the role of telecommunications in human communication, and in particular the potential for electronic meetings.

It was against this backcloth that the Communications Studies Group was born—funded jointly by the Civil Service Department, the British Post Office (now British Telecom), and American and British industry. Under the leadership of Alex Reid and Brian Champness, the Group pioneered research on electronic meetings. Other research groups around the world—especially at Bell Telephone Laboratories, Australia Telecom, the Canadian Department of Communications, and elsewhere, soon followed suit. Much of the work was written up as a book by Short, Williams and Christie (1976), and some of the original research papers can still be obtained from British Telecom, 88 Hills Road, Cambridge, England.

Research peaked during the period 1972–76, but in very recent years there has been a resurgence of interest as companies like Atlantic Richfield in the USA have once again decided to look seriously at the scope for electronic meetings. This new interest has been associated somewhat with the rise of satellite communications, local area networks, bandwidth compression, and other technical developments which promise to reduce communication costs.

Key issues

The key psychological issues addressed in this chapter are:

— What sorts of meetings are there? What are the key psychological dimensions that need to be considered in distinguishing between different kinds of meetings?
— What sorts of meeting tasks are affected by the medium of communication used, and what sorts are relatively unaffected? What are the important distinctions to be made between different kinds of meetings?
— What can be said about users' attitudes toward electronic meetings?
— Can an optimal type of electronic system for holding meetings be identified?

Types of electronic meeting systems

In addressing these issues, it is necessary to consider the various types of systems that are available for holding meetings electronically. The main distinction is between systems which provide a visual channel as well as audio, and those that provide only audio.

Audio–video systems. Figure 5.1 shows one of British Telecom's Confravision studios. Based on two-way broadcast standard monochrome television, Confravision is a service that links studios in several major cities on the mainland UK, including London. One monitor shows the people in the remote studio (or pictures such as graphs if they choose to use their overhead camera for that purpose), and a second monitor shows the outgoing picture.

Figure 5.1: British Telecom's Confravision. (Reproduced
by permission of British Telecom.)

Confravision was the first major audio–video teleconferencing service to be provided to the public. Figure 5.2 shows a more recent studio-based system, 'Arcovision', used by Atlantic Richfield in the USA for corporate communications nationwide. This system uses optical boxes to provide voice-switched images of the speaker. A further recent trend has been towards movable audio–video systems (e.g. Compression Labs Inc.'s, Mini Conference system and NEC's Rollabout family) which are considerably cheaper than studio-based systems to set up. These systems are used mainly for pairwise linkages between studios. In contrast, the MRC-TV system operating in the tri-state area of New York, Connecticut and New Jersey links several studios at one time; a controller at the MRC-TV headquarters in the World Trade Centre, New York decides who or what should be 'on screen' at any given time. Various other audio–video systems operate in the USA, Canada, Australia, and elsewhere. The Commission of the European Communities is reported to have decided recently to implement an audio–video system between its offices in Brussels and Luxembourg, to replace the existing audio-only link, as part of its INSIS programme.

Figure 5.2: Atlantic Richfield's Arcovision. (Reproduced by permission of the Atlantic Richfield Company.)

Audio-only systems. Figure 5.3 shows the Remote Meeting Table. Devised by Alex Reid and Barry Stapley at the Communications Studies Group, this system now links civil service offices in many different parts of the UK. The people in the studio sit around a circular table. Loudspeakers arranged around the table represent the people at the remote location; when one of

Figure 5.3: The Remote Meeting Table. (Reproduced by permission of the Central Computer and Telecommunications Agency.)

the remote participants speaks, his or her voice is automatically directed to the appropriate loudspeaker. The separation of voices around the table adds to the realism of the effect and aids in the identification of who is speaking at any given time. The Remote Meeting table was designed to link two or three studios at a time. Some other audio-only systems—e.g. British Telecom's Orator, and various systems offered by Televerket in Scandinavia—can be used to link several different locations together.

Audio plus 'papers'. Where audio systems are used, some facility for exchanging papers is often provided as well. The system set up in ICI in the mid-1970s uses a remote writing facility so that someone in one studio can write notes or make a drawing that is reproduced simultaneously in the remote studio. An audio system set up in the Union Trust Bank in Connecticut uses a facsimile link to allow papers to be exchanged.

Computer mediated systems. These systems, also referred to as 'computer conferencing systems', first appeared towards the mid- and late-1970s. FORUM, a system developed at the Institute For The Future in California, was one of the first and was a predecessor of PLANET, the first such system to be offered on a commercial basis (by Infomedia, in California). These systems typically operate in one of two modes: 'synchronous' and 'asynchronous'. In a synchronous conference, everyone involved (typically up to about thirty people, at widely separated locations—possibly in different countries) sits at a computer terminal and types in questions, answers to questions, comments, and other contributions to the conference. Public messages of this sort are routed by the computer to everyone; private messages are sent only to those for whom they are intended. Various other facilities may be provided; for example, the person leading the meeting may choose to distribute a questionnaire which is then analysed online and feedback provided during the course of the conference. An 'asynchronous' conference is similar except that people come online and offline as convenient, such conferences often lasting for weeks or months. Asynchronous conferences are often used for extended project coordination to supplement other forms of communication.

TYPES OF MEETINGS

Some general distinctions

A widely accepted general taxonomy of meetings does not exist. Instead, individual theorists have focused on a multitude of theoretical distinctions, of which the following are among the more important to consider here.

Meetings can be described as either formal or informal, forming part of the processes of formal and informal communication. Formal communications are those which emanate from official sources and carry official sanctions. They usually flow through formal channels, acquiring legitimacy and authenticity. Formal groups are created in order to fulfil specific goals and carry on specific tasks which are clearly related to the total organizational mission. In contrast, informal communication channels are not specified rationally. They develop through accidents of spatial arrangement, friendships, and the varying levels of ability in the organization's 'boxes'. Informal groups may develop because men and women have needs which go beyond those of simply doing one's job, and they seek fulfillment of these by developing relationships with other members of the organization. It is important that electronic systems are not used in such a way as to preclude the possibility of sufficient informal communication.

Communication can also be related to what Thorngren (1972) has called programmed, planning and orientation activities.

Programmed activities are routine, repetitive and standardized. The subject matter is relatively specific and involves regular contacts between individuals who are well acquainted with one another. In contrast, orientation activities are novel, unstructured and complex. The subject matter is of a more general nature and often involves individuals who have not met before. The orientation process allows the organization to adjust to changes in its environment, e.g. new markets, new products or new legislation. Between these two extremes there are planning activities which are concerned with the development and realization of alternatives suggested by the orientation activities.

At a finer level of analysis, Bales (1955) and others have demonstrated that it is possible to code the interactions which go on during a prolonged encounter. Bales was concerned principally with group discussions and showed that the following twelve categories can be coded reliably: shows solidarity; shows tension release; shows agreement; gives suggestion; gives opinion; asks for suggestion; shows disagreement; shows tension; shows antagonism. Bales suggested that four broader categories may also be defined: positive reactions; problem-solving attempts; questions; and negative reactions. We could also suggest that these could be divided into two even broader categories which might be of some significance. Bales' 'positive reactions' and 'negative reactions' could be classed together as person-oriented communications, implying that they reflect an attitude of one group member towards another. 'Problem-solving attempts' and 'questions' could be regarded as non-person oriented since they are concerned more directly with the task in hand and do not reflect the attitudes of the group members towards one another. The distinction between these two categories may be important for two reasons. First, people may vary in terms of which of these they choose to focus on, and this may be partly a matter of temperament. Secondly, evidence presented later in this chapter suggests that some telecommunications media which can handle non-person oriented communications adequately may not be suitable for person-oriented communications.

Three major facets

The Communications Studies Group at University College London under-took to develop its own taxonomy of meetings based on the verbal descrip-tors which management personnel in business and the civil service used to describe their face to face meetings (see Short, Williams and Christie, 1976). The taxonomy is based on a three-dimensional theoretical frame-work in which every meeting can be described in terms of its purpose or aims, the particular interpersonal interactions which occur, and the general atmosphere of the meeting (e.g. friendly or hostile). The Group conducted a study to determine how many kinds of aims, interactions and types of atmosphere could be identified.

In this study—called DACOM (for 'Description and Classification of Meetings')—311 management personnel rated their most recent business meeting on 104 seven-point scales (based on a series of 65 open-ended pilot interviews). For each of the 104 items, the respondent gave a score from 0 to 6 to indicate how well the item applied to the meeting being consid-ered. The items were grouped into 'functions of the meeting' (aims) 'what went on' (interactions), and 'atmosphere'. Each section was submitted to a factor analysis to see how many different kinds of aims, interactions and atmospheres could be described.

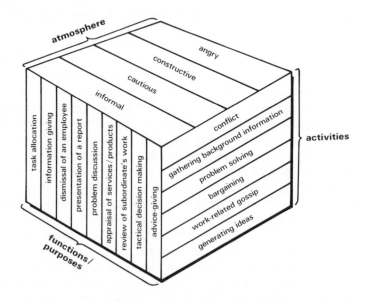

Figure 5.4: Three facets of meetings

Respondents seemed to use a wider range of aims than any other category of descriptors. Nine factors were required to account for 45% of the variance for the scales in this category, compared with only 6 factors to account for 47% of the variance for the scales describing interpersonal interactions, and only 4 factors to account for 50% of the variance for the scales describing the atmosphere of the meetings. This seems reasonable when one considers that what the meeting is meant to achieve is perhaps the single most important thing a manager would want to know before attending it. The particular aims, interactions and types of atmosphere identified are shown in Figure 5.4 which provides the basis for a taxonomy of business meetings based on the verbal expression of managers' perceptions.

DACOM was conceived as a pilot study and more systematic research is still required to confirm and elaborate on the findings. However, a survey of 1791 face to face meetings (see Short, Williams and Christie, 1976) using a classification derived from DACOM findings suggests that meetings concerned with problem solving and exchanging information are amongst the most common kinds of meetings to be held in business and government.

EFFECTS OF MEDIA

The role of vision in interpersonal communication

Visual contact is an obvious feature of all normal face to face groups whatever their purpose. It is expensive to provide vision electronically, very much more expensive than to provide sound, and so it is important to ask whether vision is always really necessary or whether it is possible to dispense with it for certain kinds of communication. Fortunately, social psychologists have made significant contributions to our understanding of the use made of visual contact during normal face to face encounters.

Argyle (1969) suggests there are six main functions served by visual contact, excluding straightforward written communication: to provide continuous evidence that each person is attending to the others; to coordinate patterns of speaking by means of head nods, eye contact, and other non-verbal cues; to provide feedback, e.g. to indicate agreement or disagreement; to introduce redundancy into the message, e.g. gestures to emphasize a point; to substitute for verbal communication, e.g. a shake of the head to mean 'No'; and to indicate interpersonal attitudes, e.g. by a formal or a more relaxed posture. Some of these, particularly the first three, serve what Birdwhistell (1970) has called an integrational function—they keep the interaction going smoothly—whereas the others are more directly concerned with the message itself.

Visual communication evidently does not serve a single function. Neither is it simply a matter of choosing to provide visual contact or not providing visual contact when it comes to designing electronic media. A range of options are available from say, life-size colour images of all members of the group down to a small monochrome head-and-shoulders view of the person who happens to be speaking at the time (an option often selected in practice). Does it matter? To answer this, it is necessary to look in more detail at some of the factors involved in visual contact.

First, there is proximity. Hall (1963) suggests there are four main levels of proximity. From near to far these are: 'intimate', 'personal', 'social' and, beyond about twelve feet, 'public'. (The labels are fairly descriptive provided it is recognized that the precise distances involved and the kinds of behaviours deemed appropriate for each level vary somewhat from one culture to another.) Different non-verbal cues may be important at different distances. For example, pupil size may be important in the 'intimate' zone (Hess, 1965) but even mutual gaze may not be important at the 'public' range where it is very difficult to be sure whether or not gaze is indeed mutual. Apparent proximity is affected markedly by telecommunication. Apart from the obvious effect which zooming a camera in can have, the quality and level of the sound can also have a marked effect, as anyone who has used a telephone will know.

Orientation is also important. For example, people communicating face to face tend to sit opposite one another for a competitive task but side by side for a cooperative one (Sommer, 1965). When telecommunicating, apparent orientation will depend on the position of the cameras. When there are two groups involved the most usual arrangement creates an impression of the groups sitting opposite one another, which might indicate a competitive or conflictful attitude if the groups were meeting face to face. In an audio-only system, orientation is usually ambiguous.

Physical appearance may have short-term effects on an interaction (Argyle and McHenry, 1971), and can be ambiguous if strangers are communicating by means of a non-visual medium. This may result, for example, in the sex of people involved having a significant effect in some situations, perhaps because when men and women communicate without seeing one another each is free to imagine the other as conforming to his or her 'ideal' (or stereotype, perhaps).

Many cues deriving from arm movements or general posture are lost if, as is often the case with videotelephones for example, only the head and shoulders are visible. Even facial signals may be distorted or lost depending on the quality of the picture and the position of the camera. Eye contact is particularly important here. Argyle and Dean (1965) have linked eye contact to the general level of intimacy which is established, suggesting that this interacts with other relevant variables such as proximity and posture and can be varied by the people involved to maintain an optimum level of intimacy. If this is so, it may be important that a two-way television system can apparently reduce the amount of eye contact. Each person will tend to look at his or her screen, not the camera. Each can look at the camera to convey 'eye contact' to the other, but even then it will be a very one-sided affair since the other person must look away from the camera (to his or her screen) in order to perceive eye contact. Both people will experience eye contact simultaneously only when each camera is positioned in the centre of the screen (or, in practice, some trick is used to achieve the same result, e.g. positioning the camera just above a very small screen, or perhaps using a mirror system so the person appears to be looking into the camera when actually looking at the screen).

Removing the visual channel entirely or replacing it with an electronic substitute could be expected to have significant effects on communication. But the effects may be subtle in view of the complexity involved. Birdwhistell (1970) has pointed out that no part of a communication has meaning in and of itself; meaning emerges from the complex interactions between the many parts (the whole is greater than the sum of the parts, as the Gestalt psychologists emphasized). The problem is further complicated by the fact that social psychologists have concentrated on describing the role of the visual channel in normal face to face communication. To discover that it is used in particular ways during communication when the parties are face to face does not necessarily imply it is essential for effective communication. Many people travel to work by car, but this does not mean they could not get to work if deprived of their cars. They might even be able to do so better, in terms of some criteria (e.g. environmental pollution).

The concept of social presence

Three different approaches to the psychology of mediated communication can usefully be distinguished: simple theories based on a concept of efficiency; theories which concentrate on non-verbal communication; and a theory suggested by Short, Williams and Christie (1976) based on a concept of social presence.

Naive efficiency theory holds quite simply that some communications media are better than others, specifically that face to face communication is best—presumably because it would seem to include all the other media in it (which view ignores special facilities that some electronic media can offer)—and all other media represent various levels of degradation. The theory has two main weaknesses. First, it is often difficult to decide which outcome to an interaction is 'best'; for example, in negotiation situations individual and collective interests often conflict so one has to ask, best for whom, the individual or the group as a whole? Secondly, there is the implication that everything is done better face to face whereas evidence suggests that some kinds of communication, e.g. simple information exchange, may be done no more efficiently face to face than by, say, telephone.

Theories which concentrate on non-verbal communication suffer from three main weaknesses. First, it is likely that people can modify their behaviour in reduced-cue situations to compensate effectively for the cues which are lost, for example by indicating agreement verbally instead of nodding. Secondly, particular cues and combinations of cues are not invariant in their meaning across different situations. Failure to shake hands when meeting someone face to face may be significant, but will have no meaning if the communication is by telephone. Thirdly, it is often difficult to formulate precise predictions based on this type of theory because of uncertainty about the relationship between use of particular cues and the outcome of the interaction. This is especially difficult when one is concerned with complex situations.

The main theory focusing on non-verbal communication to explain the differences caused by media is 'Cuelessness' which is illustrated in Figure 5.5. Rutter, Stephenson and Dewey (1981) define the degree of Cuelessness in terms of the aggregate number of social cues available in the interaction. According to this theory, the few social cues available over audio cause the content to become more task-oriented which in turn causes the style to become more depersonalized, and changes the outcome of the discussion. This model can be criticized on both theoretical and practical grounds. The theoretical criticism centres on the causal link between content and style; since both are measured at the same time it is difficult to argue that one causes the other in a sequential causal chain. The criticism of the practicality of the theory focuses on the difficulty of measuring cuelessness— since many social cues are redundant the aggregate number is impossible to measure. The theory, therefore, lacks the ability to predict effects.

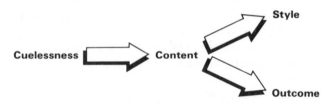

Figure 5.5: The theory of the Cuelessness, based on
Rutter, Stephenson and Dewey (1981)

The theory proposed by Short, Williams and Christie suggests that what is important is the 'social presence' of the communications medium. Social presence is conceived as a cognitive synthesis of many factors. The capacity of a medium to transmit information about facial expression, direction of looking, posture, dress and non-verbal vocal cues, all contribute to its social presence. However, Short *et al.* believe that what the person perceives is the most important variable and they see social presence as a subjective, rather than an objective quality of the medium. It is the person's perceptions which are important, even though these will normally be influenced by objective differences between media.

Their definition of social presence as a subjective quality of communications media has led them to measure it by means of verbal ratings. They claim that high social presence is indicated by a person: rating a medium as being sensitive, not impersonal, and warm rather than cold; endorsing items such as 'one can easily assess the other people's reactions to what has been said', 'it provides a great sense of realism', 'one gets a good feel for people at the other end', and, 'it was just as though we were all in the same room'; and not endorsing items such as 'one gets no real impression of personal contact with people at the other end of the link', or 'people at the other end do not seem real'.

The concept of social presence is related to Argyle and Dean's (1965) concept of intimacy. Argyle and Dean proposed that when two people communicate face to face they will behave in various ways—such as adjusting their posture, moving closer or further apart, smiling, or adjusting the amount of mutual gaze—to maintain an optimum level of intimacy. Short, Williams and Christie propose that the social presence of the communications medium is one of the factors contributing to intimacy.

Social presence is also related to the concept of immediacy. Immediacy (Wiener and Mehrabian, 1968) is a measure of the psychological distance which a person puts between himself or herself and the object of the communication, the addressee or the communication itself. Negative affect, low evaluation and non-preference for any of these things are said to be associated with non-immediacy in communications. The concept was originally applied to speech but can be extended to include non-verbal communication, e.g. immediacy conveyed by physical proximity. It has even been suggested (e.g. Heilbronn and Libby, 1973) that immediacy or non-immediacy can be conveyed by the medium of communication used. This is 'technological immediacy' rather than 'social immediacy' which is conveyed through speech and associated non-verbal cues. Technological immediacy and social presence may be linked closely in some situations. For example, if a person can choose between using a voice telephone or a videotelephone, both technological immediacy and social presence will be higher if (s)he chooses the latter. However, technological immediacy can vary when social presence does not. For example, assume a person makes two telephone calls, one to Mr. A who is fifty miles away and one to Mr. B who is in the next office. Social presence is the same for both calls, but immediacy is much lower for the second call than for the first because the person could easily have talked to Mr. B face to face so using the telephone had a significance it did not have for the call to Mr. A. Even if both calls had been over a long distance, the call to Mr. B may have been lower on immediacy if the caller chose to adopt a relatively formal style and convey an impression of aloofness. But in all cases the person making the call would have seemed just as 'real'—the social presence afforded by the telephone would have been the same. Therefore, immediacy and social presence, although related, are not identical; they can vary independently of one another.

Early results reported by Short, Williams and Christie suggest that the largest differences in social presence are found between visual and non-visual media and that differences within each of these classes are much smaller. This is congruent with studies which have obtained similar results for the influence of media on effectiveness.

Some postulated effects of social presence

The social presence theory suggests that some media are perceived as 'warmer', more 'personal', and more 'sensitive' than other media. These differences might be expected to be more important for some tasks than for others, and for some aspects of communication than for others. An important result of the empirical research conducted in this area is that media are indeed specific in their effects. A non-visual medium, for example, does not affect all types of communication equally, as might be expected from a simple efficiency theory, but has much more marked effects under conditions where 'warmth' and 'sensitivity' could be expected to be important. However, before pursuing this further it may be useful to summarize briefly some of the social psychological research which was done prior to work on telecommunications as such and before the emergence of the theory of social presence.

A relevant area of research concerned the effects of the mere presence of another person or persons. It was discovered very early on (e.g. Allport, 1920; Pessin, 1933) that a subject's performance can be affected by awareness that another person is present. But physical presence is not required. All that is necessary is that the subject is aware that another person is 'present' in some sense, e.g. working on the same problem in another room (Dashiell, 1935). Wapner and Alper (1952) found that subjects were affected even more by being told they were being watched through a one-way vision screen than by the audience being in the same room. This is an important result in the present context because a telephone link also creates a situation where it is obvious another person is 'present' and yet physical presence is missing. It is because 'presence' in some sense does not depend on physical presence that the concept of social presence is defined in terms of the subject's own perceptions of his or her environment.

Physical presence or separation has also been found to affect the likelihood of a subject conforming to a group's judgement (e.g. Deutsch and Gerard, 1955). But, again, it is not physical presence that seems to be the key variable because conformity is also reduced if the group is physically present but the subject does not have to give his or her name (making the situation more impersonal?) (e.g. Mouton, Blake and Olmstead, 1956).

Variation in social presence could also be used to explain results obtained by Milgram (1965). Milgram created a situation in which an experimenter instructed a subject to give electric shocks to a confederate. The situation varied from the subject not seeing or hearing the confederate, through a condition where the confederate could be heard, to one where the subject and confederate were face to face, and even to one where the subject had to hold the confederate's hand down on the shock plate. Increasing social presence decreased the amount of cooperation with the experimenter's instructions. (These results could also be explained in terms of immediacy, once again demonstrating the close link between this concept and that of social presence.)

The concept of person-orientation

Social presence may be an important key to understanding the effects of different communications media. Another concept which promises to be of use in this connection is that of person-oriented versus task-oriented interactions. The idea here is that some kinds of interactions, e.g. sensitive negotiation, require the people involved to attend to one another as people—to take account of each other's personal style and idiosyncrasies—much more than some other kinds of interactions, e.g. conveying a simple message (where, in the extreme, the person receiving the message could almost as well be replaced by a tape-recorder or a pad of paper).

In practice, most communication episodes will involve both kinds of interactions to some degree or another. For example, in a recruitment interview the participants will be concerned partly with exchanging factual information (which could be done and to some extent may already have been done by the potential employer providing written information about the job and the potential employee completing an application form) but partly also with learning about one another as people. Even in presenting the factual information each party may be concerned with creating a good impression and will need to attend carefully not just to the message but to how the other person is receiving the message and responding to it.

The distinction is similar in some ways to the distinction between interparty and interpersonal exchange (e.g. Douglas, 1957; Morley and Stephenson, 1969). Douglas proposed that in any communication episode the individuals involved are concerned both with acting out certain roles and with developing or maintaining some personal relationship. Morley and Stephenson proposed that the balance between these two could be affected by the medium of communication used (in their view, the 'formality' of the medium, but social presence may be equally if not more important than formality). They proposed that during a telephone conversation, for example, there would be more emphasis on interparty exchange, and less on interpersonal exchange, than during a face to face conversation.

There is a difference between the two concepts because a person may have to attend very carefully to the other individual as a person in order to carry out his or her role effectively. One example would be a social worker talking to a client. The emphasis may be very much on interparty exchange (with clear definition of roles) as the client describes a problem and asks for help, but the social worker will need to be very person-oriented in this situation (attending, for example, to how the person asks for help) to really understand the person's needs.

Whether or not a particular communications medium is satisfactory will depend, according to this theory, on the interaction between the social presence afforded by the medium and the balance between the person-orientation/task-orientation of the communication episode. This is shown in Figure 5.6. As can be seen, the theory predicts that most media will be satisfactory for most communication episodes. Only when the communication episode is predominantly person-oriented does the choice of medium become crucial. Here, media low on social presence will not be satisfactory. Examples falling in this category include cases where there is some emphasis on conflict, person perception, interpersonal attraction or group cohesion.

| | | Social Presence of Communications Medium | |
		Low	High
Nature of the Communication Episodes	Predominantly Person Oriented	Not Satisfactory	Satisfactory
	Predominantly Task Oriented	Satisfactory	Satisfactory

Figure 5.6: Postulated effects of social presence and person-orientation on acceptability of mediated communication

Effects on cooperation and conflict

Wichman (1970) measured the amount of cooperation shown in a Prisoner's Dilemma game (in which two subjects can either cooperate for medium gain by each or compete for higher individual gain for the 'winner') under four communication conditions: face to face; visual contact but no talking permitted; talking allowed but a screen interposed to prevent visual contact; and no communication at all. The amount of cooperation shown was highest in the face to face condition and least when there was no communication at all; the other conditions were intermediate. Wichman interpreted his results in terms of the number of cues available to the subjects, but the results are also consistent with social presence theory since a person-oriented task was used.

Dorris, Gentry and Kelley (1972) obtained somewhat similar results in a different kind of game. Their game required subjects to work in pairs to divide a number of points between themselves, two minutes being allowed to discuss this and decide on an acceptable split. Each subject's winnings were related in a complex way to the number of points gained by the split. Subjects could either see one another or not (a screen being interposed for this condition). The average joint payoff was higher when subjects were in visual contact, but only for the first of the four games they played. In all four games, lack of visual contact resulted in a longer time to reach agreement and fewer settlements. Again these results are consistent with social presence theory and, as the theory would predict, all the effects were more marked when the subjects were competitively oriented (resulting, in terms of the theory, in greater person-orientation and therefore greater sensitivity to the social presence afforded by the communications medium).

The medium of communication has also been shown to have an effect in more complex situations involving conflict; for example, a simulated negotiation between defence council and plaintiff's attorney (Smith, 1969), and a simulated negotiation over an industrial wage dispute (Morley and Stephenson, 1969).

The Morley and Stephenson findings were replicated in a carefully controlled experiment by Short (1971a). In this experiment, one subject in each pair represented 'management' and the other 'the union'. One side was given a stronger case than the other. This side did significantly better when communication was by telephone than when it was face to face. (Morley and Stephenson had observed a trend in this direction but it was not statistically significant; they did find that allowing or not allowing the subjects to interrupt one another had a statistically significant effect, but Short did not repeat this part of the experiment.) Seven of the thirty pairs of subjects in Short's experiment failed to reach agreement. A very interesting finding was that six of these were in the telephone condition, only one was in the face to face condition. In a sense, then, this experiment suggests not only that the medium of communication affects situations involving conflict but where it is possible to define a 'better' outcome, a high social presence medium such as face to face communication is 'better' than a low social presence medium such as the telephone.

In a subsequent experiment, Short (1974) compared three media: face to face; two-way television; and two-way audio with no visual contact. The outcomes in the television and face to face conditions were very similar and significantly different from the audio-only condition, indicating that the absence of a visual channel had an effect beyond that of mere physical separation (because the subjects were physically separated in the television condition as well). However, Short also measured the amount of time the subjects spent looking at their television monitors in the television condition, i.e. how much use they actually made of the visual channel. It is very interesting that he found there was no significant correlation between this and the outcome of the negotiation. This suggests it was the subjects' awareness of the fact that visual contact was possible which was important and which largely accounted for the difference between this condition and the audio-only condition. One is reminded again of Dashiell's (1935) experiment in which the subject's awareness of another's presence, even though the other person was in a different room, was sufficient to affect his or her behaviour. Short's finding provides strong support for the social presence theory and very weak support for the theory that what is important is visual contact *per se*.

The evidence reviewed so far indicates that conflictful situations are sensitive to the effects of communications media. Before going further it may be useful to consider what is meant by conflictful situations. Realistic conflicts tend to be rather complex. There is a conflict of interest: what one side wins another loses (the type of conflict considered so far); there is also the possibility of ideological conflict between the two sides; and, finally, the possibility of personal conflict between the individuals involved. The latter two seem to be rather different from the simple conflict of interest. It is interesting in this connection that the taxonomy of communication episodes suggested by the DACOM study (see Figure 5.4) includes two different types of interpersonal interactions which seem relevant here. One is 'conflict' and in the study this was associated with reports of 'disagreement' and 'arguing'—it seems somewhat similar to the ideological and personal clashes just noted. The other is 'bargaining' and was associated with reports of 'compromising'—it seems more related to the simple conflict of interest usually studied in the laboratory.

Short (1972a) conducted an experiment to examine the effects of media in a conflict situation where personal clashes of opinion were more important than any simple conflicts of interests. In this experiment, subjects (in pairs) discussed controversial issues such as 'public transport in large towns should be heavily subsidized out of taxation to discourage the use of the private car' and 'foxhunting should be forbidden'. The purpose of the discussion was to reach agreement in terms of an eleven-point scale ranging from 'strongly agree' to 'strongly disagree'. In one condition, the two subjects were in real conflict (the issue for discussion being chosen on the basis of a pre-test questionnaire to ensure this). In a second condition, the two subjects really agreed but one was asked to play 'devil's advocate'. The subjects were not aware they had been treated differently. Each pair of subjects discussed two items (for fifteen minutes each) in each of the two experimental conditions. Medium of communication (each pair used the same medium throughout, either: face to face; two-way television; or audio-only) had no significant effect on the rating agreed by the pair but did affect the amount of personal opinion change shown by the individual subjects. (Opinion change was measured by asking the subject to rate his or her real, personal opinion after each discussion and comparing this with his or her rating on the pre-test questionnaire.) There was significantly more opinion change in the audio condition than face to face, and an intermediate amount in the television condition. This was true whether or not one subject was playing 'devil's advocate'.

The finding of more personal opinion change under conditions of audio-only communication rather than visual contact has been replicated several times (Short, 1972b, 1972c, 1973b; Young, 1974). The effect is not large and although it was observable in all the experiments cited it attained statistical significance in only two (Short, 1972c, 1973b).

In one experiment (Short, 1972b) the subjects were asked to rate one another on twenty-two seven-point rating scales. Subjects in the audio condition generally rated their partner more favourably, e.g. more 'pleasant', 'reasonable', 'trustworthy', than in the television or face to face conditions. It seems that subjects not only are more likely to be swayed by another's opinions if social presence is low in the communications medium, they are also more likely to judge the person favourably. This is consistent with the hypothesis that under these conditions subjects are more free to imagine their partner as being like their 'ideal' (or, perhaps, stereotype) man or woman. In this case, one would expect the amount of opinion change to be affected by the interaction between the sex of the individuals, the topic discussed (whether 'masculine' or 'feminine') and the subject's attitude toward his or her partner's sex.

Effects on person perception, interpersonal attraction and group cohesion

Getting to know people is an important aspect of many business contacts. Strangers are often present at face to face meetings. Collins (1972) found that 20% of the 6397 civil service meetings he surveyed involved one or more strangers being present. It is important to know what kind of effects, if any, electronic media are likely to have on the process of getting to know another person.

One important aspect of perceiving another person is the degree of mutual liking or attraction which is established. Another aspect worth studying in group situations is group cohesion, i.e. the extent to which telecommunication is likely to affect the likelihood of a group splitting up in some way.

Users of telecommunication systems certainly appear to feel this general question to be important. Comments such as the following (from Short, 1973b; and Christie, 1973b) are not at all uncommon:

'The picture helps if you don't know the other.'

'I didn't get to know the other as well by video as face to face.'

'Noting facial expressions is an important guide to the other's feelings.'

'... if he had come over here, we wouldn't have got the same impression that it was us and him; it would just have been a sort of cosy chat. And this, I suppose, is inevitable; so long as you have got two locations, you are bound to have two sides.'

'Around a table we are five or six; we feel on equal terms. Over the TV we felt the four of us were against him.'

Accuracy. The effects of communications media on the accuracy of person perception have been examined experimentally in a number of studies. Several early experiments (e.g. Geidt, 1955; English and Jelenevsky, 1971; Maier and Thurber, 1968; and Berman, Shulman and Marwit, 1975) failed to find any effect. However, although the total range of media examined in these experiments was wide—from written transcripts to audio–video film— the situation investigated was always unidirectional, no meaningful two-way interaction being allowed. Other experiments have looked at the two-way interaction situation.

Laplante (1971) had pairs of students play a game (the Prisoner's Dilemma) in which both cooperative and uncooperative moves were possible. At various points during the game one student (the confederate) would speak to the other (the subject). The message was either friendly or unfriendly. The following media were compared: face to face; television; telephone; written message. If the message was friendly, the subject rated the confederate most favourably when communication was face to face or by television (the non-verbally rich media). If the message was unfriendly, the subject rated the confederate least favourably when either of these media was used. The medium of communication seemed to modify the effect of the message.

Klemmer and Stocker (1971) compared the videotelephone with an audio-only system. Each subject had a ten-minute conversation using each medium of communication; each conversation was with a stranger. In the first replication of this experiment, subjects were rated significantly more submissive by their partners when using the videotelephone, but there were no differences on several other scales (e.g. friendly–unfriendly). In a second replication of the experiment there were no differences at all.

Confidence. The evidence that medium of communication can affect the accuracy of person perception is weak but there is stronger evidence that it can affect confidence.

Reid (1970) ran groups of three subjects in which one was the speaker and the other two listeners. One listener could see the speaker, the other could not. The listeners did not differ in how accurately they could tell if the speaker was lying or telling the truth, but the listener who could see the speaker was significantly more confident in his or her judgements. Reid obtained similar results in a second experiment in which interviewers rated an interviewee on a series of scales (e.g. serious, mature, responsible). Interviewers who saw the interviewee face to face were more confident in their judgements than those who talked to him or her by telephone but the judgements themselves did not differ.

The general finding has been replicated in a more elaborate experiment conducted by Young (1974) in which care was taken to avoid some of the methodological weaknesses of the earlier experiments. However, in this experiment a few differences in the interviewers' ratings of the interviewees' personality also emerged. Interviewees seen face to face were rated less broadminded and less rational than those seen by television or talked to by telephone. Those on the telephone were judged more dominant. Interestingly, the interviewees' ratings of the interviewers were not affected by the medium of communication.

Evaluative Biases. Young's experiment and several others (e.g. Klemmer and Stocker, 1971) suggest that the medium of communication can affect subjects' evaluations of one another under some conditions. Williams (reported in Short, Williams and Christie, 1976) designed an experiment specifically to examine this question in more detail.

One hundred and forty-four civil servants took part. Individuals were paired on the basis of them being strangers or only slight acquaintances. Each person had two 15-minute conversations, each with a different partner and each using a different medium of communication. Three media were compared in all: face to face, television, and telephone. One of two tasks was carried out, either: a) free discussion on the 'Problems of Modern Life', or b) a 'Priorities' task in which the subjects had to agree on the three most important problems of modern life, each subject having written down the four which (s)he considered most important just prior to the discussion. After both conversations were over, subjects ranked their partners and the conversations they had had on a number of scales.

Factor analysis of the scales resulted in three factors for the conversations ('conversation evaluation', 'conversation interest', and 'conversation argumentativeness') and three for the people ('person evaluation', 'person intelligence', and 'person domineeringness'). Analysis of variance of the factor scores revealed an interesting interaction (most of the effects being concentrated on the evaluative factors): In the free discussion condition, face to face conversations and people were preferred to television ones which were preferred to audio ones; but in the priorities condition, the television conversations and people were preferred over the face to face and audio ones, these two not differing significantly.

An arousal theory interpretation. Williams' results are consistent with an arousal theory interpretation. According to this interpretation, arousal potential (Berlyne, 1960) increases both with the addition of sensory stimulation by means of a visual channel and with the increase in the person-orientation of the discussion in the priorities task due to the element of competition and conflict of opinion. Thus, the experimental conditions can be rank ordered from non-visual communication system with low person-oriented discussion (very low arousal potential) through to a visual system with a high person-oriented discussion (very high arousal potential). The optimum level of arousal potential in Williams' experiment appears to have been when face to face communication was used for a relatively low person-oriented discussion or when the television system was used for a high person-oriented discussion, but arousal potential was unpleasantly high when face to face communication was used for the high person-oriented task.

The general model suggested by these results is shown in Figure 5.7. According to this model, both the person-orientation of the task and the social presence of the communications medium (especially visual contact) contribute to arousal potential but in most situations the social presence of the medium will contribute more than the person-orientation of the task.

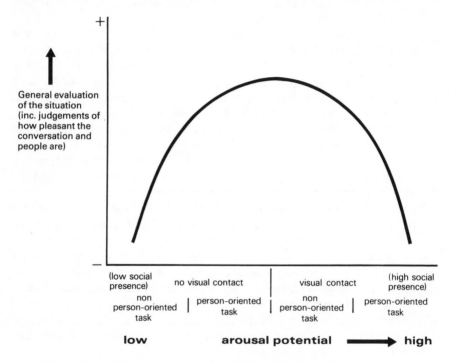

Figure 5.7: Postulated relationship between subjects' evaluations and the arousal potential of mediated communications

Results obtained by Christie and Kingan (1975) are also consistent with this model. Sixty-five civil servants participated in 4-, 5-, and 6-person groups, half of the group in one studio and half in another. The two sub-groups could communicate by telecommunication, with visual contact for half the groups and without for the others. Two tasks were used: a) discussing and agreeing on a suitable market for a car which was described in detail (low person-orientation), and b) the two subgroups arguing for different markets and having to come to an agreement (high person-orientation).

Following their discussion, the subjects were each asked to use two 7-point scales (ineffective–effective; unenjoyable–enjoyable) to rate each of the following: the discussion; the medium of communication; the other participants (separately for the two rooms); and how the other participants perceived the subject (separately for the two rooms). In addition, each subject was asked to consider eight different cases in which (s)he might choose to telecommunicate or travel to a meeting. The number of times out of eight that subjects chose to 'telecommunicate' was the thirteenth dependent variable. Factor analysis of the 13 variables resulted in two clear factors: a) a general evaluation of the whole teleconferencing environment, including the discussion, the communication system, and the people involved; and b) a self-judgement in terms of how the subject felt (s)he was perceived by the other participants. Neither factor loaded the measure of preparedness to use telecommunications in hypothetical instances. For both factors, the most favourable judgements were made when the visual system was used for the low person-oriented task.

Neither of the two experiments described was designed to test the model in Figure 5.7. Rather, the model represents a plausible generalization that can be drawn from the results of the experiments. What is needed now is an experiment specifically designed to test the model. Also needed is an experiment to demonstrate that psychophysiological indices of arousal follow the arousal potential continuum suggested by the model.

Metaperceptions. A metaperception is what A thinks B thinks of A, and could be regarded as a measure of understanding. Weston and Kristen (1973) found there was considerable uncertainty at this level. Television communication was judged more favourably than audio-only communication on several relevant scales, e.g. 'I had quite a bit of trouble knowing how the people at the other end were reacting to the things I said'. Young (1975) found a similar result when subjects were asked, 'How well did they understand your views, in regards to agreeing or disagreeing with them?' Those at the other end of audio links were judged to understand less well than those at the end of video links or face to face. However, Young (1974) found no significant differences between the media when he compared A's ratings of what (s)he thought B thought with B's own ratings reflecting what (s)he actually did think. As with simple perceptions, medium of communication may affect confidence but does not seem to affect accuracy.

Coalitions in small groups. Williams (1973) set up 48 three-person groups. In each group, two members sat together and communicated with the third by telecommunication (visual for half the groups, audio-only for the others). The game used involved two members of the group having to vote down a third in order to complete the task. It was hypothesized that the 'lone' person at the end of the telecommunication link would be voted down most often. Actually, this did not happen—perhaps, Williams suggests, because the other two felt sorry for the lone person.

In a subsequent experiment reported in Short, Williams and Christie (1976), Williams used four-person groups to overcome any tendency for two members to feel sorry for a third. In this experiment, subjects had to propose ideas. Each idea had to be seconded to be recorded and subjects could also dissent if they wished. The medium of communication (face to face, television or audio-only) had no significant effect on the number, quality or originality of ideas generated. However, it did affect the patterns of seconding and dissenting. For both telecommunication conditions, it was significantly more frequent for an idea to be seconded by the other person in the same studio than by a subject from the other end of the telecommunication link. In the audio-only condition, but not the television condition, it was also more common for someone at the end of the telecommunication link to dissent than for someone in the same studio as the person proposing the idea to do so. The experiment therefore provides some evidence that telecommunication can sometimes result in the formation of subgroups, creating an 'us–them' situation.

Person perception, opinion change and social presence

Recently, the view that media low in social presence cause more opinion change has been challenged. De Alberdi (1982) collected social presence ratings, opinion change and perceptions of the other participant in an experiment that compared the effects of visual communication and physical separation.

In a conflict-of-opinion experiment similar to those carried out by Short (see above) student participants discussed social issues.

The two media conditions in which subjects could see one another (face-to-face; video and audio) were rated higher on social presence than those without visual contact (audio; curtain separation). Opinion change was similar in all four conditions, and individual social presence ratings did not correlate with opinion change scores. More opinion change did, however, correlate with more positive ratings of the other participants.

Opinion change seems, therefore, to be related to what one thinks of the other person, rather than in any simple way to social presence. In support of this line, it should be noted that opinion change and evaluation of the other were correlated in Short's (1972b) experiment. This suggests that the 'more opinion change over audio' result may be a pseudo-process (Hyland, 1981), or sometime correlate of positive person-perception effects of media.

USERS' ATTITUDES

Attitudes toward electronic meetings

The evidence on attitudes towards meetings held over teleconferencing media generally suggests that users tend to judge these less favourably than face-to-face. Meetings held using a video system tend to be judged more favourably than those held using an audio-only system. For example, Weston and Kristen (1973) compared Carlton University tutorial groups communicating over audio-only and television systems with face-to-face tutorial discussions. Students were asked to rate their feelings towards the discussions on 14 rating scales. The audio discussions were rated significantly less favourably than the television discussions.

The differences between audio-mediated and audio/video-mediated discussions are not necessarily very large, however, and do not necessarily apply to all aspects of the meeting—as is illustrated by a study by Christie (1973a). Christie gave 54 American business executives the opportunity to try four different communications media in groups of 6 (3 at either end). The media used were:

- Speakerphone (loudspeaking telephone)
- A high fidelity monaural audio system
- An audio system in which each speaker was represented by a different loudspeaker
- A two-way monochrome television system with monaural sound.

In a fifth condition the groups met face-to-face. Immediately following each discussion the groups rated the media on nine rating scales. It is clear that media are grouped in different.ways depending on the attribute considered. The results on these scales are shown in Figure 5.8.

Quite small differences in the technical performance of a system can be important. Phillips and Treuniet (1978), report for example that delays in satellite transmission (over 170 msec) result in lower subjective ratings of the other participant and increases in the amount and duration of mutual silence.

Not all aspects of electronic meetings are rated less favourably than face to face meetings. Champness (1973) in a study of business people who had used the Post Office's 'Confravision' system found that the majority agreed the system would increase how business-like a meeting was and decrease how heated it was. Similar results were obtained by Williams and Holloway (1974) in their study of the Bell Canada intercity video-conferencing system. Some of the effects of teleconferencing therefore may be viewed as beneficial.

Tomey (1974) reports a study of an audio-only conferencing system used for bank committee meetings. The majority of users agreed that teleconferencing had the following advantages:

- Discussion of particular items tended to be shorter.
- It was easier to get a point across without lengthy debate.
- Committee members were more attentive.

Users agreed that the meetings were more impersonal, and one tended to lose the impression of personal, emotional contact, but were also agreed that this was not important.

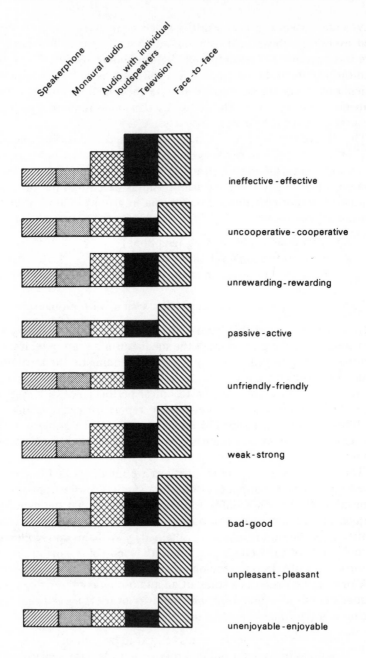

Each step represents a statistically significant difference but does not indicate the size of the difference

Source: Christie 1973

Figure 5.8: Judgments of discussions held by different communications media. (Reproduced by permission of the Controller of Her Majesty's Stationery Office. Crown Copyright.)

Evidence suggests, therefore, that whilst electronic meetings may be evaluated overall less favourably than face-to-face, there is the advantage of a more business-like atmosphere. This provides further support for the view, based on the effects of media on task outcome, that electronic meetings are particularly appropriate for very task-oriented meetings, but less appropriate for meetings where interpersonal relations are especially important.

Attitudes toward electronic meeting systems

Early studies of user attitudes, using the semantic differential technique, (e.g. Champness, 1972a, 1972b) proved useful for ordering communications media according to very general criteria. Using this technique, video telecommunication systems were found to be rated more favourably than audio-only systems.

Other studies examined what advantages and disadvantages potential users perceived various sytems to have. Using this technique, Christie (1973) found that his subjects—a sample of American managers—considered stereophonic audio conferencing systems to have several distinct advantages, but also some disadvantages, compared with monaural systems. The possibility of a system's advantages and disadvantages tending to cancel each other may help to explain why differences in general attitudes towards different systems (as measured by the semantic differential technique) are not larger.

Other studies (e.g. Champness, 1972c; Christie, 1974b; Williams and Holloway, 1974) showed that the acceptability of any given system is not fixed but depends upon the kind of meeting involved.

Williams and Holloway (1974) present comparisons between three telecommunication systems used for real business meetings on a trial basis. The results are presented in Figure 5.9. The three systems were all designed for group-to-group conferencing. They include two interactive television systems (the Bell Canada system and the British Confravision system), and the Remote Meeting Table (an audio-only system with automatic identification of the participants by individual loudspeakers and illuminated nameplates). As would be expected on the basis of the other research results discussed, the three systems were judged most satisfactory for relatively 'task-oriented' activities such as asking questions, giving or receiving information, or exchanging opinions, and rather less satisfactory for relatively 'person-oriented' activities such as bargaining or getting to know someone. It is interesting how similarly the three systems were judged in view of the technical differences between them and the fact that they were used by different groups of people. It seems that attitudes towards the use of electronic meeting systems depend more upon the nature of the meeting involved than the type of system.

Noll (1977) conducted a somewhat similar study of users of the Bell Telephone Laboratory video conferencing system. Users were asked to rate

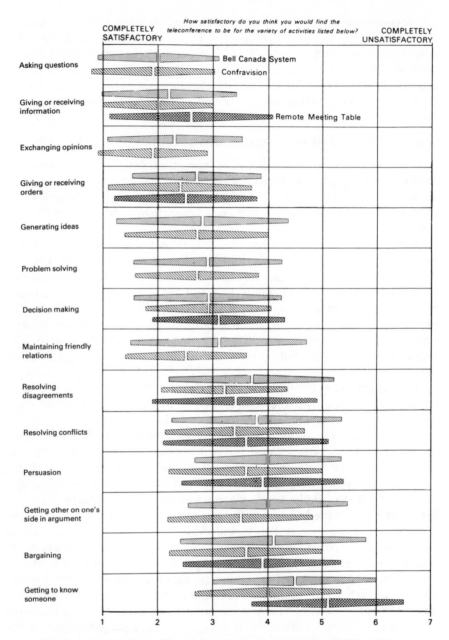

The scales show scores for selected activities carried out over the teleconference system. They are listed according to satisfaction with the system, and each result is compared with scores collected from the data of British Confravision and, where applicable, with scores from the Remote Meeting Table data (an audio-only system). Each horizontal bar indicates both a mean score and its standard deviation.

Figure 5.9: The acceptability of different electronic meeting systems according to the type of interaction involved. (Reproduced by permission of the Controller of Her Majesty's Stationery Office. Crown Copyright.)

the system on 34 scales measuring satisfaction with the system for various purposes such as 'communicating with strangers', 'discussing confidential matters', 'bargaining', 'coordinating efforts', and so forth. The respondents were also asked to rate face-to-face communication and telephone conversations on the same scales. Once again, satisfaction varied according to the purpose of the communication. Interestingly, differences between the media themselves were more marked in this study than in the Williams and Holloway study, probably because the media involved spanned a broader range than in that study.

It is not, however, only the type of meeting that influences the acceptability of meeting electronically. Social factors also have to be taken into account. A study by Christie and Holloway (1975) illustrates the importance of acquaintance in the decision to hold a meeting over a teleconferencing system or to travel.

In this experiment, 96 business managers and senior civil servants held five- or six-person discussions using one of six different telecommunication systems. They were then asked to imagine they would be involved in a business discussion. The nature of the discussion was described in detail, and in different ways for different subjects. The subjects were asked to decide, hypothetically, whether they would travel to meet the other discussants face-to-face or use the telecommunication system they had just experienced. Significantly more subjects chose to telecommunicate: when the discussion was said to involve asking and answering of questions, exchange of information and simple problem solving, than when it was said to involve bargaining, negotiation and personal conflicts; when subjects were told the discussion involved acquaintances rather than strangers; and when the travel time associated with holding the discussion face-to-face was long. The subjects' choices did not depend upon the type of telecommunication system available to a degree which was statistically significant.

In summary, whilst electronic meetings—especially those held using audio-only systems—tend to be less favourably evaluated than face-to-face meetings, there is some indication that there is the advantage of a more 'business-like' approach to the meeting. Attitudes towards electronic meeting systems, and the decision to telecommunicate, are very dependent on the type of meeting that is to be held: electronic meeting systems are generally considered more suitable for relatively 'task-oriented' meetings than for relatively 'person-oriented' meetings.

CONCLUSIONS

The research which has been done on electronic meetings suggests the following conclusions concerning the key issues identified above.

Very many different sorts of meetings take place in organizations but there are three major facets that are especially important in considering the potential for electronic meetings. These are: the aims of the meeting, what goes on at the meeting, and the atmosphere associated with the meeting. The research done has helped to identify the major dimensions which discriminate between meetings in regard to each of these facets.

Meetings which are relatively 'person-oriented' (e.g. where the emphasis is on persuasion, and other instances where interpersonal relations play an especially significant part such as when meeting people for the first time) are more affected by the medium of communication used than are meetings which are relatively 'task-oriented' (e.g. where the emphasis is on simple exchange of information). The most important distinction is between face to face communication and telecommunication. The differences between different telecommunication systems are relatively small, but where there are differences video meetings tend to be more like face to face than are audio meetings.

Users usually judge video systems more favourably than audio systems in general terms, and there are noticeable differences between different systems in each of these categories. However, their willingness to use electronic meeting systems depends much more on the type of meeting involved than the type of system. This is probably because the advantages and disadvantages of different systems tend to cancel out to some extent. For example, whilst audio systems tend to provide poorer 'social presence' than video systems (a disadvantage) they often result in a more 'business-like' approach to the meeting (which can be an advantage).

For many purposes, an optimal electronic meeting system would be one that provides high 'social presence' at low generalized cost (i.e. money spent, inconvenience involved, and other sorts of costs). In many circumstances this would probably be based on an audio system of some sort rather than video, but this situation is changing as satellite communications, local area networks, and other technical developments are making video a more affordable option. It is certainly the case that video is not as important as has sometimes been assumed and that audio systems can be effective for many purposes.

In the next two chapters we turn our attention to Type B communication and in particular to the emerging problem of how to help users cope with increasingly large amounts of information.

Chapter 6

Personal Information Systems

BRUCE CHRISTIE

INTRODUCTION

The office worker and the office are closely associated. The office is 'base', and many office workers spend a large proportion of their time in their offices, mostly at their desks. An important question in systems psychology therefore is how the office worker organizes this environment, and how electronic systems can help.

The single most important aspect of this question is how the user organizes the 'papers' in the office. This is a major information management task for the average office worker—who controls an estimated 20,000 pages of information (and, even worse, adds 5,000 a year whilst disposing of only 3,000) (Jarrett, 1982). Yet people by and large do manage their papers fairly effectively. It is vital to the success of electronic office systems that they should be at least as effective as the paper-based systems of the conventional office, and preferably should provide significant advantages.

In this chapter we consider: some general psychological principles; the psychological advantages of the paper-based office compared with conventional computer systems; some recent approaches to developing more effective electronic environments; and a number of hypotheses concerning the psychology of personal information management that should guide the development of future systems.

Key issues

The key psychological issues addressed in this chapter are:

— What general psychological principles are particularly relevant to the design of personal information systems?

— What are the psychologically salient characteristics of paper-based information systems?
— What are the positive psychological features of paper-based systems that should be incorporated into electronic systems?
— What can be learned from early examples of electronic systems for personal information management?

SOME GENERAL PSYCHOLOGICAL PRINCIPLES

The human is an extremely complex system and can make use of a wide variety of psychological mechanisms in working with its environment. It is unfortunate that the user interfaces to conventional computer systems typically do not capitalize on this fact; indeed, they typically force the user to use procedures which are among the less powerful in the user's psychological repertoire.

Recognition versus recall

Traditional user–system interfaces (USIs) often require the user to recall information about an item the user wants to retrieve from the computer. This is often the file name or some similar piece of information. That this is difficult for users to do is evidenced by the frequency with which users fail to recall the exact name and resort to looking through the directory—it is easier to recognize the name than to recall it. (Unfortunately, even then the computer makes life unnecessarily difficult for the user. The filenames which make up a conventional directory are usually so similar—e.g. all alphanumeric, all the same colour, all the same size characters—that searching through the directory can be quite slow.)

The superiority of recognition over recall can be demonstrated experimentally. A classic experiment by Wolford (1971) illustrates this, and how it can be modelled mathematically. His subjects learned a series of 'paired associates' such as 'girl– 62' and later had either to recall them or recognize them. In the recall conditions they were either given the first item (e.g. girl) and had to recall the second (e.g. 62) or were given the second (e.g. 62) and had to recall the first (e.g. girl). These were called the 'forward recall' and 'backward recall' conditions, respectively. In the recognition condition they were presented with two possibilities in a form like this 'girl (71 or 62)' and had to recognize which number went with the word.

Wolford developed a mathematical model of this situation which hypothesized that remembering depended on associations between the stimuli (words and numbers). According to the model,

> Probability of correct recognition = (probability of a forward association) + (probability of no forward association but a backward association from the correct item) + (probability of no

forward or backward association involving the correct item but backward association from the incorrect item indicating a stimulus other than the probe, e.g. a backward association from '71' to an item other than 'girl') + (probability of no associations involving either item but a correct guess).

The two recall probabilities (forward and backward) obtained by Wolford were 0.38 and 0.21, respectively. Substituting these in the formula above yields a predicted probability of correct recognition of 0.81, much better than either of the recall scores. The efficacy of the mathematical model is attested to by the fact that the observed probability of a correct recognition turned out to be very close to the predicted figure, in fact 0.80.

User-oriented USIs to personal information systems should not require the user to recall information (such as an exact file name), which suits a computer and is relatively easy to program but which users are relatively poor at. User-oriented USIs should allow the user to make more use of recognition, which may not be so easy to program but which is better suited to human psychology.

Incidental learning

Wolford's experiment, and the mathematical analysis of its results, simplifies what occurs in everyday complex situations. The analysis of Wolford's experiment assumes a very small number of possible associations between a very limited number of elements (e.g. 'girl', '71' and '62'). In most everyday situations, things are much more complex. I may recognize a document in my office by its title, the names of the authors, its position on a shelf, or any of a number of other features.

A question arises as to what elements or features ('prompts' or 'cues') a user might be able to use to retrieve a wanted item from a personal information system. The answer is very many, and furthermore they do not have to be features the user has deliberately tried to associate with the wanted item. It has been known for about twenty years (see Postman, 1964) that whether one intends to remember something or not does not matter; what matters is that one processes items when they are presented. This is illustrated in an experiment by Hyde and Jenkins (1973).

In the Hyde and Jenkins experiment, subjects were presented briefly with each of 24 words and asked to process them in some way—either by checking whether the words had an 'e' or a 'g' or by rating the pleasantness of the words. Half the subjects were told the true purpose of the experiment was to learn the words, the others were not.

The subjects who were told the true purpose of the experiment was to learn the words performed no better than those who were not (69 compared with 68 per cent, and 43 compared with 39 per cent, depending on the type of processing done by the subjects).

User-oriented USIs should therefore capitalize upon the facility humans have for forming incidental associations (incidental learning) by encouraging processing of items at the time the user puts them into the electronic store. It is not necessary for the user to feel that he or she is having to try and remember anything, so long as some processing of the items is done.

Level and elaborateness of processing

How the user processes the items as they are filed away is important. We need to consider two concepts:

— levels of processing
— elaborateness of processing.

In the Hyde and Jenkins experiment mentioned above, half the subjects were asked to process the items by checking whether the items (single words in that experiment) had an 'e' or 'g'. This was called 'phonemic processing'. The other subjects were asked to rate the pleasantness of the words. This was called 'semantic processing'. Semantic processing was considered to be a deeper level of processing than phonemic processing.

This distinction between levels of processing was proposed originally by Craik and Lockhart (1972) as part of a model for guiding research. According to the model, perception/cognition of words (or other items) proceeds through various stages from shallow to deep. The deeper the processing, the better (longer) the items are remembered. Processing of the physical features of the item is the shallowest level, phonemic processing is a moderate level, and semantic processing is the deepest level.

According to this model, the subjects in Hyde and Jenkins' experiment who were asked to rate the pleasantness of the words (semantic processing) should have remembered the words better than those who were asked to check their spelling (phonemic processing). This was exactly what was found. The 'phonemic' subjects recalled only 39 per cent (if not told the purpose of the experiment was to learn the words) or 43 per cent (if they were). The 'semantic' subjects recalled many more, 68 per cent (not told to learn) or 69 per cent (told).

A general principle suggested by this model is that a user-oriented USI should encourage the user to process items to be filed deeply rather than shallowly. This will help the user to retrieve the items when they are needed later.

In practice, it is difficult to predict what sorts of tasks will involve 'deep' processing as opposed to 'shallow' processing, except intuitively or circularly (by seeing what results in better memory). For this and other reasons, the 'levels of processing' model does not by itself provide an adequate explanation of behaviour, and it is necessary to take account of other theoretical constructs, especially the concept of elaborateness of processing.

So as a pragmatic 'rule of thumb', the USI should be designed to encourage depth of processing, but as a theoretical construct this has limited value and we need to consider the concept of elaborateness.

The distinction between depth of processing and elaborateness of processing is illustrated in an experiment by Craik and Tulving (1975). In this experiment, subjects were asked whether words such as 'watch' made sense when presented briefly in various sentences such as:

'He dropped the ...   .'

and

'The old man hobbled across the room and picked up the valuable
... from the mahogany table.'

Both sentences require semantic processing, but the elaborateness of the processing varies. The second sentence requires more elaborate processing because it involves a richer, more complex context. The results of a memory test, given without any warning, showed that the words were remembered better in the more 'elaborate' sentences.

A second guideline therefore is that a user-oriented USI should encourage elaborate processing at the time the user files an item away.

More work is needed to develop the theory of processing further. It is difficult to define 'elaborateness', as it is difficult to define 'depth', but certainly the way in which items are processed does seem to affect memory for the items. As pragmatic concepts both 'depth' and 'elaborateness' have some value in the design of the USI.

One way in which research needs to proceed is to extend experimentation to include items that are more complex than those used in the early experiments—e.g. memos and letters rather than single words. These more complex items involve many features (e.g. colour, size) that are potential prompts which the user can use later to find the items needed. The depth and/or elaborateness of processing done by the user when the item is filed should affect the number and kinds of potential prompts that are converted into actual prompts which the user can use later.

Multiple prompts

Traditional USIs do not provide many features that users can use as prompts later. An item may be retrievable only through its filename—made up of letters and numbers in monochrome (probably all in upper case) in a straight line, standard size. This is neat and tidy and convenient for programming, but deprives the user of the possibility of forming rich incidental associations.

User-oriented USIs should provide users with more opportunity to capitalize on their facility to form many incidental associations that can be used to find items later. User-oriented USIs should provide the possibility of using many different sorts of prompts, such as: colour, shape, size, typefont, orientation, texture, and many of the other prompts that people use in their conventional office environments.

The importance of context

Prompts such as the size of a book, or other features can be used when the person trying to find the book has incidentally learned the associations involved, probably without realizing that this learning has taken place (until there is a need to find the book). There may be many other potential prompts to do with the context in which the book was last seen or the document filed. These can be grouped into several categories, such as:

- physical: the physical context
- emotional: e.g. anxious, bored, in a panic, angry, or other emotional state
- time-related: e.g. a document received last year as opposed to last week
- event-related: e.g. 'I remember writing it at about the time we were also working on the Exeter Project'.

Any of these sorts of information can be helpful as prompts in helping a person to remember where an item was placed or filed away.

A technique which some hypnotists use in helping a person (e.g. witness to an accident) recall information is to help the person reconstruct or 'relive' in imagination the whole context, including emotional and other aspects. The multiple prompts thus provided can help to recall details that might otherwise not be available to the person concerned. A similar technique may be helpful in locating an item that has been lost.

The general principle can be demonstrated experimentally. For example, an experiment by Smith, Glenberg and Bjork (1978) illustrates the importance of physical context. In that experiment, subjects learned lists of paired associates in two different physical settings. On the first day, they sat in a windowless room and saw the paired associates presented on slides by an experimenter who was smartly dressed in a coat and tie. On the second day, they sat in a different room which had windows and listened to the paired associates played on a tape recorder by the experimenter dressed sloppily in a flannel shirt and jeans. On the third day, they were tested for their recall of the items presented—half the subjects in one of the settings used for learning, the remainder in the other. Subjects could recall 59 per cent of the list learned in the same setting as tested, but only 46 per cent of those learned in the other setting, so the physical context of the test was an important influence on recall.

A second example illustrates the importance of the emotional context. Bower, Monteiro and Gilligan (1978) had their subjects learn two lists. For one list, the subjects hypnotically reviewed a pleasant episode in their lives in order to induce a positive emotional state; for the other list they hypnotically reviewed a traumatic event in order to induce a negative emotional state. The subsequent recall test was done under conditions of either a hypnotically induced positive state or a hypnotically induced negative state. Recall was better when the test state matched the learning state, illustrating the importance of emotional context on recall.

The relative ease with which people can remember information when they can return to the same physical and emotional state they were in when they learned the information, and the relative difficulty they experience when trying to remember it when in a different state, is called state-dependent learning.

A lesson to be learned for USI design is that users can be expected to find it easier to retrieve items from an electronic system (items they themselves have previously filed) to the extent that the USI can help the user to reconstruct the context in which the item was filed.

SOME CHARACTERISTICS OF PAPER-BASED SYSTEMS

The paper-based conventional office provides a very good environment in which users can make use of the various psychological mechanisms they have available in their psychological repertoire for the management of personal information.

A systematic study of personal information management in a paper-based (or other) environment based on adequate methodology and an adequate sample remains to be published. We shall base our comments here on everyday observation and on the findings from a small and relatively informal exploratory study by Malone (1983).

Malone interviewed ten office workers for one hour each. The interview was split into three main parts:

— a 'tour of the office' in which the interviewer attempted to elicit a description of how the office was organized and why;
— a part of the interview in which the interviewee was asked to find certain pre-determined 'probe' items in the office (not included in all the interviews);
— a set of standard questions covering organization of the office and problems experienced.

The following are among the key findings reported:

Neat and messy offices

Malone describes two of the cases in some detail. One illustrates a 'neat' office, and one a 'messy' office. The difference seemed to relate to the type of work done. The neat office belonged to a purchasing agent whose work was based largely on standard forms. The work was routine and well-defined with information being processed according to clear cut procedures. The messy office belonged to a research scientist whose work involved using a variety of different sorts of items—magazines, journals, papers, computer listings, and personal notes. In this case there was very little routine paper flow. The 'mess' was in fact highly structured and personalized to the needs of the person concerned.

These cases support everyday observation that people differ in how they organize their personal space in an office environment. Malone relates the differences between the two cases reported primarily to differences in the type of work done. It is likely that differences between offices also to some extent reflect differences in the personalities of their users. It is interesting to note that when Malone divided his cases into 'neat' and 'messy' he found that several of the people in neat offices said they 'couldn't stand clutter'. It may be of interest to some psychologist readers to speculate to what extent such attitudes reflect broad personality characteristics. (Non-psychologist readers may like to refer to Kline, 1983, for an easy to read good overview and discussion of personality theory.)

There was evidence that, whether because of differences in the type of work done or for other reasons, some of the people with messy offices appeared to have more difficulties with the management of their information (e.g. somewhat greater difficulty in finding items).

Files and piles

As everyday observation suggests, Malone found that people kept both 'systematic' files and 'unsystematic' piles (e.g. on their desks). He makes the interesting point that conventional computer systems allow only files and that, since people seem to have a need for piles as well, perhaps more user-oriented systems should also allow for the possibility of piles.

Finding and reminding

People organize their personal information space in ways that help them find items they need, but Malone also found evidence for a 'reminding' factor as well. Five of his ten interviewees made explicit remarks about the importance of reminders. It seemed that about two-thirds or so of the piles observed appeared to serve a reminding function.

One of the difficulties identified in filing rather than piling information was the difficulty in deciding on the appropriate categories in which the items should go, but Malone suggests additional reasons why people keep piles: the mechanical difficulties of actually creating files easily and quickly; the need to be reminded of tasks to be done; and the wish to have frequently used information easily accessible.

ADVANTAGES OF PAPER-BASED SYSTEMS

We shall consider three key advantages that paper-based office systems have compared with conventional computer systems. Compared with conventional computer systems, paper-based systems are:

— image-rich
— personal
— familiar.

The need for an image-rich environment

Imagery is a key aspect of human information processing. Imagine trying to do your work if you were blind. You would be able to manipulate only auditory and tactile space, all the things you can do visually would be gone.

Imagine trying to do your work if books and other items did not exist—if all that did exist were endless monochrome lines of text and numbers (in upper case only if you were unlucky enough). You would not be able to place a letter on your chair when you left the office to remind you to deal with it as soon as you returned, nor dog-ear a page in a book to remind you of where you had reached, nor find a leaflet in a pile on your desk by remembering it was on blue paper.

Books and papers and other items in a typical office provide very rich images. Colours and shapes fill space in a highly complex way that often gives the impression of complete chaos. Often it does get somewhat out of hand but, for the most part, is it not much more difficult to work with this complexity than with a 'tidy office'—and is it not irritatingly difficult to find papers and so forth once someone has been 'kind' enough to reduce the complexity by 'tidying up'? The reason is that the complexity is only chaotic to an outside observer. To the user it is a complex network of associations that provide a basis for effective information handling.

Traditional computer displays are grossly image-impoverished compared with a typical office environment. Traditional displays restrict the user to monochrome lines of text and numbers with perhaps a few very simple graphics. This is a very poor substitute for the richness that paper affords.

The need for personalization

The value of personalization is evident in everyday examples. For example, consider a typical 'untidy' desk. It is not simply untidy, it is personalized.

It may appear chaotic to an outside observer, but the user can make sense of it—it is in fact highly organized, but in a way that is idiosyncratic to the user. This becomes evident if by some mishap the items on the desk are swept onto the floor in a heap and then put back on the desk in an arbitrary layout. The result may appear no more or less complex than before, but now the desk's owner may find it as difficult to find particular items as the outsider does. The typical desk only appears chaotic because its organization is highly personalized to the preferences of the particular user.

Some office workers do keep very tidy desks, and this again reflects their personalization of their information environment.

The particular schemes people use to organize their personal files are also highly personalized, as anyone who has any files for their personal use will know.

Electronic office systems should provide at least as much opportunity for personalization as is provided by the traditional office environment. This can be a matter of personality to some extent, and to some extent a reflection of the type of work done. Malone (1983) reports contrasting cases he came across in his interviews. In the first case—that of a purchasing agent—the person worked almost exclusively with standard forms that needed to be processed in a very rigidly defined way; his desk was 'neat'. In the second case—that of a research scientist—the person had very little routine paper flow; his desk was 'messy'.

The need for familiarity

The paper-based office is a familiar environment to most office workers. They know the principles on which it is organized and they know how to manipulate it. The typical office worker is expert in handling information, especially in the form of text, but is not familiar with the intricasies of traditional computing. In psychological terms, the 'schemas' or 'schemata' (framework of understanding) which the office worker brings to the information-handling task are very different from the schemas which computer specialists use.

It would be possible in principle to teach the office professional appropriate psychological schemas for interacting with computer systems using traditional dialogues. In practice this has not happened, and is most unlikely ever to happen. Professional office workers have shown themselves by and large unwilling to learn the necessary skills, partly because they do not believe the returns would warrant the effort, and partly because they simply do not have the time.

An alternative, more practical approach is to design the environment to fit in with the user's existing schemas as far as possible. In this way, the new situation will appear familiar and the user will not feel at a loss or need nearly as much training.

Conventional computer systems compared with paper-based systems

Compared with the image-rich, personalized, familiar environment created by paper-based office systems, conventional computer systems typically have been

— image-impoverished
— impersonal
— unfamiliar to their intended users.

EARLY STEPS TOWARDS MORE EFFECTIVE ELECTRONIC ENVIRONMENTS

Recognizing that conventional computer systems have been deficient in important respects from the user's point of view, several different centres have been developing approaches to electronic environments that are more satisfactory. We shall look at four examples of such developments and consider what general principles can be discerned.

The MIT Spatial Data Management System

The Spatial Data Management System (SDMS) at MIT was developed a few years ago as a prototype system that demonstrates how the spatiality inherent in how people typically organize their personal information space can be interpreted into an electronic environment that incorporates concrete and familiar images. SDMS has had an important influence on human factors work in the area of office systems.

The prototype system uses a wall sized visual display, projected from a full colour digital television. The display presents a virtual spatial world, over which the user can 'helicopter' with joystick control. The user can zoom in on items of interest and interact with them. The items, or 'data types', include: maps, text, book-like items, letters, photographs, slides, movies, sound, and television. Sound is provided by a stereophonic system synchronized with the visual information.

The items available to the user for inspection are arranged in a two-dimensional 'Dataland'. The whole of this Dataland is shown continuously on a small monitor—the 'world view monitor'—to the user's right. The items on the world view monitor are about one-half to three-quarters of an inch tall. Some of the items are located by themselves against a darkish background; others are grouped together on coloured backgrounds. A small part (about one sixty-fourth) of Dataland is shown in fine detail on the wall-sized screen. An indicator on the world view monitor shows the region of Dataland currently being presented on the large screen.

To 'move to' a new area of Dataland, i.e. to change that portion of Dataland shown on the large screen, the user can manipulate a joystick to traverse Dataland continuously or can touch the relevant part of the world view monitor to jump to the new area.

The spatiality of the interface between the user and the system is enhanced by the use of sound as well as the visual images. Some of the items in Dataland are associated with particular sounds. These sounds can all be heard all the time through the stereophonic sound system, but their relative loudness depends on how near the relevant items are to those in view on the large screen. For example, as the user traverses Dataland in the direction of an image of a telephone (which is one of the data types used) the sound of the telephone ringing becomes louder.

Spatiality in SDMS. The emphasis on spatiality in SDMS is explicit in descriptions its designers give of it (e.g. Bolt, 1979). Spatiality is certainly one of the most striking features of the imagery used in the design of the interface, being conveyed visually (on the screen), auditorily (through the sound system), and kinaesthetically (through manipulation of the joystick). But there are other important aspects of the SDMS imagery that deserve note, as follows.

Concretization. Information is abstract in essence but in the external world (outside a human or computer memory) is always embodied in a concrete form. The images of items in Dataland are not labels pointing to abstract computer files; they are images of concrete objects such as letters, maps and so on.

Familiarity. The images are not just concrete, they are also familiar. The letters look exactly like we expect letters to look, including their shape, typefonts, layout, letter-headings, and so on. We know how to relate to them because, in psychological terms, they activate existing cognitive schemas.

The Panorama Project

Spence and his colleagues at Imperial College of Science and Technology (e.g. Spence and Apperley, 1982) have taken some of the ideas suggested by the work at MIT, including SDMS and other projects, and applied them to the office of the professional in a project they call Panorama.

The user has available a wall display of 'icons' arranged spatially in two dimensions. The icons are concrete, familiar images (e.g. journal covers, picture of an in-tray, and other objects). These are either fixed to the wall or are projected onto the wall from a slide projector, in which case the particular set of icons can be changed easily at will.

The user can select a particular icon by pointing to it. Items relevant to that icon are then displayed on a screen at the user's desk. For example, if the user points to the 'in-tray' icon, then the contents of the in-tray will be displayed on the screen.

The user's screen presents items in what is called a 'bifocal display'. This means that all the relevant items (e.g. all the items in the in-tray) are always on the screen, but only the item 'in focus' in the centre of the screen is shown in detail—all the others are shown squashed up in less detail in the outer regions of the display. This allows the user to read the item of interest whilst retaining context information about the location of the other items. Again, the emphasis is on using spatial information in a two-dimensional array of concrete images of familiar objects (letters, telexes, reports, journals).

The user can bring other items into focus by touching the relevant parts of the screen.

Spatiality, concretization and familiarity are important aspects of the imagery used, just as they are in SDMS. In addition, Panorama introduces the following notions.

Hierarchy. There is no 'world view' as such. Instead, there is a high-level set of icons which provide clues to what lies behind them. The spatial context information provided by the outer regions of the bifocal display does not provide a world view, but only an overall view of a particular set of items selected by the user. Spence and Apperley (1982) discuss a whole range of what they call 'information levels'.

Use of the office environment. Panorama uses the office environment more than does SDMS. The office wall is used for the high-level icons; the bifocal display unit is incorporated into the ordinary desk (along with other Panorama devices, such as a speech recognition unit); a monitor in another part of the office is used to provide a 'public' display. The Panorama system becomes a part of the office, whereas SDMS is based in a special 'media room'.

Queen Mary College's NAN System

NAN is an office information system developed at Queen Mary College in London (Lamming, 1979). It provides a set of general purpose information handling facilities that can be oriented towards specific applications (e.g. document preparation, filing).

The user at any one time normally will be interested in only a relatively small number, e.g. twenty, of the total set of activities. These activities are treated specially by the system in that they are arranged like pieces of paper across a visual simulation of a desk top. The user can move activities back and forth between the 'desk top' and the 'filing cabinet' at will.

Comments. The imagery used in NAN again includes spatiality (a two-dimensional array of representations of activities) and familiarity (the 'desk top', 'filing cabinet', and 'pieces of paper'). It also includes the notion of hierarchy—only some activities (those on the 'desk top') are of interest at any given time, others (in the 'filing cabinet') are not visible.

The Xerox Star and the Apple Lisa and Macintosh

The Star (or '8010 Star Information System') was announced by Xerox in April 1981 as a new personal computer designed for offices. Consisting of a processor, a monochrome display, a keyboard, and a cursor-control device (the 'mouse'), it is intended for business professionals who handle information (Smith *et al.*, 1982). The Star created a lot of excitement in the human factors world when it was first launched and has had a significant influence on subsequent developments in USI design. The Apple Lisa and Macintosh conceptually owe much to the Star, and they all incorporate some of the ideas demonstrated in the earlier SDMS.

Every user's initial view of the Star is the 'Desktop'. This is a grey area which entirely fills the screen. Icons on the Desktop represent the working environment, current projects and accessible resources. They are images of familiar office objects, such as documents, folders, file drawers, in-baskets and out-baskets. They are visible, concrete embodiments of corresponding physical objects and users are encouraged to think of them in physical terms. The intention is that users will be able to intuit things to do with the icons, and that those things will indeed form part of the system. The designers felt this would happen if

— the Star were to model the real world accurately enough, its similarity with the traditional office environment preserving familiar ways of working and existing concepts and knowledge; and
— the system were to incorporate sufficient uniformity, allowing principles and commands learned in one area to be applied in others.

The same principles of spatiality, concretization and familiarity are embodied in the imagery used in the other systems reviewed above. The Star also embodies the notion of an office environment, although its representation of this is grossly distorted as all the objects represented are shown on the 'desk top'. The imagery is limited further by the use of the same shaped icons to represent all objects of a given type, the absence of colour, and in various minor ways (e.g. constraints on how items can be arranged on the 'desk top').

Literal versus functional representation

The imagery used in 'Dataland', the Star, and Lisa tends to be literal. That is, it tends to recreate the physical appearance of the office and typical office objects (e.g. folders, 'in' and 'out' trays). The trend can be taken further than in those systems. Commodore's 'Magic Desk', for example, presents a very literal picture of the office to the user. It includes a desk, a typewriter, a filing cabinet, desk drawers, and so on. The user can use any of these by pointing a 'hand' shown on the screen at the object which then behaves like a conventional piece of office equipment. For example, if the user points to the typewriter, the screen changes to look like the view the user would see of a conventional typewriter—and the user's keyboard behaves like a conventional typewriter keyboard. The same kind of literal interpretation of the conventional office environment into the electronic medium applies to the 'filing cabinets'.

This kind of literal interpretation is a two-edged sword. Its advantage is that the images shown activate cognitive schemas that are well-learned. So long as the 'equipment' depicted then behaves like conventional equipment, say a typewriter, the user can make use of that aspect of the system without having to learn any new procedures. The disadvantage is that if the item on the screen behaves like its conventional counterpart, then the user is deprived of many of the advantages that computer power can offer, e.g. the 'typewriter' will not offer the user sophisticated word processing facilities— if it were to it would be confusing because conventional typewriters do not, and the user would have to positively 'unlearn' some things about the 'typewriter' as well as still having to learn the new word processing procedures. To accommodate word processing in the 'literal' approach, the USI would have to display a picture of a word processor—but what should this look like? One very quickly lands in a circular situation.

The same applies to personal filing. Many personal computers offer personal filing systems based on a more or less literal translation of physical card index systems or filing cabinets into the electronic medium. The result is the systems are easy to learn (because they are consistent with existing cognitive schemas) but fail to make full use of the computer power available.

The 'literal' approach may well be very suitable for users who may need to use a filing system (or typewriter, etc.) occasionally, who keep only small amounts of information on file (or type only a few letters, etc.), and who do not wish to learn new procedures. Such an approach may have some application in the home, for example. It is much less likely to be very suitable for the user who needs to use the system often for reasonably high levels of work.

For the latter type of user, a 'functional' approach may prove of more value than the literal approach. The 'functional' approach uses an image-rich USI, and will incorporate familiar elements in the imagery, but will not present the system simply as a 'filing cabinet', 'card index system' (or 'typewriter', etc.). it will provide the user with many opportunities to make full use of the computer power available.

Conclusions on the systems reviewed

All of the systems reviewed above recognize the value of providing users with an image-rich interface to the system. The following aspects of this imagery are used in all the systems:

— spatiality
— concretization
— familiarity.

In addition, there is some recognition of hierarchy or levels of information, e.g. some items in view, others 'put away' somewhere or not visible.

All the systems base their familiarity to a greater or lesser extent on providing images of items that are found in traditional offices. The desktop is given special prominence in most of the systems.

Despite this general recognition of the value of imagery, all the systems are still *relatively* image-impoverished:

SDMS provides the user with vivid visual, auditory and kinaesthetic images, but the items portrayed seem to exist not exactly in a vacuum but in a rather nebulous 'Dataland'. There is no definite environment, just a two-dimensional array.

Panorama adds to the cues associated with the various images (the richness of the imagery) by integrating the images into the ordinary office environment but not entirely successfully. The images seem to emanate from an abstract system that exists 'somewhere in the computer' and are merely projected onto or into the 'alien' office environment, not really belonging to it.

NAN presents items that are supposed to look like pieces of paper but otherwise are not necessarily very familiar, and does so in a two-dimensional array that is meant to represent a desktop but does so only if one first accepts the imagery of the pieces of paper—there is no environment around the 'desktop' to reinforce the imagery.

The Star attempts to portray more of the office environment but does so in a relatively distorted and image-impoverished way because it:

— ignores the uniqueness of real, individual objects by giving all objects of a given class the same shape
— reduces the richness of the imagery by reducing everything to monochrome
— removes much of the structure of a real office by distorting and breaking down the relationships between different objects by making them all more or less the same size and putting them all on top of the 'desk'
— makes the interface rather impersonal by standardizing everything except the particular positions (within a standard grid) of the various icons used.

The various systems considered have been important in demonstrating that the user interface to a computer-based system does not have to look like a conventional computer. They have demonstrated the possibility of designing image-rich interfaces, and have had an important influence on trends in USI design.

They have also served to point up a distinction between a 'literal' and a 'functional' interpretation of the office into the electronic medium. The literal approach may have some benefit for casual users but the functional approach allows the user to make more use of the computer power available.

HYPOTHESES AND CONCLUSIONS

The systems considered are all complex and amenable to different interpretations by different people. There is a need for systematic research to submit the various hypotheses they suggest to controlled tests. The following are some key hypotheses suggested by the systems and by the general psychological principles considered earlier.

— recognition	— the USI should allow the user to use recognition as well as recall to retrieve items from the system;
— incidental learning	— the USI should encourage the formation of incidental associations at the time the user puts an item into store, without requiring the user to make a deliberate effort to remember anything;
— depth and elaborateness of processing	— the USI should provide opportunities for the user to process items deeply and elaborately at the time they are filed;
— multiple prompts	— the USI should facilitate incidental associations by using size and many other features to provide the user with the possibility of multiple prompts;
— context	— the USI should be designed to help the user to reconstruct the context in which an item was filed when it is needed later;
— imagery	— the USI should be image-rich, allowing the user to develop multiple associations and use other psychological mechanisms that are so useful in a paper-based environment;
— spatiality	— the imagery should incorporate a virtual space of at least two and preferably more dimensions;
— concretization	— the imagery should be as concrete as possible, giving an impression of 'reality';
— familiarity	— the imagery should be familiar so the user can bring established psychological schemas to bear, and learning can be minimized;

— hierarchy	— the imagery should incorporate 'levels of information', reflecting the psychological constraint that the user cannot give equal attention to everything all at the same time;
— environment	— the imagery should incorporate a virtual environment in which items are located, and this environment should be image-rich, not merely the physical boundaries of the VDU screen;
— personalization	— the imagery should be capable of being personalized to fit in with the user's own particular preferences;
— literal versus functional interpretation	— the imagery should not normally be a literal interpretation of the conventional office environment except perhaps for casual users with a low workload; this applies to personal information management and to other services provided by the systems.

In the next chapter, we turn our attention to the world of information that lies beyond the user's personal information space, and how the user can be helped to search through that world to find the particular items of information that are needed.

Chapter 7

Shared Information Systems

BRUCE CHRISTIE

INTRODUCTION

In the previous chapter we considered the professional office worker's need to be able to access his or her personal files quickly and efficiently. Very often, however, the information that is needed does not exist in one's own files. What happens then? What happens when the user needs to seek information from the world that lies beyond his or her personal files—in public, organizational, or other shared databases. That is the subject of this chapter.

It is not possible simply to extend the whole range of techniques that are potentially available to help in the management of personal files. One cannot, for example, 'dog ear' or otherwise tag pages in order to be able to locate them easily in future unless one has already accessed those pages—in which case, by definition, they form part of one's personal files.

The person engaging in 'information seeking' as the term will be used in this chapter is searching an ill-defined range of information sources for items of information that often may not exist in exactly the form required, even if the user is capable of specifying that form—which is not always the case. If the information does exist it may be in the form of a graph in a magazine article, a table of numbers in a textbook, in a colleague's head, in a videotex page—the user often does not know. Maybe the information does not exist at all.

Meadow (1970) has suggested the user typically:

— has only partial knowledge of the information (s)he wants
— may have no knowledge at all of the way the information is structured in whatever information systems may be available

— may not know the appropriate language for dealing with the available systems, especially if they are computerized, and may not know how to formulate his or her requirements
— does not know what penalties may be imposed on his or her search strategies.

It may seem surprising that people ever manage to find what they are looking for, given these uncertainties, yet they do. In this chapter we shall consider the psychology of what is involved and some of the possibilities for improving the interface between the user and the world of information beyond the user's personal files.

A possible sequence of steps that might occur when someone seeks information is as follows, and we shall look at each of these steps in turn:

— The person (or system) identifies a need (or possible need) for information.
— The person formulates the need more clearly in his or her mind.
— The person selects a source of information.
— The person attempts to find the relevant information by entering into a dialogue with the source, which may be an electronic system or some other type of source (e.g. a colleague).
— The system attempts to identify relevant information and to present it to the user.
— The performance of the system is evaluated.

This is not the only possible sequence of steps but it is convenient for organizing the remainder of this chapter.

Key issues

The key psychological issues addressed in this chapter are:

— What psychological factors underlie and influence the decision to seek information?
— What do people need in order to help them formulate their information requirements?
— What factors influence a user's choice of a source of information, e.g. whether to use an electronic system or some other means of obtaining the information needed?
— How well do conventional retrieval systems support users' needs?
— How can 'relevance' be assessed and used by an intelligent system?

IDENTIFYING A NEED FOR INFORMATION

People only seek information when there is a need to do so. Perhaps the first point to be appreciated in designing the USI for information seeking is that there are two broad categories of need that must be considered. Any given system in principle can address either or both types of need, but the psychological requirements on the USI are different for the two. The two types are:

— diversive information seeking (or 'diversive exploration', after Berlyne, 1960); and
— specific information seeking (or 'specific exploration').

Diversive information seeking occurs when a person browses through a set of information without having any specific requirement in mind.

Specific information seeking occurs when a person is seeking a specific piece of advice, an answer to a specific question, or other information that (s)he has specifically in mind (even if it is rather ill-defined).

In practice the two often occur together. For example, one may consult a dictionary for the spelling of a particular word but become distracted by reading other interesting words one comes across. Or one might consult a television magazine to check on the time of a programme and become distracted by an advertisement or start reading an article.

Diversive information seeking

Diversive information seeking is most likely to occur when one is 'feeling bored'—one is likely to explore one's environment. It may be a physical environment, or it may be an information environment (e.g. library, on-line system, or videotex), depending on circumstances. In any event, it is important to acknowledge that people do get bored and that they can alleviate their boredom by browsing 'aimlessly' through domains of information. It is also important to recognize that the same factors that direct where a person goes next when in browsing mode, can also cause the person to be distracted from specific information seeking. The following provides a very brief overview of the key factors involved.

Arousal. A popular version of general arousal theory holds that people can vary in their state of general arousal (e.g. from coma, or psychophysiologically quiet sleep to a state of hyperalertness or activity). People vary in terms of this continuum according to circumstances and according to temperament. The theory also holds that there is an optimal level of arousal, and that people actively try to attain that optimum (e.g. by drinking coffee or alcohol, or putting music on, or in any of a number of other ways—including diversive information seeking).

Temperament and stimulus hunger. The theory holds that if subjected to a sufficiently impoverished environment, everyone will experience unpleasantly low arousal or 'hunger', but how much will depend on one's temperament. Eysenck (e.g. 1970) has postulated that introverts tend to be naturally more aroused than extraverts, so according to this hypothesis introverts should experience less stimulus hunger than extraverts when subjected to an impoverished environment. This was supported in an empirical test by Gale (1969).

The subjects were 36 undergraduates, selected on the basis of their scores on the EPI—a standard psychological test assessing major aspects of a person's temperament, including degree of extraversion/introversion. Each subject sat in a dimly lit corner of a dark room, facing a black console from which protruded the knobs of four morse keys. The subject could cause four different sounds to be produced by pressing the keys. During a period of 14 minutes in this situation, extraverts maintained a higher rate of key pressing than did introverts and varied their choice of keys (and therefore sounds) more often.

Arousal potential and the collative properties of stimulation. The stimulation afforded by pressing the button in the Gale experiment was minimal. In Berlyne's terms (Berlyne, 1960), it had low arousal potential—it would not be expected to raise a person's level of arousal very much. Other stimuli have greater arousal potential, depending on their 'collative' properties (Berlyne, 1960). The collative properties of stimulation include its degree of complexity, variety, incongruity, novelty, and other characteristics. One would expect the subjects in the Gale experiment to push the button less often if doing so resulted in a complex, incongruous and novel stimulus than if it resulted (as was the case) in a simple, non-incongruous and familiar stimulus. Under conditions of very high arousal potential (e.g. very complex, incongruous, novel stimuli) one would expect subjects might even work to switch off such stimulation. The author is not aware of any direct tests of this, but a number of experiments (e.g. Berlyne and McDonnell, 1965; Christie *et al.*, 1972) have indicated that EEG indexes of arousal can be affected by the collative properties of stimuli, which is consistent with the theory.

Implications for systems design. The primary point is that users of office systems from time to time have a need to engage in diversive information seeking—a need which a system can be designed to address, or not as the system's designer chooses. A secondary point is that the psychological factors which influence the course of diversive information seeking operate continuously at the USI and will sometimes affect the course of specific information seeking. For example, if a user's job is fairly boring (s)he may prefer to use a more 'interesting', more 'entertaining' way of finding, say, the time of a train. If the user is likely to be very aroused already (e.g. a 'high pressure' job with very little time and lot of stress), the user may prefer a simpler, less stimulating system. The use of colour, graphics, animation, speech output, and other features of the user–system dialogue can all be expected to be important in this connection.

Specific information seeking

Specific information seeking by definition only occurs when the user has a particular requirement. However, 'objectively' having a requirement—even subjectively acknowledging that a requirement exists—is not sufficient to ensure that information seeking will occur.

Janis and Mann (1977) have discussed the antecedents to specific information seeking at some length. They propose that they are more complex than was originally thought. Prior to the mid-1960s, psychologists generally believed that people censored the information they were prepared to accept in a highly biased way in order to protect their existing beliefs and decisions. People were believed to seek information that supported their position and to avoid information that did not. Deviations from this 'selective exposure hypothesis' were usually regarded as unimportant exceptions to the general rule, and the hypothesis became incorporated as a fundamental part of various attitude theories, especially cognitive dissonance theory. The hypothesis was called into question by a number of studies done during the 1960s which showed that it was by no means as general as had been assumed. Janis and Mann suggest that the problem should be looked at a different way: to specify the conditions under which people display different sorts of preferences for information.

They discuss information preferences in terms of seven main categories. These are summarized in Figure 7.1. When there are no serious risks and no deadline pressures, interest in seeking information is low. Interest in information increases as pressures increase. Eventually, however, if the pressures become too great—high risk and little hope of a satisfactory solution—then the person concerned is likely to cope by what Janis and Mann call 'defensive avoidance', and interest in seeking information becomes low again.

Implications for systems design. People cannot be relied upon to be entirely 'rational' about when they seek information, or what information they are prepared to look for. If control is left entirely to the user of an information system, the user may fail to be exposed to appropriate information at the optimal time. There is therefore a case for giving the electronic system some degree of control over what is presented to the user and when. For this to be possible, however, it is necessary for the system to be able to evaluate what information is appropriate to present to the user. This problem is considered later. First, let us consider the need the user may have to formulate his or her information requirements more precisely before attempting to explain them to an electronic system or satisfy them in some other way.

FORMULATING THE NEED

Before an intelligent electronic information system can provide relevant information, it must have some understanding of the user's requirements.

	Coping pattern	Dominant information mode	Characteristic information preferences	Level of interest in information
A.	Unconflicted adherence	Indifference	Nonselective exposure	Low
B.	Unconflicted change	Indifference	Nonselective exposure	Low
C.	Defensive avoidance			
C-1	Procrastination	Evasion	Passive interest in supportive information avoidance of all challenging information	Low
C-2	Shifting responsibility	Evasion	Delegation of search and appraisal to others	Low
C-3	Bolstering	Selectivity	Selective exposure search for supportive information and avoidance of discrepant information	Medium
D.	Hypervigilance	Indiscriminate search	Active search for both supportive and nonsupportive information, with failure to discriminate between relevant and irrelevant, trustworthy and untrustworthy	Very high
E.	Vigilance	Discriminating search with openmindedness	Active search for supportive and nonsupportive information, with careful evaluation for relevance and trustworthiness; preference for trustworthy nonsupportive information if threats are vague or ambiguous	High

Figure 7.1:   Preferences for information. (Reproduced by permission of Collier MacMillan.)

Typically this will come about at least partly by the user specifying the requirements to the system (although the system could infer requirements in other ways, e.g. by monitoring the user's efforts to produce, say, a report or research article and then volunteering information from time to time). For the user to be able to specify his or her requirements to the system, (s)he must have some understanding of what they are.

Preliminary briefing

People use a variety of different strategies in developing an understanding of what it is they need. Some of these strategies were identified in a study by Wolek (1972) of scientists and engineers. He was interested in what such people do when they need to brief themselves on a technical subject prior to a meeting.

Thirty scientists and engineers from three different establishments were interviewed about the most recent incident in which they had searched for and obtained useful information through interpersonal channels. The 'critical incident' method was adopted to tease out information about all the relevant events leading up to the meeting.

Three main strategies were identified, as follows.

Piggybacking. Especially if there were higher priority demands on his or her time (the 'opportunity cost of time'), the person concerned would delay any specific search for information in the hope that between the time of identifying a need for information and the time when action would be required (the meeting) the information would become available through the normal course of events, e.g. routine information gathering, chance encounters with colleagues, and so on. The need for information was 'piggybacked' on routine activities. This seems to correspond roughly with Janis and Mann's 'procrastination' pattern which is associated with no deadline pressures and results in a low level of interest in information.

Friendly consultation. This involved contacting friends for references to likely sources of information, and the persons concerned would be guided in this by previous experiences ('experience with similar needs'). They preferred to go to friends first even when an expert was available who was an acquaintance. It seemed to serve three main functions:

— reducing the possibility of embarrassment by asking an uninformed question
— improving the precision of the questions that needed to be asked so as to avoid ambiguities or misunderstandings
— keeping friends in touch with the work being done.

Professional peripheration. This was a deliberate attempt to learn about areas of work peripheral to the person's special field of expertise, and was used especially to aid in the understanding of the context of the question or problem ('context appreciation'). It usually involved reading research abstracts, review articles, and similar materials. The people interviewed seemed to assume it was their responsibility to educate themselves sufficiently to be able to ask appropriate questions and understand the replies.

Special communicators

The strategies identified above indicate that contacts with other professionals can be important in helping to formulate questions. Some professionals are more valuable to their colleagues in this respect than are others. Holland (1972) studied these 'special communicators' to see what characterized their use of information compared with that of their colleagues.

Two hundred and nineteen professional researchers in three different organizations were sent questionnaires, and 143 completed questionnaires were returned. The 'information potential' of the researchers was defined as the number of times they were cited by the others as likely sources of technical information. They were asked various questions regarding their exposure to information, such as the number of journals read and meetings attended.

The researchers with high information potential made somewhat different use of external and internal information, as follows. With regard to external information, they

— either contacted an unusually large number of acquaintances or read an unusually large number of journals and reports
— tended to contact a more diverse set of information sources
— in contacting personal sources, tended to prefer face-to-face and telephone communication rather than written communication.

With regard to internal information, they

— were more frequent users of personal contacts
— read only about the same number of internal reports as the low information potential researchers (except in one of the three organizations, where they read more)
— tended (in two organizations) to contact a more diverse set of information sources
— tended to prefer telephone communication rather than face-to-face communication for internal technical communication.

Implications for systems design

The studies cited above illustrate the point that professionals use multiple strategies in briefing themselves on particular topics they need to address, and that contact with other professionals plays an important part in this. In this regard, contacts with other professionals are not spread evenly across all the other professionals available—it is concentrated on friends and on 'special communicators' who are regarded by their colleagues as having high information potential. For these special communicators also, contacts with other professionals are very important.

There would seem to be three alternative strategies for the design of intelligent information systems, derived from this analysis:

— acknowledge the need for professionals to contact other professionals, probably before using an electronic system, and do nothing about that; or,

— design the system to facilitate that by using the system to identify the 'special communicators' in particular areas whom the user may wish to consult. This could be done by the system automatically, by the system

    — monitoring professionals' use of external and internal information and identifying those who fit the profile of the 'special communicator' (above)

    — monitoring which professionals are contacted most often by their colleagues; or

    — researching the details of what goes on when a professional contacts a 'special communicator' and then designing the system to mimick some or all aspects of this, thereby reducing the load on the 'special communicators' and doing some of this work using electronic intelligence.

SELECTING A SOURCE

Having acknowledged a need for information and having formulated what that need is, the person concerned then needs to select a possible source of information. This could be a library, an electronic system, a colleague, or other source. How does the user go about selecting a source? Some of the factors involved are illustrated by the following two studies, by Rosenberg (1967) and Christie (1981).

Rosenberg was able to show that ease of use can be an especially important factor—more important than the amount of information expected.

Rosenberg asked 96 professionals in industry and Government (52 in research, 44 not) to consider three possible situations, described as follows:

— You are working on a design for a procedure or experiment and wish to know if similar work has been done.

— You are preparing a proposal for a new project either to the management of your organization or to an outside agency. You wish to substantiate the proposal with a thorough bibliography. The proposal involves approximately $60,000.

— You wish to gather information in order to write an article in your area of specialization for a trade or research journal.

The subjects then ranked eight possible sources of information according to their preference in each situation. The possible sources were:

— search your personal library
— search material in the same building where you work, excluding your personal library
— visit a knowledgeable person 20 miles away or more
— use a library that it not within your organization
— consult a reference librarian
— visit a knowledgeable person nearby (within your organization)
— write a letter requesting information from a knowledgeable person 20 miles away or more
— telephone a knowledgeable person who may be of help.

The subjects also rated the sources, using 7-point scales, according to a) their ease of use, and b) the amount of information expected.

A statistically significant correlation was found between order of preference (averaged over the three situations) and ease of use (Spearman rank correlation coefficient greater than 0.86 for both groups), but not between preference and the amount of information expected (coefficients of -0.17 and -0.11 for the 'research' and 'non-research' groups).

The experiment suggests that ease of use can sometimes be the crucial factor in determining which source of information a user chooses. This does not mean that ease of use is so important in every situation or for all categories of user, and a later experiment by Christie (1981) suggests that there can be interactions between the various factors involved, the important factor depending on the situation.

Christie was particularly interested in the choice of Prestel (the British Telecom videotex service) compared with other sources of information available in the home. In view of Rosenberg's findings he hypothesized that convenience would be more important than perceived effectiveness. The experiment was designed to test this hypothesis and the relative influence of several other factors.

Prestel was demonstrated to a group of 20 women. Each subject was then given a set of five sheets, in random order. One of five possible situations (selected to be diverse) was described at the top of each sheet:

— The children need new clothes. You want to find out the best place to go to get some.
— You want to catch up on the national news.
— A pipe has just started to leak badly. You need a plumber urgently.
— You are bored cooking the same old things. You want to try something new. You need to get an interesting recipe that you will feel confident attempting.
— You are thinking of buying a car. You are not sure what sort. You need information.

The subjects were asked to consider nine possible sources of information in relation to these, as follows:

— Prestel (= videotex)
— turn on the TV or radio
— make one or more telephone calls
— visit somewhere (e.g. a shop, Post Office or Agency) to make enquiries
— go and see a neighbour about it
— look in a newspaper, magazine or directory (e.g. yellow pages)
— look in some other kind of book or magazine you already have
— get a book or magazine from the library or a shop
— other (please describe briefly).

The subjects were given 15 minutes to rate each source of information on the following six rating scales for as many as possible of the five situations:

— How likely it is you would use this material?
— What you think it would probably cost you, taking everything into account?
— How convenient you think it would be?
— How effective you think it would be?
— How speedy you think it would be?
— How good or bad for you in very general terms (taking all the above factors into account and any others felt relevant)?

The scales each ran from 0 (e.g. 'extremely unlikely') to 5 (e.g. 'extremely likely').

The product–moment correlations between the scales were calculated across the methods and subjects, separately for each situation. The results showed that which factor was the most important in influencing the likelihood of choosing an information source depended upon the situation. For most situations it was the perceived effectiveness of the source, but for the 'recipe' and 'car' situations, convenience was equally important.

The experiment therefore did not support the hypothesis that convenience would generally be more important than effectiveness. In fact, the relative importance of convenience varied over quite a wide range. It had most influence in the 'recipe' situation ($r = 0.83$), and was least important in the 'clothes' situation ($r = 0.25$).

Interestingly, the likelihood of selecting a source was positively related to its perceived cost (not negatively, as one might have expected), i.e. the more costly sources were more likely to be chosen than the less costly. Presumably this was because the more costly sources were felt to have various advantages over the less costly, and the subjects were willing to pay for these added benefits. In line with this hypothesis, the effect was greatest for the 'car' situation ($r = 0.65$), where presumably a relatively small investment might influence a big decision.

The two studies together, Rosenberg's and Christie's, suggest that no single factor is of overriding importance in determining the likelihood of a particular source of information being selected in all situations. Ease of use, effectiveness, generalized cost, and other factors are all important to greater or lesser degrees depending on the situation.

A corollary of all this is that as different sources become equalized with respect to some factors, so the importance of the remaining factors will increase. For example, as the cost differences between using various computer-based and non computer-based information sources become smaller, so the importance of factors such as ease of use and effectiveness becomes more critical.

USING ONLINE SYSTEMS

Once the user has selected a particular source of information, how well can (s)he expect that system to meet his or her needs? In this section, we concentrate on the use of online retrieval systems.

Over relatively recent years it has become increasingly possible for professionals and others to use large computerized databases as sources of information. The Commission of the European Communities has done a lot through its Euronet Diane program to foster the development of such 'online' services throughout Europe. Euronet Diane provides an international network through which users can access databases held on host computers in many different European countries. The databases contain information on many different areas of scientific and technical information (STI), including engineering, chemistry, agriculture, energy, psychology, and many other topics.

Retrieval systems were used prior to the days of online systems, but were manual. Nevertheless, much of the research done on the manual systems applies to current online systems.

The major focus of research into retrieval systems during the last twenty years has been on 'documentary' systems—systems designed for the storage and retrieval of information in and about research reports, published and unpublished papers and articles of all sorts. Although 'information science' as such is at best described as embryonic, a number of useful generalizations have emerged from the research done which we shall review briefly.

Precision and recall

Put simply, what a user hopes to get from an information system is relevant information. The nature of relevance is discussed later in this chapter but let us assume for now that we can decide whether or not an item of information is relevant. One way of providing relevant information is to dump the whole database on the user. Of course the user would be disappointed in such a performance because it would put all the load on the user to sort through the various items. Alternatively, the system could find the item that best fits the user's needs and present that. That would also be disappointing, for obvious reasons. Ideally, one would like the system to present all the relevant items that are in the database, and only those. Unfortunately, both manual and online systems fall far short of this ideal. How well they actually manage to perform is measured by precision and recall.

Precision is defined to be the proportion of retrieved items that are relevant. Recall is defined to be the proportion of relevant items that have been retrieved.

Retrieval system performance is typically about 40 to 60 per cent in terms of each of these (Sparck Jones, 1981a), far short of the ideal.

The situation is even worse than that statistic implies. Not only are precision and recall generally less than one would like, they are generally inversely related, i.e. one cannot normally expect to have both good precision and good recall. As precision improves, recall generally deteriorates.

Since there is this very basic trade-off, it would be interesting to know which users prefer in various situations, but the author is not aware of any systematic study of this.

One reason for the apparent lack of such a study is the lack of an adequate taxonomy of information seeking situations, a state of affairs that applies to the psychology of situations more generally. Some progress has been made in relation to the classification of social situations (see Argyle, Furnham and Graham, 1981) and it is possible that some parts of the methodology used could also be applied to the classification of situations from the point of view of information seeking.

Full text search. The advent of cheap memory and high speed processing has made it possible for some office system manufacturers to offer 'full text search' for limited domains of information—e.g. departmental files. The idea is that as long as the user can specify something about the item (e.g. it probably includes the word 'consumer') then the computer will search the whole domain until it finds an item containing somewhere in the text the words specified. This is sometimes referred to as if it were the answer to the problem of designing an effective information system, but this is far from the case. There are at least two reasons for this:

— There is every likelihood of high recall with low precision if common words are specified (e.g. it probably contains the word 'the'— a ridiculous example, but it illustrates the point). For even moderately large domains, the user would be overwhelmed with irrelevant information.

— The system puts the onus completely on the user to think of all the synonyms that might be used. For example, someone interested in papers on mushrooms might not think to use the terms 'fungus' or 'toadstool'.

This leads on to the question of indexing items so they can be found when needed by users with differing needs.

Indexing systems

In order for the retrieval system to find a particular item, it must be able to identify the item according to some system of 'tags' or labels. This is the indexing system. It could be a simple alphabetical ordering of items, it could use keywords, or it could use other tags. As explained above, it could even be every word or word combination that appears anywhere in the text of the item.

One might suppose that the indexing system used would be the key to the successful performance of a retrieval system. Curiously, this does not seem to be the case. A classic series of experiments which pointed to this somewhat surprising conclusion is usually referred to as 'Cranfield 1', started in the late 1950s (Sparck Jones, 1981b.)

Cranfield 1 was a laboratory test focused on comparing four widely differing indexing systems:

— the Universal Decimal Classification (UDC)
— alphabetical subject catalogue
— a faceted classification scheme
— the Uniterm system of coordinate indexing.

This selection was made in order to include indexing systems which differed as fundamentally as possible and which represented the principal types of retrieval systems which were felt to have any significance at the time. The UDC system was chosen because it was the most widely used 'enumerative' type of system, based on a 'tree of knowledge' and using decimal notation. Facet classification specifically lacks those characteristics. The alphabetical index is deliberately anti-classificatory and relatively uncontrolled in language. The Uniterm system is also uncontrolled but allows all permutations and combinations of individual terms to define items.

The indexing systems were systematically compared by using all of them to index 18,000 research reports and periodical articles in the general field of aeronautical engineering. Half of the items dealt with the specialized subject of high speed aerodynamics. A set of 1200 questions was then obtained by soliciting questions based on sample document lists from those involved in the indexing.

A complex series of tests was then conducted which showed that whilst the indexing systems did vary to some degree in terms of retrieval performance the differences were much less important than had been thought and not especially important for practical purposes when considered in the context of other factors operating in practical situations.

The results reported for all searches show recall values of 74 per cent for facet, 76 per cent for UDC, 82 per cent for alphabetical, and 82 per cent for Uniterm. One reason for the unusually high recall scores overall is probably that the questions were based on items that were known to exist in the database being searched. A test of precision yielded values ranging from 7 per cent for UDC to 13 per cent for alphabetical but the researchers showed that quite different figures for precision could also be demonstrated and that little credance should be attached to any particular set.

The general comparability of different indexing systems in terms of their performance has been substantiated by other studies, but this does not mean that there is nothing to choose between them. For example, in discussing the findings from an INSPEC study, Evans (1981) found that controlled-language and free-language boolean profiles were comparable in terms of their overall retrieval performance but differed in other ways. The controlled-language profiles were produced more quickly (given a knowledge of the controlled language) and comprised fewer search terms, but a) they would be unlikely to appeal to the general user (i.e. someone who is not an information specialist), and b) in rapidly evolving subject areas are likely to lag behind and be inadequate until updated.

Automatic indexing

Indexing of items traditionally was done manually but by the early 1960s was beginning to be done automatically. The simplest programs would scan document texts or text excerpts (e.g. abstracts) and assign as descriptors words that occurred sufficiently frequently.

Linguists were quick to point out the theoretical inadequacies of such simple systems and identified a number of linguistic techniques that were felt to be very important or even essential for an adequate system. These included, amongst others:

— the use of term hierarchies, so that descriptions could be expanded by adding hierarchically more general or more specific terms
— the use of synonym dictionaries or thesauri so that individual terms could be replaced by a broader set of terms, broadening the original context description
— the use of syntactic analysis to supply specific content identifications and avoid confusion between composite terms such as 'blind Venetian' and 'Venetian blind'
— the use of semantic analysis to make use of context knowledge not contained in the documents themselves, often specified by preconstructed 'semantic graphs' and other related constructs.

Salton (1981) describes a programme of work using the 'Smart' system which, amongst other things, compared search results obtained by using term hierarchies and thesauri with other simpler systems based on the use of single, frequency-weighted terms extracted from the items of text concerned. He reports how the early results in 1964 and 1965 failed to demonstrate any value in the use of syntactic analysis procedures to construct syntactic content phrases, or the use of concept hierarchies, under any circumstances. The most helpful content analysis seemed to be based on the use of weighted word stems extracted from titles and abstracts, and the use of a thesaurus to recognize some synonyms and related terms.

Other studies, including the Cranfield tests, added support to the Smart findings and at the present time, according to Salton (1981, p. 231):

> there is an understanding among retrieval experts that an over-specification of document content normally produced by the more refined indexing methodologies can be just as detrimental as an underspecification.

It often takes a long time for the results of research to filter through to practice, and Salton notes (p. 231) that:

> this evidence does not, however, prevent many people from still clamouring for more sophisticated linguistic analysis procedures to be incorporated into automatic indexing systems, or indeed from incorporating such methodologies into newly designed retrieval systems.

Indexing exhaustivity. The use of computers makes it easier to base the indexing of an item on more of the item's content (e.g. title, abstract or whole document)—the extreme is illustrated by the notion of 'full text search' discussed briefly above. One might suppose that the more of the item that is used in indexing the better but this is not so. It turns out (Keen, 1981) that for any given level of required recall there is an optimum level of indexing exhaustivity. Below the optimum, recall will be reduced; above the optimum, precision will deteriorate.

Indexing specificity. Precision can be improved by making the indexing of items more specific. Specificity can be increased by using index terms that are semantically more specific or by increasing the levels of term combination that are available in searching. For a given level of required precision there is an optimum level of specificity (Keen, 1981).

Summary of key findings

Sparck Jones (1981a) has lamented the relatively slow progress made in research into retrieval systems during the last twenty years, but some general conclusions can be drawn.

We have seen that,

— Precision and recall are typically fairly low, and that there is a trade-off between them.
— Natural language indexing is competitive with artificial indexing languages.
— Simple descriptions of items are competitive with more complex descriptions.
— The formal properties of items may be turned to advantage, as in weighting schemes.
— The characterization of queries is more important than that of the items to be searched.
— The number of searching keys is more important than their individual quality.

THE EMERGENCE OF VIDEOTEX

The systems considered above have been used primarily by information specialists on behalf of other professionals. At the close of the 1970s, British Telecom (then the Post Office) embarked upon an exciting venture, now called videotex, aimed at making information directly accessible by non-specialists.

The venture has met with mixed success. It has been successful in so far as videotex systems of one sort or another have sprung up all around the world and videotex has become an established and well known approach to supporting information seeking. On the other hand, the market projections forecast have yet to materialize.

Videotex is a family of electronic systems, of which British Telecom's 'Viewdata' was the first. Viewdata is the system that supports British Telecom's 'Prestel' service in the UK and International Prestel. The other main videotex standards are:

— Bildschirmtext (FR Germany)
— Antiope (France)
— Telidon (Canada)
— Captain (Japan).

All these systems share the following characteristics. They:

— provide an interactive link between the user's television set and the host computer
— present information on the television screen primarily page by page although other possibilities exist
— allow the user to work through a hierarchy of menus as the primary means of finding a page of interest (although, once again, other possibilities exist).

System performance

It is not known for sure to what extent the conclusions cited above in relation to retrieval systems generally apply to videotex. Relatively little has been published that is directly relevant.

One would expect, prior to any further evidence, that the findings concerning recall and precision would apply to videotex as much as to other retrieval systems.

Recall does seem to be low, as a study by Bush and Williams (1978) illustrates. They conducted an experiment in which 24 people who had never used Prestel before each searched for specified target pages which were known to exist in the database. Each subject searched for 10 of the 20 possible targets, five via one tree structure and five via a second. The order of using the two trees and the order in which questions were asked was varied.

Success was only 59 per cent (only 43 per cent with no backtracking) for the tree structure based on an arbitrary but sensible structure. There was some improvement when a more 'natural' structure was used, based on an analysis of how people naturally formed categories and super-categories from the pages in the database (see Bush, 1977). Success rose to 79 per cent (56 per cent with no backtracking)—still far short of the 100 per cent that would have been ideal.

The results of the Bush and Williams test are not directly comparable with more traditional tests of recall and precision because the test was not conducted in an exactly analogous way, but it is clear that system performance fell significantly short of the ideal.

Videotex and the user's personal information space

An important selling point concerning videotex is that its pages can be standardized with respect to a topic but very frequently updated. Thus, a particular page number can be dedicated to, say, stock prices, but the actual prices shown can be updated as frequently as desired.

A key psychological corollary of this is that some pages 'whilst physically in the external world' can be psychologically incorporated by the user into his or her personal information space. Knowing the page number becomes analogous to knowing the place in the office where, say, a particular reference book is kept. The remaining pages which the user does not 'know' in this sense (i.e. most of the database) remain psychologically part of the ill-defined external world from which the user cannot retrieve information in the sense that information can be retrieved from one's personal information space but in which the user can seek information as the occasion requires.

BEYOND CURRENT SYSTEMS: INTELLIGENCE APPLIED
TO RELEVANCE

Conventional retrieval systems born out of a library-oriented psychological environment focused on the unintelligent retrieval of functionally independent items (e.g. abstracts of technical papers) according to criteria made as explicit and as complete as possible by the user.

British Telecom, with a different perspective on how information could be bought and sold as a marketable commodity, broke away from that tradition slightly with videotex by

- deliberately fostering the development of a highly heterogeneous database covering as many aspects of life as possible, and by
- acknowledging the functional dependencies between different items in the database—e.g. having identified an area where there might be better employment prospects, a user might wish to see what the housing situation is like there, see what schools there are, and maybe take a train journey to visit the place.

In connection with the latter point, the videotex user can perform calculations on data received through videotex pages and can respond to information directly using 'response pages' to place orders for goods, book tickets, state requirements, or respond in other ways.

Videotex is, however, still unintelligent and it still puts the onus on the user to specify the information requirements (by selecting from menus). As in the conventional type of system, the videotex system still responds to the user more or less passively.

As we move into the era of machine intelligence and the use of knowledge bases, it is becoming more and more realistic to imagine an intelligent system that could take a more active role in helping the user, either by:

- monitoring what the user is doing when not actively seeking information (e.g. when preparing a report) and deciding what information if any it could offer that might help the user; or by
- responding in a more intelligent way when the user actively seeks information from the system.

Relevance

If an intelligent electronic information system is to take any degree of initiative in presenting the user with appropriate information it must have some method for deciding what information it is appropriate to present. Clearly, the information must, by definition, be relevant. However, that does not get us very far, for at least two reasons.

First, consider the following two sets of statements and then decide which statement is the appropriate one to start with in the following kind of interaction.

a. 'Mrs Brown, your son is dead.'
   'He was killed in a car accident.'
   'I am a police officer.'

b. 'Mrs Brown, I am a police officer.'
   'I am afraid there has been a car accident.'
   'Your son is dead.'

Both lack the degree of tact one might hope for but the second is clearly more appropriate than the first. It is better to start with 'Mrs Brown, I am a police officer', than 'Mrs Brown, your son is dead.' However, both statements are relevant—in fact the second is arguably more relevant than the first, even though it is less appropriate.

Secondly, also illustrated by the example above, it is necessary to decide, relevant to what? Clearly, the information 'I am a police officer' is relevant. But to what, and why? It is not given in response to any question that Mrs Brown has asked. Its relevance comes from the implicit need to provide credibility for another item of information presented concurrently ('Your son is dead'). The latter item is clearly relevant to Mrs Brown in the sense that she almost certainly would want to know, but she has not asked.

Multiple criteria for relevance. Information can be relevant in different ways. The example above illustrates how it may be relevant factually or relevant because it affects the credibility of facts presented. The latter, translated into a research environment, fits category (iv) below—one of five ways in which Kent (1974) suggested an item of information might be relevant (although he preferred the term 'useful'):

(i) The item itself is cited.
(ii) The information in the item is reported.
(iii) The item causes the user to take some action other than citing or reporting.
(iv) The item has some other effect on the user, e.g. changing his or her confidence in a research finding.
(v) The item is used in some other way.

These five ways in which information may be relevant fit readily into a research environment where researchers are seeking information from libraries or online systems for use in writing research reports or in other research related activities. However, they also apply in general management and other professional situations, and even in the home. The following sequence illustrates the categories applied to a domestic situation (the items are numbered according to the scheme above):

  (i) 'Look at this.'
 (ii) 'It says they are going to build a road through our house.'
(iii) 'Get on to that solicitor friend of yours. He'll know what to do.'
 (iv) 'I knew we shouldn't have trusted that estate agent.'
  (v) 'And you know what you can do with this.'

Implications for systems design. An intelligent system must take account of the user's 'state of readiness', over which it may have some control (e.g. by informing the user there has been some unexpected business activity and an unusual dip in the price of some particular commodity before informing him he has been wiped out). It must also take account of the different ways in which information can be relevant. Even if it cannot, for example, contribute to the user's decision-making in a particular area, it may be a useful item for the user to be able to cite; or, even though the item does not provide any new facts as such, it may modify the user's confidence in the facts that have been presented. Finally, it must take account of how the user is likely to behave following presentation of particular items. It is often the case that users terminate searches following receipt of one or a few items that are sufficient for immediate purposes even though other items still in the system may in some sense be even more relevant. To help the user fully, the system must have some understanding of user psychology, and adapt its own behaviour accordingly.

The binary choice unit or bcu

Yovits and Abilock (1974) have proposed a measure of the value of information which in principle could be used by an intelligent system in the special case where information is being presented in order to aid a user in choosing between various possible courses of action. Their measure is called the 'binary choice unit' or 'bcu', which can be explained as follows.

They are concerned solely with situations in which information is being used as a basis for action of some sort. They regard information as that which changes the probability distribution representing the user's overall inclination towards the possible courses of action available (his or her 'decision state'). They propose a measure, I, of the amount of information 'contained in' a decision state, as follows:

$$I = M \sum_{i=1}^{M} \left\{ P(a_i) \right\}^2 - 1$$

where M is the number of possible courses of action available to the user, and $P(a_i)$ is the probability of each course of action.

The value of I is zero when all the $P(a_i)$'s are equal to I/M, i.e. a state of complete uncertainty. It reaches its maximum value of $M-1$ when one of the $P(a_i)$'s is unity and all the others are zero, i.e. a state of complete certainty.

When there are only two possible courses of action, I varies from zero (complete uncertainty) to unity (complete certainty). Yovits and Abilock choose this situation to define a unit of information which they call the 'binary choice unit' or 'bcu'.

In general, the maximum amount of information for any given decision state will be $M-1$ bcu's. Any given item of information can therefore in principle be assigned a bcu value. This is the amount of reduction of uncertainty (measured in bcu's) that results following receipt of the item.

The bcu is a psychological, not a physical or logical measure. This is because, as Yovits and Abilock point out, the bcu value of any given physical piece of information (e.g. piece of text, or graph) is time and situation dependent. The item will have different significance to different users at any given point in time, or to the same user at different points in time.

Implications for systems design. An intelligent system could use the bcu values of items of information to decide which items to present to the user. This applies only to situations where the user's need for information can be related to a meaningful and definable set of possible courses of action. It also depends upon the system having estimates of bcu values available. It is possible that in some circumstances these estimates could be derived from parallel previous instances where analogous users with analogous needs have sought analogous information.

Types of information

The bcu is a quantitative measure of information, potentially useful but not sufficient for selecting items of information. It is like calories as a measure of food intake. The calorific value of an item of food may be of some use in connection with maintaining a healthy diet but by itself it could lead to some very peculiar diets indeed (e.g. eating nothing but baked beans in the appropriate quantity to maintain a given level of calorific intake, or nothing but small amounts of cream cake).

The bcu measures overall reduction in psychological uncertainty but uncertainty can be reduced by an equal amount in different ways, depending on the type of information.

There are five main types of information that it is useful to consider here. These are as follows, derived from expectancy-value theory (e.g. Fishbein and Ajzen, 1975):

(i) Information that affects the strength of beliefs about the outcomes of particular courses of action. (*Bi*); for example, research articles that demonstrate a link between smoking and a variety of cancers.

(ii) Information that affects the evaluation of the outcomes of particular courses of action (*ei*); for example, a visit to a hospital ward full of people dying from cancers of various sorts, or the death of a close relative from cancer, or a horrific television documentary that shows the symptoms of various kinds of cancer.

(iii) Information that affects the person's beliefs about the expectations of others (*NB*); for example, the prevalance of 'no smoking' signs, the attitudes expressed by other people in various situations.

(iv) Information that affects the weighting given to (i) and (ii) above (*W₁*); for example, admonitions from parents, teachers and others to 'be rational', and 'think'.

(v) Information that affects the weighting given to (iii) above (*W₂*); for example, admonitions from parents, teachers and others to take account of other people's feelings and 'when in Rome do as the Romans'.

According to Fishbein's model, the strength of 'behavioural intention' (*BI*) for any given course of action is given by the following expression:

$$BI = (Bi.ei) * W_1 + NB * W_2$$

where (*Bi.ei*) is called A_{act} (or 'attitude toward the act') and *NB* is called 'normative belief'.

For a further discussion of types of information, the reader is referred to Christie (1981).

Relationship to bcu's. The value of $P(a_i)$ in Yovits and Abilock's model is conceptually related to the value of *BI* in Fishbein's model. What the exact relationship is between the two is not entirely clear, but one could start by assuming a simple linear relationship. The bcu therefore provides a measure of the overall reduction in the user's level of uncertainty, and the analysis by type of information given above explains how that overall reduction was achieved.

Implications for systems design. The expression for *BI* could in principle provide the basis for a set of heuristics the system would use in deciding what information to present to the user. The system could learn from previous experience (with other users in similar situations or the same user in similar situations) that, say, W_2 is greater than W_1 in a given type of situation.

This would mean that the user is likely to consider other people's expectations to be especially important, so the system might decide to concentrate on providing information about other people's expectations. Or it could decide that this predisposition on the part of the user is best regarded as a 'bias' and that it should compensate for the 'bias' by concentrating on presenting information about the likely consequences of particular courses of action.

Depending on the amount of 'intelligence' one wishes to build into such an information system, one can imagine the system being able in principle to adjust the flow of information to the user in terms of any of the variables in the expression for *BI*.

There are three key points to appreciate about this:

— As electronic systems gain access to larger and larger amounts of information, so it becomes more and more necessary for the system to make some decisions about what to present and what not to present to the user, and in what order. There will often be just too much information available to expect the user to be able to cope with a complete, unintelligent dump.
— The decisions made by the system cannot be 'value free' or 'objective'. What could these terms mean? Certainly, there is nothing about, say, a random selection of items that makes that any more 'objective' or 'value free' than any other system—it is no more or less automatic, and it simply means that every type of information is valued equally (which may be desirable, but may not be).
— It is necessary for the designer of such systems to be very clear about the value judgements that are made in programming the system.

CONCLUSIONS

The research on information seeking reviewed suggests the following conclusions concerning the key issues identified above.

Two broad categories of information seeking can be defined on the basis of their psychological antecedents. These are 'diversive information seeking', which can be related to general arousal theory, and 'specific information seeking'. This chapter has concentrated on the latter. People cannot be relied upon to be entirely 'rational' about when they seek information, or what information they are prepared to look for, so there is a case to be made for the electronic system to take a more active role in information seeking than has been the case conventionally.

People often need to brief themselves before engaging in information seeking proper, so as to formulate their needs more clearly. They use and in an electronic environment would continue to use multiple strategies in this. Electronic systems might possibly play a part, either by identifying 'special communicators' the user may wish to contact for advice, or by mimicking what goes on when a person talks to a special communicator.

Once the user has formulated his or her needs, (s)he must select an appropriate source of information. Ease of use, effectiveness, generalized cost, and other factors are all important to a greater or lesser degree depending on the situation. Unfortunately, an adequate taxonomy of situations does not exist for predicting which factors will be important and when.

Conventional retrieval systems perform significantly below the ideal in terms of recall and precision. More important even than this, they are passive and unintelligent, putting the onus completely on the user to tell the system exactly what is needed.

Future systems will be more intelligent, have an understanding of psychologically different types of information and their relevance to user needs. They will also be more active, taking more of an initiative in finding and presenting the user with relevant information. This poses new kinds of problems for the systems designer. It will not be possible to programme future systems to be entirely 'objective' or 'value free' in what information they decide to present to the user. It will be necessary for the designer to be very clear about the value judgements that are made in programming the system.

In the next chapter we conclude examples of emerging types of electronic systems by looking briefly at another area in which it might prove especially fruitful to apply artificial intelligence: that of decision-making.

Chapter 8

Decision Systems

MARCO DE ALBERDI AND JAMES HARVEY

INTRODUCTION

In the previous few chapters we illustrated some of the ways in which electronic systems are emerging in support of Type A communication (where we discussed the psychological aspects of 'electronic meetings') and Type B communication (where we focused on personal and shared information systems). In the case of shared information systems especially, we noted an apparent scope for the application of artificial intelligence. In this chapter we conclude our sampling of product trends by looking briefly at another area where artificial intelligence seems likely to play an increasingly important role in the future—that of decision systems. This chapter, then, is concerned specifically with the emerging area of Type C communication—communication between human and 'intelligent' machine, a new type of human communication that by definition could not have occurred until the technology of artificial intelligence had been developed.

For the purposes of this chapter we view decision making as a knowledge rich activity combining the skills of the 'decision analyst', the 'domain specialist' and the 'organization analyst'—vested, notionally, in the 'decision maker'. What these terms mean is explained later in the chapter. We briefly review the contributions of those working in the field of decision support systems and in particular the contribution made by psychology to an understanding of the human limitations on decision making. We conclude by outlining an approach to the use of artificial intelligence in decision systems.

DECISION TYPES AND PROCESSES

Types of decisions

Simon (1960) distinguished between two general categories of decisions, which he called 'programmed' and 'non-programmed'. Programmed decisions are repetitive and routine, and a definite procedure can be worked out for handling them so that they do not have to be treated uniquely each time they occur. Non-programmed decisions are novel, unstructured and consequential. There is no well-defined method of handling the problem because it has not arisen before, or because its precise nature and structures are elusive or complex, or because it is so important that it warrants separate, individual treatment.

Keen and Scott Morton (1978) have made a further distinction which is useful in the present context—between two sorts of programmed decisions, 'structured' and 'unstructured', where structured decisions are those that do not involve a manager—the decision is well enough understood to be suitably given to clerks or to be automated in a relatively straightforward way. Unstructured decisions lack these characteristics. In between the two extremes are semi-structured decisions, for which neither managerial judgement alone nor a simple data processing model alone will be adequate. The problem may be too large, or the computational complexity and precision needed to solve it may be too great. The solution is likely to involve some judgement and subjective analysis as well as sufficient data. Under these conditions the manager plus the system can provide a more effective solution than either is capable of alone.

Examples of computer support for programmed, especially structured, decision making can be found from the field of management science in the realms of linear programming, inventory scheduling, and discounted cash flow techniques.

Examples of computer support for semi-structured decision making can be found in purpose built decision making support systems such as described by Keen and Gambino (1980), and in packages providing high level language tools such as Prosper from ICL.

There are few examples of computer support for unstructured decisions and indeed there are conflicting views about the possibility of doing this. The 'unstructured decision' is unstructured both from the point of view of the decision maker and in terms of relevant theory. Under these circumstances, it could be argued that computer tools are inapplicable—that development of decision support systems provides a means for going beyond the traditional use of computers in structured situations, but should avoid ineffectual efforts to automate inherently unstructured decisions (cf. Keen, 1980). Contrary to this view, our interest in this chapter lies in the possibility of providing computer support for unstructured decisions within an organizational environment.

One view of decision making in such an environment is provided by Thompson and Tuder (1959), and is shown in Figure 8.1.

		Preferences about possible outcomes	
		Agreement	**Disagreement**
Belief about causation	**Agreement**	Computation	Compromise
	Disagreement	Judgement	Inspiration

Figure 8.1: Decision types, based on Thompson and Tuder (1959)

'Computation' can be roughly equated with programmed or structured decision making, and the other decision types with non-programmed or unstructured decision making (semi-structured being somewhere at a boundary).

Thompson suggests that beliefs about causation become relevant when the decision makers involved are unsure about the outcomes associated with different projects or uncertain about costs or undesirable effects. Preferences about possible outcomes provide a value dimension to the decision-making process.

Thompson uses the categorization to propose 'structures' for organizing decision making within each quadrant. These structures are essentially concerned with the routing of information pertinent to the decision and the protocols to be observed by the decision unit. For example, a structure for judgement requires:

— fidelity to the group's preference hierarchy
— inviting all members to participate
— routing pertinent information about causation to each member
— giving each member equal influence
— designating as the ultimate choice the majority decision.

Thompson views the elements of any decision as comprising alternative courses of action where there is a need to explore the differential consequences of the alternatives and to make an evaluation of different potential outcomes in terms of their desirability.

Hopwood has developed these ideas further (Hopwood, 1974, 1983), as summarized in Figure 8.2. The central tenet of this view is the importance of information in supporting the decision maker. In the case of the 'answer machine' it exists to provide the result for a specified inquiry; in the case of the 'learning machine' it exists to provide an insight into the decision problem by exploring the domain; in the case of the 'ammunition machine' it exists to provide justification for a decision; in the case of the 'rationalization machine' it exists to provide a mass of information to enrich the domain and to support the decision maker in eliminating alternatives.

These views of the types of decisions involved in organizational decision making provide a framework within which we can now consider the decision process itself.

| | | Preferences about possible outcomes | |
		Certain	Uncertain
Beliefs about causation	Certain	Decision by computation ('Answer Machine')	Decision by compromise ('Ammunition Machine')
	Uncertain	Decision by judgement ('Learning Machine')	Decision by inspiration ('Rationalisation Machine')

Figure 8.2: Hopwood's model of decision making, based on Hopwood (1983)

The decision process

The decision process is established by the recognition of the need to make a decision; this may be implicit or explicit. In either case the pre-eminent need is for the decision maker to articulate the assumptions behind the need: that is, there is a need to answer the question 'why'.

Hopwood (1974) develops a view of the decision process which is summarized in Figure 8.3.

The environment in which the decision takes place is an 'open systems' one. This means a system which is characterized by being open ended, incremental and undergoing continual evolution. It is very difficult to determine what exactly exists at any particular point in time—for example, a query might never finish looking for possible answers. One can imagine a situation in which a query to find a bargain-priced used refrigerator in good condition might reference information stored in any number of personal and organizational computers. Enormous amounts of effort and time could be expended processing the query to find such a refrigerator without being certain that the best buy had been located. This might happen in an 'open system'. In contrast, systems based on the assumption of a 'closed world' assume that they can find all the instances of a concept that exist—typically by searching their local storage.

Within this general framework outlined above, one can postulate at least four roles (and information sources) contributing to decision making. These are illustrated in Figure 8.4. The degree to which these roles are present in any one individual is unimportant to our theme, although it does affect the tactics of decision making. Our intention here is to examine the contribution

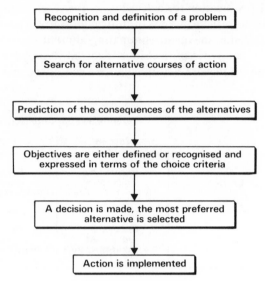

Figure 8.3: The decision process, based on Hopwood (1974)

Figure 8.4: Roles (and information sources) in decision making. (Reproduced by permission of ITT Industries Ltd.)

made by each of the roles and gain an understanding of their importance. Then we will identify opportunities for providing computer support.

KEY ROLES IN DECISION MAKING

The decision maker

Clearly the decision maker operates at the focal point of the decision making process. As Hopwood (1974) points out, in many large organizations the decision maker is concerned with 'finding occasions for making a decision and recognizing a problem, searching for and analysing alternative courses of action, choosing among the alternatives'. The essential questions seem to us to be:

— How are these alternatives generated? What is the motivating force? What governs the richness of the alternatives available? Our premise is that these are not givens and that they result from specialist domain knowledge.
— How are the consequences predicted? What is the limiting factor? Our premise is that these are not givens.
— What are the limits imposed by human cognitive capacity in decision making?

The theory of economic rationality requires that all alternatives are exhaustively reviewed in order that the result be maximized within a defined cost; this may have been possible given a closed world system but seems impossible given an open system model. In any case such a possibility defies the limits of human cognitive capacity within any reasonable timeframe or with any reasonable assumption of supporting computer processing power.

Simon (1982a) has pointed to a clash between economic rationality and what he calls 'bounded rationality'. He points out that classical economic theory is wrong in that 'rational choice (is not) a choice among objectively given alternatives with objectively given consequences'. Rather he formulates 'the principle of bounded rationality' in which he suggests that

> The capacity of the human mind for formulating and solving complex problems is very small compared with the size of the problems whose solution is required for objectively rational behaviour in the real world—or even for a reasonable approximation to such objective rationality.

He notes two consequences that result from this view; firstly that the decision-maker must construct a simplified model of the decision domain, in order to make the problem tractable, and secondly the decision-making process must take account of the fact that there are practical limits to human rationality. The key to the simplification of the choice process in Simon's (1982a) view is the replacement of the goal of maximizing with the goal of 'satisficing', of finding a course of action that is 'good enough'.

He argues that if the alternatives in a choice situation have to be discovered or invented, and if the number of possible alternatives is large, then a choice has to be made before all of them have been looked at. This means that criteria have to be defined to determine that an adequate, or satisfactory alternative, has been found. This is what is meant by 'satisficing'. It is a concept that relates closely to the psychological concept of setting 'aspiration levels'. In a satisficing model search terminates when the best offer exceeds an aspiration level that itself adjusts gradually to the value of the offers received so far (Simon, 1982b).

Satisficing can be seen to be one strategy for coping with the degree of 'cognitive effort' required in decision making. The marginal economics of effort suggests that, within given constraints of time and cost, managers will limit their cognitive effort to the perceived value of improving the quality of their decisions. Figure 8.5 summarizes the categories of cognitive effort described by Keen. (1980b).

Further limitations on the decision maker are in the ability to compute— to combine variables and values in such a way as to predict numerical outcomes intuitively—and in limitations on drawing inferences, i.e. drawing useful conclusions from available information.

Possible biases affecting the human decision maker. One way in which humans limit cognitive load when making inferences is by adopting 'rule of thumb' strategies or heuristics. One school of thought compares these heuristics or intuitive strategies with normative statistical models (cf. Tversky and Kahneman, 1974; Nisbett and Ross, 1980) and interprets the observed differences as indicative of sub-optimal information processing strategies or biases. Further motivational biases are posited, based on needs or desires to enhance or defend the ego, and retain effective control (cf. Kruglanski and Ajzen, 1983). The picture that emerges is of human judgement prone to error, through the use of sub-optimal or inappropriate strategies necessitated by the limitations of human processing capacity and human motivation.

This view has increasingly come under attack. Humphreys *et al.* (1980) suggest that whilst the view is arguable for known data generated in the laboratory, no such claims can be made for the real world. In real world contexts, it is difficult, if not impossible, to specify a model of the task environment with which intuitive models can be contrasted (Humphreys *et al*, 1980, p. 30). Kruglanski and Ajzen (1983) go further in their criticism of the 'biases' view. They argue that the literature on heuristics and biases assumes that:

— There exist reliable criteria of inferential validity based on objective, 'true' modes of information processing.
— Motivational and cognitive factors bias inference away from these criteria and enhance error likelihood.
— The layperson's epistemic process is pluralistic, consisting of a diverse range of heuristics which are selectively invoked.

They reject all these assumptions and suggest that there are no objective veridical models that intuitive inference may be matched against, and that the lay epistemic process is unitary: in their view people only take account of information that is subjectively relevant or salient given the judgement

CATEGORIES OF COGNITIVE EFFORT	COGNITIVE STYLE	OBSERVATIONS
1. *Computation* arithmetic calculation; enumeration	mental calculation is constrained by intelligence, speed and accuracy	avoided by most managers and delegated to calculators
2. *Specification* conscious articulation of weights, priorities, preferences, point estimates of probabilities or utility functions	individuals are inconsistent in specification and do not find it easy or desirable	analystics rely on it
3. *Search* scanning; creating alternatives	although a comprehensive search of alternatives is rationally necessary, it is generally avoided by 'satisficing' and heuristics	no significant differences among individuals
4. *Inference* deriving conclusions from data; statistical reasoning; generalizing from specific instances	most people are poor statisticians who make predictable errors, rely on oversimple rules of deduction and fail to make adequate use of the information they have, especially in updating probabilities	highly constrained by ability; most people do it badly
5. *Assimilation* responding to numerical, graphical or verbal information; assessing results of analysis; 'making sense' of data	most people don't like reading long reports	depends on ability, particularly in respect of cognitive complexity; MBTI may be useful; 'sensing' people rely on concrete information, 'intuitive' on conceptual; feelers validate analysis in terms of emotional response, thinkers in relation to external logic´
6. *Explanation* justifying solutions; explaining conclusions and/or methodology that leads to them	cost depends on the gap between the user's own mode of analysis and the organization requirements, traditions and norms	

Figure 8.5: Marginal economics of effort, based on Keen (1980)

to be made. They therefore suggest that it is unreasonable to call deviations from normative models biased or flawed since these are congruent with the individual's view of the world. It is, however, important to note the title of the theory—'Lay epistemology'—since it would be unreasonable to equate the inference drawn by an expert in a knowledge rich domain with a lay person in that domain. It is, therefore, suggested that expert judgement in a knowledge rich domain will provide a more accurate, veridical inference, based on a more useful model of the domain, than is available to the lay decision maker.

Taken as a whole the limitations on the human decision maker considered above gives a view of the unsupported decision maker as being able to choose some of the potentially relevant and salient information from the environment, but with a limited model of the domain, and therefore limited in identifying all the relevant information. The decision maker is also limited in the use to which that information can be put computationally, and therefore, subject to less than optimal strategies.

Opportunities for computer support. Our approach is to recognize the psychological limits imposed on human capabilities yet at the same time recognize that information systems increasingly provide an open systems model of the world. It is to recognize that computer programs may provide the mediating force between the information system and the user applied to a decision problem. If the computer program can recognize the limits of cognitive effort it may be able to optimize where a human will satisfice. Such programs will be knowledge rich not only of the domain but of the human decision maker.

This idea of a supportive information system is not new; one idea was presented by Churchman and West (1968) over fifteen years ago. They suggested that the information system presents different views of the problem domain to stimulate the user:

> The idea behind these 'dialectical' information systems is to make clear to the user the basic assumptions that go into the support of any proposal.

This analysis of the decision maker's limitations suggests the areas of decision making that necessitate computer support within knowledge rich domains, as follows:

— search   — techniques for eliciting from the user what categories are important. The support tool searches and decides the best selection of information from the available database; (but see Chapter 7);

— inference   — develop decision analysis as a knowledge rich domain and build in corrections for biases.

The organization analyst

The 'organization analyst' of our model in Figure 8.4 can be likened to a personal assistant, putting the decision in the organization's terms. Such a role depends on organizational knowledge; that is, knowledge about the organization promoting the decision, its preferences, goals and motivations. Such knowledge forms the basis for strategic planning and underlies (either implicitly or explicitly) all organizational decision making.

To the extent that this role is satisfactory it is in providing the goal that the decision is designed to meet. To the extent that it is unsatisfactory it is in failing to elaborate the decision by knowledge of the domain.

Opportunities for computer support. These seem to be limited, although it would be possible to articulate the goals and objectives of an organization perhaps to a sub-unit level and relate these as a hierarchy of objectives. The application of these in support of decision making seems to require extensive real world knowledge.

The decision analyst

The 'decision analyst' in the model can be likened in some ways to a Rogerian psychotherapist, probing the decision maker, reacting to a key phrase, seeking an articulation of the decision in the decision maker's terms.

The purpose of this articulation is threefold:

— It provides scope for all subsequent decision making activities; this scope sets the boundary of significance of all factors thought to be relevant to the decision; it excludes as well as includes. This 'decision scoping' declares the parameters of the decision domain.
— The parameters can be challenged in terms of their adequacy and completeness, and in terms of their accuracy and relevance, either by other members of the decision unit (before or after the event) or by a domain knowledge specialist.
— The parameters can be developed through an evolving process as other information becomes available or as alternatives illuminate features of the decision not previously perceived.

The skill of the decision analyst is in being able to analyse or assess the decision in formal terms independent of the domain in which the decision is located. Again this is specialist knowledge and is considered to consist of a methodology for structuring the decision, monitoring and guiding the progress towards some conclusion. To the extent that it is satisfactory it is in providing a structuring of the decision, ensuring that the format of the decision is supported. To the extent that it is unsatisfactory it is in failing to elaborate the decision by knowledge of the domain.

Opportunities for computer support. The expertise used by the decision analyst to structure the decision can be described in terms of the method used and the heuristics applied. One noteworthy approach to capturing these heuristics and applying them in an interactive model is given by Humphreys *et al.* (1980) whose work is concerned with 'generation theory'—the body of knowledge and techniques that must be assembled in order to understand and aid structuring. The intent of this work is to provide a set of heuristics embedded in an interactive computer system which will be used to improve the quality of access to the information contained within the decision maker's semantic memory.

The success of this approach seems to us to depend upon the degree to which the decision domain is subject to decision analysis independent of the knowledge required to operate in the domain. This seems particularly true where such knowledge is critical to probing the assumptions made by the decision maker in generating the alternatives to be considered.

There exists some question concerning the possibility of applying such knowledge, however. To quote Humphreys *et al.* (1980):

> It is our contention that aiding techniques which require knowledge of the world ... should not be automated since any device doing so would have to be programmed with an enormous data base which would have to be constantly updated in the light of new information from every conceivable source.

It is our contention that the decision maker operates in a knowledge rich domain and it is knowledge of the domain that gives the decision maker greatest advantage. Increasingly, techniques are becoming available for modelling knowledge rich environments.

The domain specialist

The 'domain specialist' in our model can be likened to a technical consultant, questioning the decision maker's assumptions and probing for fallacies, errors and weaknesses in the decision maker's knowledge of the domain. In doing so, the domain specialist applies 'domain knowledge', i.e. knowledge about the particular field of interest in which the decision is located. This is by its very nature specialist knowledge and encompasses knowledge about the significance and content of the parameters as well as particular knowledge about objects which can satisfy those parameters and which are candidate solutions.

The approach of the domain specialist is complementary to that of the decision analyst. Where the decision analyst seeks an articulation of the decision in terms of the underlying assumptions, the domain specialist rationalizes in terms of their significance and appropriateness to the domain. To the extent that it is satisfactory it is in providing a solid knowledge of

the technical content of the decision. To the extent that is is unsatisfactory
it is in failing to appreciate the decision in the organization's terms.

Opportunities for computer support. The knowledge field of the domain specialist is of two kinds; firstly, a body of heuristics representing rules of good
practice and good judgement derived from experience of operating in the
domain; secondly, a body of data, both generic and specific, defining particular features and specific instances of objects in the domain. From a computer
standpoint, the body of data is typically held in an information database in-
dexed by some preconceived, and usually fixed, view of how the data are
likely to be accessed. As such it may perform inadequately in situations
where the required view is different from the preconceived view. A further
likely inadequacy is that the data contained are available only in some gross
form, that is, there are limited opportunities for filtering the data to meet
the needs of any specific access request. The body of heuristics is typically
not available as a computer aid; however as has been noted earlier, techniques are becoming available from the field of artificial intelligence which
can capture and apply such heuristics.

A COMPUTER BASED MODEL OF KNOWLEDGE
RICH DECISION MAKING

In the remainder of this chapter we outline a prototype computer based
model of decision making which integrates the various roles involved in
decision making in a specified domain. There are a number of enabling
technologies which we draw on to support us in this task, and these are:
knowledge based systems (using a production systems representation); and
knowledge acquisition through machine induction.

Knowledge Based Systems

A Knowledge Based System (Michie, 1982):

> ... embodies in a computer the knowledge-based component of
> an *expert skill* in such a form that the system can offer *intelligent
> advice*, and, on demand, *JUSTIFY* its own *line of inference*. The
> style best suited for machine inference is *RULE-BASED PRO-
> GRAMMING*.

Since such systems capture expert skill they are also known as Expert
Systems; we draw a distinction between the two terms to recognize that
Expert Systems are used by experts rather than Knowledge Based Systems
which we intend for use by non-experts. The implications of this distinction
are in terms of the nature of the user interface, terminology, and the extent
and kind of explanation provided by the system.

There are four main components to a Knowledge Based System, and these
are illustrated in Figure 8.6. The 'knowledge base' contains the heuristics

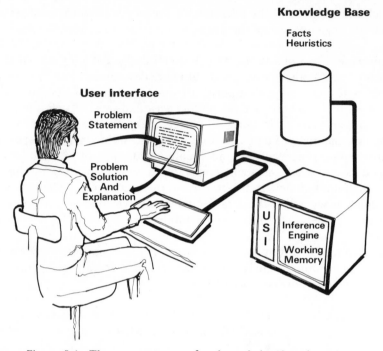

Figure 8.6: The components of a knowledge based system.
(Reproduced by permission of ITT Industries Ltd.)

representing the domain specialist skill; a number of representation for-malisms exist and our approach is based on a 'production system' rule form first suggested by Post (1943). The rules are expressed in terms of conditions, actions and certainties, that is:

IF condition THEN action WITH certainty value.

The advantages of a production rule representation are in terms of the intelligibility of the rules, and the flexibility to modify the rule base.

The 'inference engine' contains the reasoning strategy which selects and applies the knowledge. Our approach is to support a number of reasoning strategies where the system determines the selection of one appropriate to the problem at hand.

A number of examples of operational expert systems can be seen, in par-ticular in R1 or XCON (McDermott, 1980) which represents a commercially operational system using a production systems representation with a knowl-edge base of over 2000 rules.

Knowledge acquisition through machine induction

One of the major bottlenecks to the development of knowledge based sys-tems is the acquisition of the required knowledge from the relevant experts.

Our approach is based on knowledge acquisition using an approach described in 'learning from examples' (Carbonnel *et al.*, 1983) which is,

> a special case of inductive learning; given a set of examples and counter examples of a concept, the learner induces a general concept description that describes all of the positive examples and none of the counter examples.

The concept produced can be in the form of a production rule expressed in a natural language form such that it is intelligible to the expert who supplied the examples. The intent is that the process be iterative with the expert refining the examples and hence the production rule. The completion of this process is a knowledge base populated with rules induced from expert supplied examples.

Michalski *et al.* (1983) provide an example of this approach which describes knowledge acquisition in the domain of soybean plant disease diagnosis.

The computer model

A schematic view of the model can be seen in Figure 8.7. The various roles are represented in the form of their associated Knowledge Bases which contain the heuristics appropriate to each particular skill. Accordingly, the organization analyst's knowledge is maintained in the Organization Knowledge Base (although as we indicated earlier it has to be demonstrated that such knowledge can be applied at a sufficient level of detail to be useful). The decision analyst's knowledge is maintained in the User Strategy Knowledge Base; this is intended to contain both the heuristics for structuring the decision and the heuristics that apply correctives in support of the recognized limitations in human cognitive capacity (which were detailed earlier under the role of the decision maker). The Domain Specialist's Knowledge is maintained in the Domain Knowledge Base which is specific to a decision domain. It contains heuristics specific to that domain which will be used to validate and enhance the decision maker's understanding of a particular decision. It operates as a filter to the Information Database, containing specific data pertinent to the domain. The integration of these roles is achieved through a user interface prompted by action from the decision maker.

One scenario we envisage being used is where the decision maker activates the computer model through a description of the decision. We draw on the ideas of 'exemplary programming' (Waterman, 1977) where

> the idea is to provide users with a set of small personalised programs, called *agents*, that can act either as interfaces to external operating systems or as assistants to perform useful but often mundane tasks for the user ... *exemplary programming* (EP)

means synthesising programs from examples of traces of the activity one would like the program to perform.

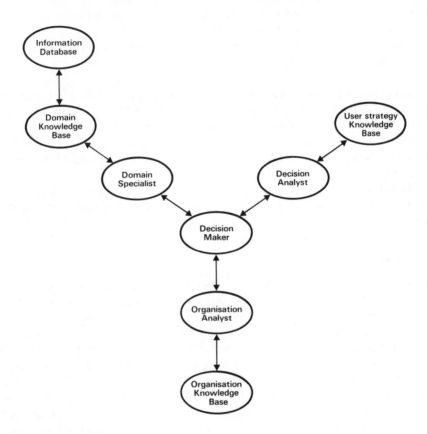

Figure 8.7: A computer model of the roles and information sources in decision making. (Reproduced by permission of ITT Industries Ltd.)

In 'exemplary decision making', the decision maker describes the features of the decision prompted by the Decision Analyst Knowledge Base and qualified by the Domain Specialist Knowledge Base. We see the result of this process as a fully articulated and structured decision reflecting the reality of the domain and the decision maker's viewpoint. The articulated decision forms the basis for interrogating the information database as it seeks information to illuminate the alternative options—that is, the reasons underlying the decision become the keywords with which any information database is interrogated.

The process of searching the information database comes closer to the notion of maximizing as it reviews the information against the desired criteria; the decision analyst's knowledge base of corrective heuristics may set some notional limit on the range of the search and the format of presentation of the output. This process may be interactive as information is returned which causes the decision maker to revise, add or delete alternatives.

CONCLUSIONS

The development of artificial intelligence techniques has made a new kind of communication possible, different from either Type A (person–person) or Type B (person–'paper'/'paper'–person); This is Type C—person to 'intelligent' machine. The technology is only just now emerging and little is known about the detailed psychology of what is involved. In the previous chapter we saw how it might have a role to play in 'information systems'. In this chapter we have considered its possible role in decision systems. This is a new area where much work needs to be done and where we can expect interesting developments during the coming five to ten years.

This concludes our sampling of product trends. We have attempted to be illustrative rather than comprehensive, to give a flavour of the kinds of developments that are going on rather than to document them fully.

In the following chapters we turn our attention to the question of 'usability'—the notion that, whatever particular kind of system we are considering, some systems are 'better' and easier to use than others. We start by considering what this concept means and what sorts of measurement techniques are available, within a 'psychophysiological' framework (Chapter 9) and then we go on to review what is currently known about how to achieve good 'usability' (Chapters 10 and 11).

Part Three
Product Usability

Chapter 9

Assessing Product Usability: A Psychophysiological Approach

ANTHONY GALE

In the last four chapters we have briefly reviewed some of the key ways in which electronic office systems can support their users. In this and the following two chapters we turn our attention to the design of the user interface to these products. The potential which the underlying electronic technology has to offer will only be realized in practice to the extent that the user–system interface is designed appropriately for the humans concerned.

How can the psychologist/designer know when this has been achieved? Sales figures for the product are not adequate because a product may achieve initial sales despite relatively poor design (although this is becoming increasingly unlikely as competition for market share increases). Alternatively, sales figures—even if quite high—may still be lower than could be achieved if the design were improved. Task performance criteria in themselves are insufficient because high task performance may sometimes be achieved at the expense of hidden costs to the users—'physical' costs such as backache, or 'psychological' costs such as reduced job satisfaction. Such 'hidden costs' are likely at some stage to become very visible, both to the user and to the organization (e.g. as days lost through sickness or absenteeism), and eventually to be reflected in sales of the product and the image of the manufacturer concerned.

In this chapter we propose that the design of a user–system interface is successful to the extent that working with the system produces optimal results—maximum overall benefits—in terms of the user's:

— behaviour, including specific task performance and broader aspects such as absenteeism
— subjective experiences, relating to use of the particular product concerned and to the work situation more generally; and
— physiological responses, including specific stress responses which may or may not result in physical symptoms—and which, incidentally, may well be less in a well-designed electronic office than in a conventional office—as well as other physiological responses.

This framework for assessing the user–system interface is called 'psychophysiological' because it includes both psychological (behavioural and experiential) criteria as well as physiological criteria.

This chapter explains the psychophysiological approach further and within this framework discusses some of the methodological intricacies involved in a professional assessment of when good design has been achieved in human factors terms. Behavioural and subjective criteria have been referred to frequently in the last four chapters dealing with experiments relating to particular types of products, and so this chapter will focus primarily on physiological aspects. The many factors involved in making a meaningful assessment of how well a user–system interface has been designed will be illustrated in the context of this discussion. Some comparisons between the relative usefulness of behavioural, experiential and physiological data will be made in order to illustrate the different, complementary contributions these different types of data can make.

INTRODUCTION

Psychophysiologists study the relationship between physiology, behaviour and experience and believe that rather special information becomes available if we monitor all these aspects of the person. For example, following the introduction of cordless switchboards—and contrary to the expectations of the designers—operators complained of job dissatisfaction and their performance deteriorated. Wastell, Brown and Copeman (1981) conducted a study of physiological and performance differences between new (cordless) and old exchanges. This enabled them to detect whether the demands of the two tasks were different without interrupting the operator's performance, since their principal dependent variables were brain evoked responses (ERPs) and heart rate, monitored continuously throughout a work period. Interviews, questionnaires, retrospective ratings, or any method which involved *interruption* of performance, would not provide information upon fluctuations of attention, task demands or the relationship between specific bodily changes and specific aspects of performance.

The relationship between mind and body constitutes the major and most complex question in psychology. Psychophysiology is not reductionist; it does not seek to reduce all events to the physiological domain. It considers the person as a system, made up of sub-systems (physiology, behaviour and experience). At any point in time, a particular system or system set may prevail. Thus, in sleep, the physiological sub-system is dominant and behaviour and subjective experience are under physiological control. During focused attention there is bodily stilling, the individual's intention to work at a particular task is seen to have consequential effects on his or her physiology, and behaviour can become relatively automatic. In a new situation and when faced with novel tasks, emotional responses may be dominant and attention become narrowed; but as learning develops and performance becomes smooth and regular, both physiology and behaviour become predictable and show less variability. During a working day, there will be fluctuations in performance partially attributable to the effects of circadian rhythms in physiological systems and the development of fatigue. It is not easy to tease out the independent and interactive effects of the various sub-systems, but by sampling the three domains of physiology, behaviour and experience in any particular context, we obtain the data needed to help us construct a model of processes occurring within the person.

Two further aspects of psychophysiology must be mentioned. The first is that changes within the person are seen to occur in response to external cues and to the past experience of the individual. Thus, *context* and the person's interpretation of what is happening are seen to be crucial. This will affect the ways in which the person's sub-systems interact; for example, bomb disposal officers experience the physiological correlates of anxiety which all of us know, but past training, esprit de corps and a sense of professionalism, enable them to ignore their fears and perform effectively. In contrast, the established habits of the stereotypical hypochondriac, make the person exaggerate the impact of physiological events upon conscious experience.

Secondly, psychophysiology recognizes the impact of individual differences upon behaviour. Certain individual characteristics, such as personality and intelligence, are known to be attributable in part to biological determination through heredity (Gale and Edwards, 1983b). The expression of inherited factors within the nervous system affects the ways in which the individual acts upon the physical and social world. One theory (e.g. Strelau, 1983) claims that individuals are born with potential differences in *information processing* capacity and in the level of *activity* which they sustain. Such characteristics are particularly salient to the operation of new technology devices. Strelau believes that individuals develop a *style of action* as they grow up, which is their particular way of matching their information processing coefficient and capacity for activity to the demands imposed on them by the environment. Strelau's experiments and field studies involve not only physiological monitoring, but the observation of *all* behaviours, not only those which are specifically work-related. Thus, in one study of foundry workers, work actions, preparatory actions (e.g. preparing tools, cleaning) and ancillary actions (e.g. smoking, chatting to fellow workers, bodily movements) were all monitored, since all are seen to be psychologically adaptive and therefore significant for the evaluation of the operator–task interface.

People differ in the ways they move between specifically work-related and other activities.

We have seen therefore that psychophysiological researchers sample: physiological events, behaviour and performance, subjective response, meaning or context, and individual differences. Given the richness of the data it is not always easy to specify causal routes or to build a model of the necessary and sufficient conditions for high grade performance. The discipline is at present very young, and given the state of our knowledge, it should be seen as an *approach* to tackling problems, rather than as a means of providing simple answers.

A very important issue is whether measurement should be complex or simple. Simple measurement may be expedient but it may well distort reality. A characteristic of systems is that they have emergent properties, that is to say, certain phenomena only appear when the systems work in combination; when studied in isolation they often reveal properties which may have little functional utility within the context for which the systems were designed. Thus, given the complexity of human beings, psychophysiology offers the promise of capturing reality in a faithful fashion.

What can psychophysiology tell us about computer–user interaction, particularly in the context of the office of the future? Our account so far makes it clear that in measuring the impact of new technology or in designing new work-systems for future use, we would be wise to sample as many aspects of the person as possible. In that way, we may capture information on emotional responses, response and adaptation to new circumstances and working conditions, and subjective reactions and performance changes which arise when new information processing demands are imposed on the individual. We will also determine how changes in working arrangements affect social relationships in the office and the ways in which the meaning of working experience shifts for the operator.

PHYSIOLOGICAL MEASURES AND CRITERIA FOR THEIR USE IN APPLIED CONTEXTS

A tremendous range of physiological measures has been used in field and laboratory contexts. The aim of this section is to set out the criteria for use in user–system studies. The measures include: electroencephalograms (EEG), event related brain potentials (ERP), pupillometric changes, eye movements, cardiovascular activity (heart rate, event related cardiac response, blood pressure, sinus arrhythmia), electromyograms (EMG), electrodermal activity (EDA), catecholamine excretion, and cortico-steriod excretion. There are a handful of psychophysiological studies using respiratory response, particularly in the field of human factors or ergonomics.

We need to ask two key questions: a) can the physiological measure be used without disrupting performance in laboratory or field contexts and/or without being itself disrupted by the task; and, b) are available theoretical constructs concerning the measure and existing research using the measure of direct relevance to user–system interaction?

Kak (1981) set out a list of criteria for using psychophysiological measures in applied contexts. These apply whether the measures are of behaviour, subjective experiences or physiological responses:

(i) *Reliability* refers to the extent to which a measure yields stable data within constant conditions, between conditions and between subjects. As crucial as this minimal requirement is for any scientific measurement, it is not easy to make a judgement on existing physiological measures, since data are rarely available from studies specifically designed to measure reliability. Variations in outcome between research studies may be due to unreliability or to other weaknesses in the research. For information processing studies ERP, EEG, heart rate, pupillometry and eye movements score relatively well for reliability. For stress studies, catecholamines and blood pressure are recommended.

(ii) *Validity*. Kak draws a distinction between conceptual validity (the rationale for linking the measure theoretically to a psychological process) and empirical validity (the demonstrable success in demonstrating systematic relationships). ERP, heart rate, pupillometry, catecholamines and EEG offer systematic theoretical accounts of their relationship to psychological constructs in the fields of information processing and stress. Authorities will differ in their interpretation of empirical validity, but on the basis of existing data, our view is that ERP, heart rate, pupillometry, EEG, and eye movements have greatest empirical salience to user–system interaction studies.

(iii) *Intrusiveness* means the ease with which the measure can be used in real-life studies. To a certain extent, satisfaction of this criterion is dependent upon technological advances; for example, radio telemetry and portable storage devices have reduced the need for trailing leads between the subject and instrumentation. Catecholamine assays can be made after extended work periods and can involve only minor discomfort. Unfortunately, many of the physiological measures involve application of transducers which disturb normal activity (e.g. EDA is normally recorded from the hand) or require the person to work in, or in fixed association with, the recording device (e.g. eye movements and pupillometry require very special recording conditions). Heart rate, ERP and EEG have been used successfully in free-moving contexts and most of our research examples will relate to the use of these measures. It is unfortunate that eye movement and pupillometric studies are currently excluded on this criterion, since they have good validity.

(iv) *Freedom from artefact* is a criterion which currently restricts the majority of physiological measures. Heart rate, ERP and EEG (in that order) are probably the measures of choice for simultaneous measurement paralleling performance, while catecholamine excretion is dependable for post-task measurement. In all cases of course, the systems under study have many functions to perform, maintaining bodily and vegetative systems which have little to do with information processing or emotional experience. To a certain extent therefore, such normal bodily activities do constitute 'artefacts' and their effects need to be excluded before changes of psychological significance can be detected.

(v) *Ease of analysis.* Kak does not consider this criterion. But the introduction of computing and microprocessors has made available not only considerable potential for data storage, but for data manipulation, both online and offline. None of the measures we have mentioned is simple to evaluate and a variety of descriptive models is available. Measurement of ERP and EEG involves simultaneous processing of several channels of data and calls for powerful computer programs and extensive computer memory. While the EDA response involves a relatively slow change in skin conductance, measures employed include: response latency, rise time, amplitude, and recovery. Heart rate measurement would seem at first glance to be straightforward, and at one time, sinus arrhythmia or heart rate variability was seen to be an attractively simple measure of mental load; however, more than two dozen different methods are available for measurement of heart rate variability. The present author's view is that any physiological measurement calls for powerful computing resources, since the methods of data interrogation available are numerous. Human behaviour is complex and physiology is complex. Meaningful relationships between the two domains are unlikely to be simple or straightforward.

(vi) *Acceptability* is important if a measure is to be used in the field. Cooperation and voluntary participation by operators is unlikely if the devices employed cause discomfort or if their application causes embarrassment. Such an emotional response is likely to affect the reliability of the measure and to disturb our detection of the processes under investigation.

We have concentrated upon physiological measures because these have been a key focus of psychophysiological study. Aspects of behavioural and subjective measurement are considered elsewhere in this volume (e.g. in relation to electronic meeting systems in Chapter 5). However, no psychophysiological study will be complete if it limits itself to the physiological domain alone.

THE PSYCHOLOGICAL SIGNIFICANCE OF SOME KEY MEASURES

Event related brain potentials (ERPs)

If a stimulus (e.g. a flash of light or a sound) is presented to a subject, a brain response can be detected which is made up of a series of negative and positive components. In order to detect this response it is necessary to average the EEG over several presentations of the stimulus, so that the 'noise' in the EEG cancels out and the 'signal' (the ERP) can be seen.

This is the most lively research area in contemporary psychophysiology, with new discoveries, new interpretations and new applications appearing monthly. It offers the best opportunity to study user–system interaction. The ERP waveform is complex and different components are related to different aspects of information processing. The topographical distribution of response patterns on the cranium gives an indication of which psychological resources are being utilized at any particular time. This richness of data has enabled ERP researchers to draw upon procedures and process models devised in experimental psychology. Such a marriage of expertise often leads to new discoveries within both fields of enquiry. A detailed evaluation of ERP data is given by Loveless (1983). What follows is an extremely abbreviated account of a very complex area of research, in which many alternative interpretations are available.

The resting background EEG is much larger in amplitude than are ERPs themselves and averaging of trials is needed to enhance the signal-to-noise ratio. Different components of the ERP are named in terms of positive- and negative-going shifts in voltage and the approximate time (in 100 ms) when they occur following a stimulus. Thus, N_1 is the first negative component of the ERP and P_1 is the first positive component.

Very early components, recorded below 50 ms, respond to systematic manipulation of the physical parameters of stimuli (i.e. frequency, brightness and intensity, etc.). The N_1 component, occurring early in the ERP (at approximately 100 ms) was once thought to reflect selective attention. When subjects ignored tones at one ear and attended to tones on the other ear, N_1 was enhanced for attended tones. This occurred at the vertex of the scalp and the effect overlapped with other brain area responses in a way which led to the view that it reflects a general non-specific process, a tonic attendance set, passively admitting material to the attended channel. This view has been challenged, because it seems that N_1 enhancement is accompanied by brain changes which reflect more active processing.

A later component, N_2 (the second negative component) is responsive to changes in stimulation within a sensory modality and seems to reflect automatic comparison of input with a stored neurological model, whether the person reports the change or not.

The P_{300} component (a positive part of the ERP, occurring about 300 ms following presentation of the stimulus) seems to reflect uncertainty and surprise, and can be manipulated by varying items in a sequence. With long runs of the same stimulus, P_{300} diminishes; with short runs of varying stimuli, it increases. Probability models have been constructed which predict P_{300} amplitude with considerable accuracy, on the basis of variation within sequences. While N_2 is modality dependent, and reflects deviation from past events, P_{300} seems to measure expected future stimulation. Recently it has been suggested that the larger the P_{300}, the greater the utilization of mental resources.

Finally, if a warning signal is presented, then slow potentials occur before a subsequent imperative stimulus calling for response, and such changes are seen to reflect a preparatory set. There are two potentials, capable of independent manipulation and the second is particularly sensitive to speed–accuracy trade-off instructions, reflecting a motor readiness potential.

All these attributes of ERP components can be seen to have *prima facie* relevance to user–system interaction and cognitive and motor patterns of response. Specifically, information can be obtained about: character discriminability, effect of character size and design, modality specific attention, general attention, voluntary and involuntary attention, rate of information processing and refractoriness, operator expectation, error identification, and motor preparation. As yet this promise needs to be fully exploited. We sample here a small group of studies and subject one study to detailed criticism, to illustrate both the power and pitfalls of psychophysiological research.

We have already mentioned the study by Wastell *et al.* (1981) of old and cordless switchboards. Their hypothesis was that the new switchboard induced dissatisfaction by reducing the attentional demands of the job and thus the level of interest it provides. They therefore predicted that early ERP components would be lower for ring tones on the cordless exchange, reflecting lowered attentional demand. Performance was certainly different for the two systems in terms of charge ticket completion and concurrent handling of calls and in terms of responsiveness to traffic fluctuations. However, no ERP or heart rate differences were found to parallel the highly significant performance effects.

A number of lessons can be learned from this study. Wastell *et al.* did not conduct a task analysis. This would have identified the different task components, their order of occurrence, their sequential dependencies, and their relative roles in maintaining smooth performance. Such an analysis is essential to help separate out components into sensory, attentional, decision, motor output, and feedback categories. Their selection of the ring-tone as the focus of interest seemed to have been more determined by the averaging requirements of ERP than by the task itself, and seems to have little to do with the presence or absence of a cord. This frequent event (positive response to dialling) rather than a rare event (dialling failure, engaged signal, sudden interruption) may be less affected by presence or absence of cord. Later ERP components would be better candidates for attentional allocation, as would motor preparatory potentials. Finally, telephone operation can be seen as a task which only rarely utilizes the full capacity of the operator; a secondary task should have been employed to test for residual capacity under the two operating conditions. However, in their desire to measure performance *in vivo*, Wastell *et al.* appear to have wished not to change the task from its normal pattern, a strategy which may not always be best at the initial stages of research.

It seems that Wastell *et al.* selected an inappropriate ERP component, used an inappropriate task/performance indicator, and employed an inappropriate testing procedure. In addition, although they measured physiology, mood, and performance, they made no attempt to relate them in any discrete fashion, nor explore the possibility that individual operators varied in the ways in which they handled their task. (This could have been achieved by video-analysis, and by a detailed exploration of relationships across the different data sets.)

Compensatory filler activities such as finger-tapping, chatting, scratching, gross bodily movements, and so on, are often used in working contexts to compensate for the de-arousing effects of monotonous tasks (O'Hanlon, 1981). Such activities can serve a) to keep the person alert, and b) to distract the person from the key task.

Attention to the points raised above could lead to more appropriate research. In summary, this implies: formal task analysis, identification of the ERP components most relevant to different working conditions, decision as to *in vivo or in vitro* measurement, monitoring of all behavioural components, monitoring of subjective response, modelling of the inter-relationships between data sets, and provision for ideographic as well as nomothetic analysis. Regrettably, this is a counsel of perfection which is rarely satisfied.

The dual task method already mentioned has provided some powerful ERP data. Israel, Wickens, Chesney and Donchin (1980) presented subjects with a radar screen to simulate air traffic control conditions. Their primary task was to monitor the screen; secondary tasks were detecting additional visual or auditory events, and under some conditions counting them. The operators had to detect flight direction/course changes in simulated aircraft. As the load of the primary task was increased (by increasing the number of moving items), reaction time increased, but increasing load on the secondary task did not affect primary task performance. ERP (P_{300}) changes to the secondary task reflected variations in secondary task performance as a function of the load imposed by the primary task and the consequential utilization of mental capacity. Their task analysis showed that the primary task employed largely perceptual and motor components while the secondary task involved perceptual aspects (detecting flashes on the screen) and, in one condition, memory aspects (counting flashes). The differential relationship of P_{300} to different task aspects helps to identify the resources utilized by the task.

A study by Wickens, Gill, Kramer, Ross and Donchin (1981) examined first and second order manual control in tracking, where the subject had to track and then 'capture' a moving target. Again, P_{300} responses to the secondary task (target flashes, auditory events) reflected the amount of loading on the primary task. Use of the dual task technique recognizes that operators are rarely working at full capacity; special techniques are necessary to estimate the degree of capacity utilized, and ERPs seem to be particularly suited to the estimation of the mental engagement of the operator.

Finally, Kutas and Donchin (1980) studied both ERP and EMG (electromyograms) while subjects prepared to respond with either hand. When the person knew which hand was to be used, contralateral motor potentials appeared. Later work in the same laboratory shows that EMG and ERP can be used to distinguish correct performance and anticipatory error.

Thus ERP procedures seem appropriate to assessment of displays, use of response devices such as keyboards and cursors, and to the assessment of mental loading associated with particular tasks (e.g. dialogue style, data entry, time-sharing, waiting etc.). One aspect of mental loading of increasing importance is the extent to which brain responses to synthesized speech tap normal speech recognition processes, interfere with the performance of tasks, or facilitate performance by reducing information processing requirements.

Electroencephalograms (EEG)

The EEG is a constantly fluctuating complex waveform of voltages varying between 10 and 200 microvolts, with a frequency range from 0.5 to 40.0 hertz. Within the frequency range measured, different characteristic waveforms have been identified, the most studied of which are theta (4.0 to 6.0 hertz), alpha (8.0 to 14) and beta (14 to 20). The EEG can be measured in time- or event-locked mode (as are ERPs) or continuously, sampling in milliseconds, seconds, hours and even longer durations. Such flexibility makes EEG sampling compatible with many tasks. Alpha activity is typically seen to diminish in amplitude and increase in frequency as attention is given to stimuli. As with the ERP, the locus of the recording electrode is seen to relate to activity in particular generally defined brain areas and thus relate to particular resource utilization. A great deal of research has been conducted in relation to hemispheric specialization, the view being that different sides of the brain are more strongly associated with different types of processing.

As with ERP research, there is controversy as to the meaning and interpretation of EEG changes. Unlike ERP studies, EEG researchers have not tended to draw upon contemporary models of information processing. However, given appropriate conditions of testing, the EEG has been related to many aspects of behaviour in a systematic fashion. Gale and Edwards (1983c) review the literature relating to reaction time, stimulus complexity, image formation, working memory, sustained attention, personality, expectancy, and interpersonal behaviour. In most of these fields, consistent and robust data have been obtained. Their working hypothesis concerning the functional significance of EEG is that it is responsive to fluctuations in the cue value of incoming stimuli, reflects mental loading, distinguishes between verbal and non-verbal processes, reflects long-term shifts in attention, and is responsive to general social conditions (such as proximity of others and eye contact).

They relate EEG changes to the allocation of resources, as follows:

— at low levels of activation, mental energy is allocated to incoming stimuli in a manner which reflects their cumulative significance
— as more resources are called upon, and mental load increases, there is more activation (reduced alpha amplitude)
— if a register of events must be maintained in working memory (memory mode) then too much activation disrupts the register
— if responses are to be rapid (throughput mode) then activation is beneficial
— theta amplitude increases when distracting stimuli need to be excluded in order to sustain focused attention.

This working model or hypothesis is imposed by Gale and Edwards retrospectively on existing data, much of which was not gathered with the hypothesis in mind. Therefore the hypothesis still needs to be tested in a systematic and parametric fashion. However, while there are no available studies which purport to relate the EEG to specific aspects of user–system behaviour, there is a *prima facie* case for work in relation to displays, icons, rapid processing, and maintenance of internal registers (working memory) during interactive work; indeed, there is a case for EEG work in most of the contexts where ERP techniques have been deployed. However, ERP has attracted more research funding and experimental and theoretical consideration over the last ten years, and continues to do so.

Heart rate change and variability

The mechanisms within the brain and the autonomic nervous system which control the heart have been studied in great detail, and are very complex. For example, there is dual innervation of the heart by both sympathetic and parasympathetic branches of the autonomic nervous system. This means that heart rate change could be the result of activity in either branch or both; there are drugs which block specific influences on the heart and thereby enable systematic study. During inspiration, vagal influence on the heart is blocked or attenuated by a gating system, inducing sinus arrhythmia. In psychological terms, the heart may be influenced by either emotional stimuli or by information processing and such specifically psychologically-determined changes need to be teased out from among the considerably larger fluctuations induced by vegetative, motor, and respiratory activity. We focus here on two phenomena, intake–rejection (heart rate deceleration–acceleration) and mental loading (cardiac arrhythmia).

The most elaborated theory of heart rate change and information processing is that of Lacey (1967) who claims that deceleration is associated with visual attention and empathic listening, while acceleration accompanies anagram solution and mental arithmetic. These two tasks represent intake (or external attending) and rejection (information processing) and Lacey presents a physiological model of the heart–brain relationships involved. While his view has been challenged (see Coles, 1983, for a balanced view) there are empirical data which support the intake–rejection distinction. Such a distinction clearly has relevance for the study of dialogue, since the operator has to shift when in interactive mode, between external attending and reference to inner data and processing. Sharit, Salvendy and Diesenroth (1982) have used heart rate to study the effects of machine-paced as opposed to self-paced working (see below), varying task type on the basis of the Lacey distinction. They compared an external-intake task (perceptual, letter scanning) with an internal-rejection task (mental arithmetic), including also an incentive condition. The Lacey hypothesis was confirmed and self-pacing was associated with deceleration (a finding which was not in fact predicted).

Heart rate data are probably the most easy to capture in field contexts, reliability and validity are high, and intrusiveness is low. Task analysis of many user–system contexts would probably reveal that the notion of the operator fluctuating between internal and external attending is of relevance. There are therefore grounds for believing that heart rate studies could assist in the classification and evaluation of tasks and in the monitoring of operator processing. Existing databases and theory are available to help model the findings. Thus heart rate change studies of user–system interaction await exploitation.

There is a longer tradition of using heart rate variability or cardiac arrhythmia as a measure of mental load in working contexts. Early work by Kalsbeek (e.g. Kalsbeek and Sykes, 1967) suggested that increasing information load and consequent pressure on the operator's information capacity is associated with decreased heart rate variability. This view has been modified considerably over the years. There is no commonly agreed taxonomy of the tasks employed, it is difficult to partial out other aspects of load (e.g. respiration and emotion), and there is controversy as to how best to estimate variability, since the various available measures do not intercorrelate or give the same results. Moreover, as we have indicated, the operator in the field devises personal and eccentric coping strategies which are rarely captured in the laboratory, where naive and ill-practised subjects are used and the experimental questions are focused on group rather than individual effects. A recent study by Hitchen, Brodie and Harness (1980) employed a continuous monitoring task; as rate of presentation increased, variability was reduced. However, mean heart rate and respiration also increased at the same time and the authors claim that reduced cardiac variability is a secondary consequence of such changes. Sharit and Salvendy (1982) obtained evidence to show that mean heart rate is more sensitive to information load, while variability is sensitive to pacing.

There clearly is some potential in this field for the study of user–system interaction. Given the relative ease with which heart rate can be measured, variability could be studied as a correlate of mental loading, so long as sound task analysis is carried out prior to hypothesis testing and the selection of independent variables. It must be said, however, that the conceptual validity for cardiac arrhythmia is nowhere as well supported as for Lacey's theory of acceleration–deceleration.

Pupillometry

Beatty (1982) put up a stalwart defence for the use of pupillometric measures in the assessment of information loading. Pupillometry distinguishes within-task, between-task and between-individual variations and is therefore a powerful and widely applicable measure. Pupillary responses are pervasive, have a short latency of response (100 to 200 milliseconds) and rapid offset, and relate systematically to processing load or mental effort. If subjects are given sequential material to process, systematic changes occur as items are presented, and the slope of change relates to task difficulty. Strategy instructions and stimulus meaning also lead to systematic change, including tasks involving long-term stored information retrieval.

Effects also occur in relation to selective and sustained attention. Beatty also claims that absolute, as well as relative, differences are meaningful; this is an advantage since some physiological measures can only give calibration for individual subjects, calling for data transformations before entry into a statistical evaluation procedure.

Beatty concludes that pupillometric levels and changes reflect an 'aggregate task-induced utilization of multiple processing resources'. Such a measure would be extremely useful when taken in association with measures like ERP and EEG, which combine general and specific resource measurement. Use of multiple measures, although highly recommended for applied contexts (Wierwille, 1979) occurs rarely; but the notion that different physiological measures help to identify different psychological processes is a powerful one, and offers the promise of data appropriate to model building. Unfortunately, pupillometric measurement currently involves a good deal of intrusive equipment and a high ratio of data analysis effort to data acquisition/duration. Nevertheless, it is highly recommended for laboratory testing of specific hypotheses. On the other hand, EEG and ERP can be claimed to have all the demonstrable advantages of pupillometry, without many of its disadvantages.

Eye movements

In contrast, data yielded from eye movement studies are unique, since they provide information concerning: direction of gaze, gaze dwell time, individual differences in cognitive style, scanning patterns (for text and other visually presented material), and emotional response. There is clear relevance for the study of the operator at work. For example, using eye movement analysis we can ask: how does gaze vary between source material, screen, and keyboard; do different modes of dialogue induce different eye gaze patterns; does formatting of text affect the pattern of fixations; what is the dwell time for different areas of the display; how much do different display materials attract relative amounts of attention; how do operators vary from each other; and, whether particular work schedules or tasks induce more or less fatigue.

During reading, the eyes move in quick jumps (saccades) rather than smoothly, and patterns of such movements can be related to specific textual presentations. Three processes are involved (Bouma, 1982): control of saccades and pauses, recognition processing during pauses (between saccades), and information integration over successive saccades/pauses. There are approximately four fixations per second. If text quality is good, the visual reading field can be 20 letter positions. In a rightward saccade eight (plus or minus four) letters will be covered. There are small leftward saccades (for correction) and of course, large leftward saccades for line change. Thus reading research using eye movement measurement can enable the construction of rules of thumb for VDU design and a means of evaluating existing VDUs. In conversational mode, operators divide source material and display viewing time on a 50:50 basis, with alternations at a rate of one per 5 to 10 seconds. For data entry, source material attracts most of the visual attention and display-directed gaze is infrequent (Grandjean and Vigliani, 1982). Eye movement patterns are different for good and poor readers, can be affected by specific instructions, and are sensitive also to drug effects. Some psychiatric groups have different eye movement patterns. Conjugate lateral eye movements are said to occur during information processing and the left–right direction of such movements has been claimed to indicate which cerebral hemisphere is engaged in the particular task (Oster and Stern, 1980).

Finally, studies of blinking show that it diminishes during focused attention, and conversely, increases during stress and some forms of mental loading (Wibbenmeyer, Stern and Chen, 1983).

Eye movement analysis has great promise. The three methods of measurement available all have their advantages and disadvantages. The least intrusive (videorecording) can deal only with gross movements. Electrooculography is intrusive, since electrodes need to be attached to the face, while optoelectronic techniques (the most accurate) involve application of devices to the eye surface, or the measurement of reflected light from ocular structures. Computer analysis can assist in locating the point of gaze upon the display. As with pupillometry there is currently little prospect of field work of any sophistication but the potential pay-off for laboratory-based work is considerable.

STUDIES OF WORK STRESS

We do not know if the offices of the future will have operators recognizable as clerical staff or indeed if current clerical jobs will even exist as we know them. However, several authorities have expressed concern over stresses introduced by the creation of the VDT workstation. The claim is that the clerical worker's style of work has been dramatically altered from self-pacing to machine-pacing.

In 1980, Jenner, Reynolds and Harrison reviewed 21 occupational groups. Catecholamine excretion (adrenaline and noradrenaline) as a measure of work stress was lowest for clerical workers in conventional paper offices. The highest excretion rates were for machine-paced assembly and steel workers. The well-known study of 2000 workers in over 23 occupations by Caplan, Cobb, French, Harrison and Pinneau (1975) also showed the highest experienced stress among assembly line, machine-paced workers. Smith (1981) claims that clerical workers are being shifted into a machine-paced category by modern technology and claims: 'Recent work suggests that pacing produced by computerization (of office work) may have an even greater effect than factory pacing'.

Will clerical workers now move up the stress–catecholamine excretion league? The difficulty is that researchers often exaggerate their results and frighten clients into providing cash for research. This is part of the new technology syndrome. O'Hanlon (1981) expresses caution over the view that repetitive work is stressful. There is no demonstrable causal link between repetitive work and stress; many factors operate and interact in working situations, and it is difficult to separate out their independent positive and negative effects. Moreover, people vary, both in their preference and in their response to alleged stressors. In one study, 30 per cent preferred paced work. We do not have any acceptable explanation of why some individuals become stressed or why they express stress in so many different ways.

Paced work has certain specifiable characteristics: a short time cycle; buffer stocks between work stations; work is broken up into small units; activity is continuous rather than discrete; pacing is by machine; work is monotonous; the operator is immobilized; social contact is reduced; and, there is little sense of personal control. In the case of VDU-computer automation, Sauter, Harding, Gottlieb and Quakenboss (1981) suggest the following characteristics are present: reciprocal 'yoking' of the operator with a 'smart' machine; machine control of task information; decision-making by machine; lack of freedom to accumulate or self-pace the work; frustration through time-sharing and breakdown; possibility of operator performance monitoring by supervisors; and, there is little understanding of the machine and its inner workings.

Task analysis is possible in such contexts on the basis of machine and operator activities. The machine can be available or not available for operation (e.g. presenting information, response delay) and the operator can be working or waiting. Under pacing, machine operation is fixed and operator work time is short and with little variance. Dainoff, Hurrell and Happ (1981) propose a taxonomy of tasks using initiation of work cycle (operator or machine) and duration of work cycle (operator or machine), giving four quadrants. Clerical workers in the past determined both initiation and duration; under modern working conditions, a terminal has control over both aspects. Work of this nature can lead to boredom and fatigue and the very strategies employed by the operator to maintain wakefulness (e.g. chatting, maintenance work, body movements) can themselves lead to errors. In immobilized conditions such strategies may not be available and considerable effort is needed to maintain work pace and accuracy. This effort expenditure is seen, in turn, to lead to anxiety and stress (O'Hanlon, 1981).

Physiological monitoring can be used to measure such processes at work. Catecholamine excretion has been used to estimate stress following paced and other forms of work, and the extent to which the stress is taken home by the operator after work. Mackay and Cox (1979) simulated a repetitive work situation and obtained evidence of boredom, loss of arousal, decreased heart rate, increased heart variability, and increased urine adrenaline. Performance deteriorated and became more variable, and ancilliary activities increased as a strategy for fighting against drowsiness. This is some confirmation of the O'Hanlon theory of fatigue outlined above. We must beware however of overgeneralization, either from one working context to another, or indeed from laboratory studies (which, however extended, rarely exceed a few hours' work) to field contexts.

The leading group for catecholamine research is the Swedish group led by Frankenhaeuser. They have shown that both under- and over-stimulation increased adrenaline and noradrenaline output, that relations with performance depended on the task (higher excretion better for vigilance, but worse for sensory-motor tasks), and that subjects found both over- and under-stimulation unpleasant, being bored with the latter (see Frankenhaeuser, Nordheren, Myrsten and Post, 1971). If subjects gain control over the task, then excretion diminishes.

Whilst such findings tend to be replicated and can show good relations between biochemistry, performance and reported mood, we should note that the effects obtained are rather gross in terms of the parallel psychological processes involved, in contrast to ERP work. Discrimination of general from specific stressors has not been good. Also, measurement during work is intrusive and interrupts ongoing activity. Some increases in adrenaline are helpful in certain tasks, although large increases are reliably associated with loss of control and physical stress. Noradrenaline appears to be related to physical rather than mental stress, and can therefore provide evidence of the efforts of constrained posture, a characteristic of sedentary VDU work.

Thus catecholamine measurement seems very appropriate for distinguishing tasks, for measuring degree of stress after work, and for estimating the period to recovery. Cox, Cox and Thirlaway (1983) set out a list of precautionary measures which must be taken into account in this field of research, including circadian effects, sex differences, and the need to partial out specific and general stress effects.

PSYCHOPHYSIOLOGY AND THE VDU

If a VDU is operating in machine-paced mode, in constrained physical surroundings and with little opportunity for social contact, then the dice may be loaded against the physical and mental well-being of the operator. It is still not clear whether early claims concerning eye-strain were actually direct effects of the screen or an indirect expression of psychological unease caused by a mixture of new technology, job redesign, organizational stress and fear of redundancy. In many situations where stress is experienced, individuals and groups seek a rational reason for their distress; the VDU screen was a good candidate and possibly a scapegoat for many ills. Some governments have passed extensive legislation concerning the design and use of the VDU. Can psychophysiological studies help us in evaluating the VDU?

Given our systems approach to the person, the physical surrounding and the social environment can both be expected to have an impact on operational efficiency as well as more specific attributes of the terminal. One needs both a 'top-down' and a 'bottom-up' approach to the analysis of user–computer interaction, otherwise there is a danger of excessive cognitivism, just as in other contexts there is an excess of socio-biological determinism.

Human factors work has identified the optimal conditions for current types of VDU workstations. These include: yellow, green or yellow/green phosphor for displays; minimal flicker; minimal luminance oscillation; separate operator controls for brightness and contrast; reduction of screen reflection; operator variable angle, height and distance of screen; flexible keyboard adjustments for angle, height and distance from VDU; variable height working surface; adequate space for papers and personal possessions; space for source documents; leg space; adjustable footrest; ambient illumination control; glare control; localized additional illumination; minimal illumination contrast between screen, source and background; control of air temperature, humidity and flow; seat adjustable for height, angles, back support and rotation; eyetests for operators, with provision for special spectacles to correct defective vision or aging effects; systematic training in using and adjusting the workstation; training in body posture; and, staff participation in the design and development of work organization, tasks and systems. In addition to all these factors, the design of the dialogue also affects the acceptability of the user interface. This is discussed in the next two chapters.

It is worth noting that few working environments satisfy all these requirements. While their introduction may be beneficial, the need to provide good working conditions does not arise specifically from the introduction of new technology. Any or all of these conditions could be associated with variable effects, and it would be difficult to associate increments or decrements in misery or happiness with any one particular aspect. As frequently happens in working situations, several changes are introduced at once and it is impossible to apply the sort of inferential logic to locate sources of effects that one can within a laboratory context.

Psychophysiological studies have been carried out in relation to ambient working conditions, eyestrain at work and after, posture strain, and job stress associated with psychosomatic complaints. Krueger (1982) used eye movements to study various aspects of screen presentation. Ostberg (1982) used a method called laser optometry to measure eyestrain in the form of temporary myopia following working periods. Shahnavaz (1982) found that workers preferred a lower level of ambient light than is recommended. This applied also to screen luminance, although older workers preferred a brighter screen. Haider, Kundi and Weissenbock (1982) used a variety of physiological, performance and subjective measures of eye irritation and myopia. While their data are extremely complex and difficult to interpret, they do show considerable changes in well-being after prolonged working periods. Data entry terminals give more reported strain and pain than do conversational terminals (Hunting, Laubli and Grandjean, 1982). Johansson, Aronsson and Lindstrom (1978) showed constrained posture to be associated with increased noradrenaline excretion.

It is worth completing this section with one study that showed a group of VDU operators to feel more autonomous, more self-directed, more mentally active and job satisfied, than other groups, and also less tired at the end of the day. But this study, just like those which show deleterious effects, has the same inherent difficulties. Those who preferred VDUs were compared with typesetters and non-VDU proof readers (Kalimo and Leppanen, 1981). The difficulty is that there are few studies where comparable tasks are compared, where VDU/non-VDU modes are compared within-task or within-subject groups, where testing lasts for an extended work period, where organizational factors are controlled for, and so on; the potential list of confounding factors and error sources is extremely long. Within a considerable literature, we have found no really conclusive study linking any specific aspect of VDUs to any specific measured effect.

CONCEPTS OF INFORMATION PROCESSING

Data, however rich they appear, mean nothing in the absence of theory. We have already indicated that the best work in psychophysiology borrows concepts and procedures from better established branches of psychology. Unfortunately, many of the procedures used by psychophysiologists were either devised for their own purpose or lag behind the related field of mainstream psychology by several years or even decades. Many of the procedures used in ERP work assume a linear, data-driven view of information processing; in contrast, the predominant view in experimental psychology is one of resource-driven or memory-driven processing.

Classic models of an additive factors logic variety, as developed in the field of reaction time, with serially laid out boxes and minimal numbers of feedback routes between them, are now seen as inadequate to account for the data. For example, the skilled performer in reaction time experiments, who has had many thousands of trials, shows little overlap either in response times or in errors with the typical laboratory subject. A reasonable interpretation is that the processes and brain resources used in the training stage are no longer in use once skill is established (Rabbitt, 1979). This rules out many laboratory findings, not only in terms of their own intellectual integrity, but because of their weakness as sources of generalization for field conditions. As we have seen, skilled workers devise many strategies (resources) of their own, to help them cope with adverse working conditions.

Psychophysiology has few explanatory concepts of its own. A well-known concept is that of arousal, and the associated view that performance is optimal under moderately arousing conditions but poor under low- and very high-arousal (the inverted-U relationship). Eysenck (1982) has conducted a detailed analysis of the concept. There are many sources of arousal, as the concept is used in the literature (i.e., as a state, as a response, as a motivator, as a consequence of stimulation, as a property of people, as a property of stimuli, as an effect of drugs, noise, incentives, sleep loss, circadian rhythms, and so on). There is little evidence that the 'arousal' measured in one of these contexts is that same 'arousal' as measured in others. While there is some evidence of additive effects under certain conditions, direct evidence of systematic physiological accompaniments to arousal effects is very sparse. Certainly, physiological measures do not intercorrelate very well under conditions in which arousal is said to be manipulated. While energy-related concepts are attractive and even seductive, they are very difficult to pin down operationally. Some authorities prefer to measure arousal as a subjective response, rather than a physiological state and claim that such measures are more reliable (Mackay, 1980). This brief paragraph is presented to serve as a caution, although the reader will have observed that the term 'arousal' has been used throughout the chapter.

The notion of mental load and cognitive strain seems to be more amenable to operational definition. We have seen that the dual task is employed to assess the amount of information processing capacity being utilized by a task. Some authorities (e.g. Brown, 1982), while they are sympathetic to the use of physiological measures in applied contexts, suggest that mental load should also be measured by other means as well, including performance measurement, subjective evaluation, protocol analysis, and of course, task analysis. Task analysis is essential because processing involves multiple resources. A secondary task may have very little effect on capacity if the resources it employs are different from those (sensory, perceptual, decision, motor, etc.) employed in the primary task. A good example of the use of the dual task in the field is a study of drivers in heavy and light traffic conditions by Brown and Poulton (1961). Drivers performed relatively badly on a secondary, auditory detection task, in a shopping area as compared with a residential area. There are a number of controversies about the type of secondary task to use, the ways in which data can be interpreted, and the means of handling eccentric operator strategies (see Rolfe, 1971). So far as user–system interaction is concerned, some of the best ERP studies have involved the dual task procedure.

In introducing this chapter we referred to individual differences in information processing, styles of work and coping. Such individual differences are pervasive and are likely to influence all varieties of activity. There is a long tradition of psychophysiological studies. Much of the work has been reviewed by O'Gorman (1977). Recently, Gale and Edwards (1983d; 1984) have provided an extensive review of individual differences research and attempts to relate psychometrically derived factors to specific aspects of physiology. There are no consistent findings which are powerful enough to lead us to make recommendations concerning working conditions. On the other hand, many of the concepts developed within the psychophysiology of individual differences (sensation-seeking, locus of control, helplessness, anxiety, extraversion, etc.) do have *heuristic* value for the modelling of working situations. One theory of considerable promise and which has been subjected to both field and laboratory studies is that of Strelau (1983) (see above).

This section may seem like a list of complaints about current progress in psychophysiology but the other side of the coin is that each complaint bears with it a recommendation for future practice. Guidelines for future studies are considered in our early appraisal of the Wastell *et al.* (1981) study and in the following discussion.

PSYCHOPHYSIOLOGY, GENERAL SYSTEMS THEORY, AND SITUATED ACTION THEORY

In the introductory section we referred to the General Systems Approach and suggested that it offered a means of understanding the complex interplay within the person, of physiology, experience and behaviour. However, we also mentioned the person's environment and the meaning which the individual imposes upon it. The individual at the workstation is part of a broader system, which includes not only psychological and organizational variables but sets of values and norms. Such norms govern the ways in which the individual worker sets personal goals, measures achievement at work, and makes comparisons with the behaviour and work of colleagues.

Jahoda (1979) identified a number of latent consequences of work, which have considerable bearing on personal well-being. They are that:

— work imposes a time structure on the day, giving predictability and certainty to life
— it provides shared experience outside the context of the family
— it links the person to goals beyond their immediate needs and a sense of contribution to the community at large
— work confers status and identity; and finally
— it allows the person to be active and to exercise skill.

Why should a psychologist be concerned with such things? The answer is that they are all likely to affect the physical and mental health of the individual. The notions of predictability, companionship, identification with higher goals and so on, may not be incorporated into future working experience, but in our view, that seems unlikely. For such goals would have been recognized in any of our past cultures. Social variables have proved to be some of the most powerful in manipulating physiological state, and some have argued that much of our emotional nervous system was evolved primarily for the purposes of interpersonal behaviour (Gale and Edwards, 1983a).

It is this sense of interacting systems within psychophysiology that may be of most use for the future design of interfaces between people and computers. So far as work is concerned, Jahoda's list could be a checklist for ensuring that the quality of working life is not adversely affected by innovation. New technology can enhance those positive aspects of work, or indeed, accentuate the less desirable aspects.

The notion of systems has implications for the nature of work research and how it is to be carried out. If much of peoples' actions is determined by response to the situation in which they find themselves, then it is essential to carry out research into human action within appropriate contexts. Argyle, Furnham and Graham (1981) set out a situated action theory in which the emphasis is not so much on persons but on actions in context. Situations are seen to have goals and goal structure, rules or shared beliefs which guide or regulate actions, roles for participants including duties and obligations, and elements and sequences, including rituals. All these elaborated actions occur within an environmental setting. Thus, the office should be seen as a special situation with its own rules, roles, rights, and responsibilities.

Such views are not dissimilar from those expressed by certain forward-looking psychophysiological researchers (e.g. Levenson and Gottman, 1983; O'Connor, 1983). Levenson and Gottman studied physiological changes and expressed affect in married couples as they engaged in normal end-of-the-working-day exchanges. Highly significant relations between physiology, quality of relationship and experienced emotion were obtained, not when individuals were treated separately, but on the basis of lag sequential analysis of the relationships between the successive physiological states of the couples. Thus physiology was given meaning, by sampling established patterns of interaction between personally significant individuals, in contexts where rules of action had been well practised. O'Connor (1983) favours a similar approach to the study of human performance; an approach which focuses not on outcome but on process. Such an approach to psychophysiology, which emphasizes changes over time, implies a retreat from conventional large-group designs and, instead, the use of statistical models (like Bayesian statistics) with which most psychologists have little familiarity.

What does this imply for psychophysiological studies of user–system interaction? It implies the detailed study of individuals at work, within normal working contexts and over extended periods of sampling. It implies the necessity to work with the subject in an effort to understand the meaning of work stimuli, rather than to observe the subject as a mere actor. Chapanis (1971), in a very eloquent paper, challenged laboratory-based research to show how its manipulations and criteria of worth could be applied within the field, where notions of usability and operator acceptability are far-removed from both the jargon and procedures which psychologists enjoy. What is required is a special testing environment, which captures the best of both laboratory control, and situational realism.

We believe that the proper testing context for user–system studies is a purpose built simulation of the real life office. At ITT Europe and Southampton University we have called this simulation THE CAFE OF EVE, a controlled, adaptive, flexible, experimental office of the future, in an ecologically valid environment. It is called a CAFE OF EVE because it is the modern equivalent of the Garden of Eden, where our ancestors strove in happiness and contentment until they were blasted into the real world by the first recorded knowledge explosion.

What are the characteristics of the CAFE OF EVE? Why is it different from laboratory studies on the one hand and field studies on the other? While laboratory studies help us to exercise control over independent variables, subject samples, subject behaviour, experimental design, and the whole testing environment, they often are more determined by theory than by problems. The results which emerge are often only statistically significant (rather than clinically significant and applicable with confidence), subjects are those who come to hand rather than genuine operators (i.e. compliant students or local housewives), testing is brief, while work experience goes on for many years, and designs focus on large samples tested only once, rather than on a handful of individuals tested over many sessions. It is thus unlikely that the laboratory experiment samples the strategies of the experienced operator or taps the value systems which operate in working contexts. Because experimenter and subject have situationally determined roles, subjects are often reluctant to identify which variables are influencing their behaviour.

Field studies also contain problems concerning generalizability of results. The experimenter can rarely arrange the variables under study in an orderly fashion. Access to workers is typically brief, selective and grudgingly provided. Experimenters are 'experts' subject to suspicion and possible dissembling. Change is introduced in working contexts, not step by step but in wide-ranging surges; thus it is difficult to identify the precise causal path from among many simultaneous changes which led to particular effects on dependent variables.

As we have indicated earlier, most field designs are subject to confounding; for example, it is rare to find within-group designs.

Thus the CAFE OF EVE captures all the advantages and benefits of both laboratory and field work, while avoiding many of the disadvantages of both. At the same time, the CAFE OF EVE provides a context for introducing, testing and developing new products, offering a working context for interdisciplinary collaboration between electronics experts, designers, engineers, systems analysts and psychologists. Finally, a living office of the future can be a showpiece, enabling public and sales demonstrations of new devices in operation.

Of course, such an establishment must have rules and roles which are different from those operating in the 'natural' office. But our argument is that whatever the emergent disadvantages, the CAFE OF EVE has a better chance of generating *applicable* data relating to usability of new systems, by virtue of its combination of realism and control, than do more traditional means of gaining research information. The CAFE OF EVE satisfies the requirements for good applied research specified by Howarth (1980) and Gale and Chapman (1984).

The CAFE OF EVE solution is to set up an experimental office of the future from the outset. It should be located within an entire section of an operating company so that work can continue within the normal functional requirements of the organization. Staff should be recruited on the explicit understanding that they are in a partnership of exploration with a view to improving the operational efficiency and usability of new office systems. They should appreciate the need to monitor their activities and to sample their behaviour over long durations of testing. Performance measurement, subjective reporting and physiological monitoring should be part and parcel of the job description. In such circumstances, new roles and rules can operate *vis-à-vis* researcher and subject, who now become partners in a joint and extended exploration of the office and its working. Attached to the main office should be standard laboratory settings, enabling the researcher to focus down on the controlled manipulation of variables whose salience has been identified. In so doing he or she should have available, 'genuine' skilled operators as subjects.

CONCLUSIONS

Psychophysiological approaches to the description of human problems and the analysis of complex systems have great potential for increasing our understanding of 'usability' and how to assess it. Regrettably, much of the promise which we have set out in this chapter has yet to be realized. Developments in new technology and the implementation of new office systems are not only intellectually exciting, but offer a challenge to the ingenuity of psychological scientists. Exploitation of the psychophysiological approach is likely to yield answers to a variety of important questions, particularly if researchers develop appropriate contexts for the study of the changes which follow the introduction of new systems.

In the next two chapters we turn our attention to the other side of the coin—from how to assess usability to how to achieve it. We begin with a consideration of guidelines for design of the user–system dialogue.

Chapter 10

Dialogue Design Guidelines[1]

IAN COLE, MARK LANSDALE AND BRUCE CHRISTIE

INTRODUCTION

The previous chapter introduced some of the issues involved in assessing product usability, illustrating these within a psychophysiological framework. This chapter and the next are concerned not with how to assess usability but how to achieve it. A key element of a system in regard to achieving usability is the design of the user–system dialogue, and it is to this that we turn our attention in this chapter.

The dialogue is the structure within which the user and the system exchange messages. The user and the system use the dialogue to interact with one another. It is the dynamic aspect of the user–system interface, or USI. The design of the dialogue is therefore of crucial importance in converting the potential benefits of the underlying technology into actual benefits for the user.

This chapter presents some key guidelines for the design of successful dialogues, based on a mixture of experimental research and practical experience. This is an area where much research is still needed. In order to be able to use the guidelines effectively it is necessary to understand how they relate to the design process, the kind of design tradeoffs which must be made, and the framework within which they are related.

---

[1] This chapter, with permission from ITT Industries Ltd., is based closely on excerpts from ITT Europe ESC's 'Dialogue Design Guidelines For Interactive Systems' written by Ian Cole and Mark Lansdale, 1982.

A top-down approach to the design of dialogues is often recommended and is practicable in most situations. This includes the following main stages:

— user requirements definition
— conceptual design
— semantic design
— syntactic design
— lexical design

User requirements should be extracted from an analysis of the tasks the system is intended to support, user characteristics and characteristics of the information and its use. From this, it is possible to determine the interface requirements. The objectives and constraints of the design should also be determined at this stage.

Conceptual design requires the identification of the key concepts of the application: types of system objects (e.g. mailbox, files), relations between these objects and actions on them. It is at this stage that a choice of dialogue style is made. The details of the dialogue are progressively refined over successive stages of the design process in conjunction with detailed dialogue design guidelines concerning input, display and users' requirements for interpretation and control of system processes. This will help to ensure that the dialogue is compatible with users' needs and their psychological characteristics.

Semantic design addresses the units of meaning which are conveyed between the user and the system, but not their form. This includes the commands input by the user and which operate on the system objects and the relations between objects. It also includes selection of information to be presented to the user in response to their actions.

During syntactic design the form of the semantic units conveyed between the user and the system are defined. This includes the precise grammar of input (i.e. structural rules) and the format and encoding of displayed output.

Lexical design is the stage where the defined dialogue is wedded to the hardware capability. Appropriate input devices and interaction techniques are designed, as are the output devices, output primitives and attributes, and the way primitives combine to form syntactic units.

Although the design process can be divided into well defined stages, in practice the process is iterative.

The design of an electronic office system requires that certain tradeoffs be made, for example:

— The final design may necessitate a compromise between accuracy, time, cost or ease-of-use requirements.
— Sometimes design guidelines may be incompatible with one another.

Therefore:

— It is always worthwhile spending some time testing a proposed dialogue on a representative group of potential users, not on other system designers.
— Human requirements must always take precedence over machine processing requirements; the system is a tool to aid users in their achievement of task objectives, especially where a significant proportion of users have low computer awareness.

In the remainder of this chapter we consider:

— user requirements as they relate to dialogue design: first, in terms of classification of potential levels of user skill or knowledge with respect to system interaction; second, by indicating the important task characteristics which must be interpreted
— a summary of the main types of dialogues used today; this includes a matrix to aid designers in their choice of an appropriate type of dialogue to suit user requirements
— specific dialogue design guidelines which relate to system input and output respectively
— the need for the user and the system to 'mesh in' with each other's behaviour, as humans do when holding a conversation.

Space precludes going into any more depth than this or being more extensive in this chapter, but a short bibliography is included at the end of the chapter for the benefit of those readers who need to do so.

USER REQUIREMENTS

Users must be able to use a system to achieve their task objectives by means of a dialogue compatible with their level of skill or knowledge. Therefore, to choose an appropriate dialogue style we must:

— classify users in terms of their levels of skill or knowledge
— analyse the functionality of user tasks, their important procedural characteristics, and the type of information use.

Important levels of user interaction

The skill a user displays in interacting with a system can be considered for the purposes of this chapter to be based on two dimensions: semantic and syntactic knowledge.

Semantic knowledge relates to the general concepts which are important to system use independent of the precise structure of effective interaction.

In the most general sense, it refers to knowledge (also called conceptual knowledge) of the application and approaches to tasks and problems. At its lowest level, semantic knowledge concerns the function a command elicits and the action initiated on system objects (e.g. files, documents, in-tray, etc.).

Syntactic knowledge is related to the precise structure of system interaction; the procedural and grammatical rules and conventions by which the user communicates with the system and the system communicates back.

Users can be divided into four general categories based on their semantic and syntactic knowledge (see Figure 10.1, adapted from Shneiderman, 1980). The categories are not mutually exclusive and the division between them is somewhat fuzzy. However, they provide a useful framework within which to think about the type of interaction a user is likely to require.

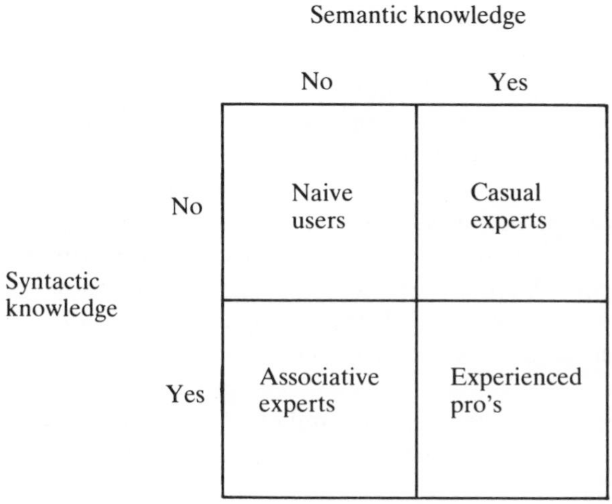

Semantic knowledge

	No	Yes
No	Naive users	Casual experts
Yes	Associative experts	Experienced pro's

Syntactic knowledge

Figure 10.1: Main categories of users (based on Shneiderman, 1980)

Naive users. Naive users have least knowledge of the system in terms of the application and the facilities and functions offered, let alone the commands available and the actions they initiate. Furthermore, this precludes any syntactic knowledge of the system. These users are not able to initiate interaction with the system and so must be led through an effective interaction by the system.

Casual experts. A casual expert is a user who has a working knowledge of the application and the various commands and functions at his or her disposal, but due to sporadic use of the system has inadequate knowledge of the exact syntax of interaction. Very infrequent use will require a full system initiated dialogue, whereas system prompts are enough for less infrequent use. With correct prompting casual experts will evolve towards taking the initiative and controlling their interaction in a flexible manner to suit their needs.

Associative experts. This category contains those users who interact with electronic systems frequently. They learn combinations of actions and specific command input sequences in order to achieve specific objectives, without understanding the system actions which are perpetrated and without fully understanding the command structure. Their use of the system is inflexible and confined to memorized procedures and grammatical structures. Unpredictable requirements and less well known approaches to a particular task will require the support of semantic information and also syntactic information; if a system syntax is consistent, however, the latter is less important.

Experienced pro's. Experienced pro's have well developed semantic and syntactic knowledge and can use the system in a powerful and flexible manner.

User skill migration. These categories are not discrete; users can progress between them, and consequently their dialogue requirements will also change. For instance, training will help naive users progress to the other, more experienced categories. Similarly, experienced pro's can regress to casual experts through lack of system use; the level of user skill is constantly changing. Also, each type of application will have a range of these user types, i.e. naive through to experienced. System design must allow for changes in user skill levels and accommodate them accordingly.

Task characteristics. There are a number of task characteristics which are relevant to choosing an appropriate dialogue. They are:

— functional structure
— procedural characteristics
— type of information use
— information characteristics.

Functional structure. The functional structure of the system is determined during the initial conceptual design stage, in line with user requirements. It is dependent upon careful task analysis, forms the basis of the system command structure and provides input into the semantic design stage.

It is important to rationalize the functional structure, and hence the command structure, so that idiosyncracies are not incorporated and an existing mess automated.

Procedural characteristics. The extent to which a task can be proceduralized is directly dependent on the degree of task structure, the ease with which the structure can be defined, and the amount of repetition implicit in the structure. A highly proceduralized task allows:

— functions to be combined into higher level functions
— information input to be embedded in structured dialogues such as form filling
— information output to be formatted and encoded consistent with user's needs.

Tasks which are not highly proceduralized, where the user's requirements are flexible and difficult to predict, and where the level of machine intelligence is low, demand a dialogue which is correspondingly flexible (e.g. a command language) and place a high information processing load on the user.

Type of information use. The way in which information is used during system interaction has an important bearing on the dialogue style. An interaction which predominently requires information input demands a compatible dialogue style, for example data forms; the formatted screen with data fields and suitable cursor control will facilitate data entry. Alternatively, an interaction predominently concerned with information retrieval will require an appropriate form of database query language. Other interactions will involve several information uses: various processing requirements will require appropriate command languages; document creation will require text entry and editing facilities.

Information characteristics. The volume and complexity of information relevant to an interaction and the optimum format in which it is used are important considerations in the design of effective systems.

SELECTING AN APPROPRIATE TYPE OF DIALOGUE

The most important decision in the course of designing the USI for an office system is choosing the style of dialogue which is optimally suited to the proposed user population and the tasks they perform. Dialogue types fall into the following main categories:

— question and answer
— form filling
— query languages
— menu selection
— function keys
— command languages
— graphic interaction

— natural language
— hybrid dialogues
— parallel dialogues.

Before looking briefly at these various categories individually, it is helpful to consider some general principles, as follows.

Initiative. Interactive dialogues can be initiated either by the user or by the system.

Inexperienced users will more often require the system to take the initiative; they will input commands, parameters, etc., in response to some form of prompting by the system (e.g. a question, or a list of alternatives).

Experienced users, however, will more often wish to exercise flexible control over the dialogue to fulfil their requirements.

Flexibility, complexity and power. Flexibility is a function of the number of ways in which a user can accomplish a given function. Highly flexible dialogues help experienced users. However, they can degrade the performance of moderately experienced (e.g. experienced non-programmers) and inexperienced users. There is evidence that the latter adopt sub-optimal interaction strategies.

Complexity is related to flexibility and is the number of options available to the user at a given point in the dialogue. There is no clear guidance at the moment as to optimal complexity. Complexity can be reduced by not displaying redundant options or by hierarchically organizing dialogue options to reduce the number displayed at any point; however, hierarchical organization can also render the dialogue structure more complex. It is reasonable to assume that high complexity should be avoided for naive users, and be tolerated only by the expert programmer. Power is related to both flexibility and complexity. It is the amount of work accomplished by the system in response to a single user command. High power is usually accompanied by either high complexity or restricted generality. The latter is a significant factor in system rejection, especially by scientific and technical users. Non-expert users may also require powerful, high-level functions. However, much effort must be spent to reduce the complexity of the dialogue in which they are embedded.

Information load. The information load imposed on the user is a major factor in the success of many systems. In principle, there is an optimum level of information load that should be sought—too high a load will cause the user problems in trying to process all the information, too low a load will tend to result in boredom and other undesirable effects. In practice, however, the problem is usually to keep the level of information load down to an acceptable level. This is especially important in the case of less experienced or naive users.

Information load can be reduced by simplifying and restructuring the dialogue, and more especially the information displayed. Major possibilities are:

— appropriate use of graphics
— reformatting displays for improved correspondence to the immediate information requirements of the user
— use of coding techniques to convey meaning or draw attention to relevant information
— use of commands of appropriate power for the immediate task of the user
— use of a language which minimizes information processing by the user (e.g. constrained natural language)
— moving clerical operations (e.g. manipulations of data before data entry) deeper into the system, beyond the USI
— structuring command language to correspond to the substructure of the user's task
— use of default values.

The following is a brief overview of the main types of dialogues in use today. In describing these it is assumed for the purposes of exposition that the system lacks any significant degree of machine intelligence. Systems with a high level of machine intelligence will be able to switch flexibly from one style of dialogue to another as appropriate for the user and task situation. It is also assumed as a further simplification that the main means of input is a keyboard and the main means of output a visual display.

Question-and-answer

In a question-and-answer (Q & A) dialogue, the user responds to questions asked by the system (i.e. system-initiated). This is the simplest dialogue type and is only appropriate for inexperienced users. It is useful in applications in which the inputs are few and easily identified, and where the task structure defines an invariable sequence of interaction.

A useful variation on question-and-answer is to allow previous answers to remain displayed, so that users can cross check subsequent answers with them and maintain consistency.

Question-and-answer dialogues have a limited number of applications, mainly those where the computer is used to elicit a profile of users (e.g. diagnosis) or to ascertain their requirements (e.g. autobank). This requires that questions are pre-defined and that users' needs can be anticipated.

Form filling

Form filling is a system-initiated type of dialogue in which the display is formatted so the required input resembles filling in a form. It is often used in situations in which the user's input is dominated by parameter values, rather than commands. Form filling allows multiple user responses in a single transaction (e.g. specifying parameters for printing a document) and provides more contextual information concerning the required input than does a Q & A dialogue. Effective use requires the formatted screen area to be 'protected' except for the input areas; the movement of the cursor between input areas should be achieved simply, e.g. by a single keystroke. It is best to complete a transaction by initiating the desired system response when the form has been completed. To prevent screen clutter however, logical sections of a form can be displayed successively, dependent upon completion of a previous section; in this way a transaction is built up on the screen step by step.

With careful screen formatting, forms can also be used for data entry. A tabular format is best because it allows easy comparison of items.

This dialogue is ideally suited for tasks which are structured, repetitious and which predominently deal with predictable input. The user is typically not required to take the initiative or use the system in a powerful or flexible manner.

Query languages

With a query language, the user interacts with a system to gain access to a database. The query language is used to express the user's specific request(s) for factual information; it does not alter the database but allows users to ask questions about the data. Users are required to ask questions in some formal manner or adhere to database access procedures. These requirements are reflected in a number of query language types, as follows:

— Keyword query languages allow the user to use a restricted form of 'natural language' to interact with the system. They employ rigid syntax to make processing easier.
— Natural languages attempt to allow the user to specify requests in 'natural language' without restrictions on the vocabulary. However, in practice only a restricted subset of the language can be processed by the system. A very high number of keystrokes or other inputs are usually needed to formulate a request.
— Hierarchical menu selection can be used to formulate database queries and is one of the easiest methods, but not without its drawbacks, e.g. tedium for experienced users, difficulties of defining menu options appropriately.
— Graphic query languages formulate requests mainly in terms of diagrams, but some typed input is usually required. Each type of object in the database is assigned a specific geometric shape.
— With query-by-example the user constructs his or her requests by giving the system an example of a reply to the request.

— A form or template which explicitly conveys the database parameters and their relationships is a useful aid to users; they can selectively input parameter values to restrict information access in accordance with their requirements.

— Multi-media languages use many different ways of interacting with a system. Text, formatted data, voice, graphics, colour, animation, and other means are used to formulate requests and present replies; e.g. higher levels of the database can be displayed using icons to reflect major information categories.

Menu selection

Menu selection is a system-initiated type of dialogue which presents the user with a complete selection of possible options on the screen at key control stages of the dialogue. It can be used for command construction and for database search (e.g. in videotex systems), where the successive presentation of menu options reflect the system command structure and/or database structure respectively.

This dialogue type is appropriate for naive users, both to lead them through a successful interaction and to familiarize them with the command and/or database structures; it is frequently used with hierarchical structures.

Function keys

Dialogues which use function keys typically have some of the properties of user-initiated dialogues and some system-initiated properties. Although the user may have to construct a meaningful sequence of key depressions without assistance from the system, the keyboard itself provides a memory aid which allows the user to rely on recognition of functions rather than recalling them from memory.

Function keys are used in command-dominated interactions and are appropriate for naive users only if the combination of command inputs is simple and/or a system-initiated form filling dialogue is used; some training is required.

Multi-function keys can be used when keyboard size is constrained. They operate in conjunction with some kind of mode switching key, the mode determining which of the keys' functions are valid. Soft function keys are software controlled. Their function changes according to the current context of the interaction. This approach is especially suitable in conjunction with touch-sensitive screens. At any stage in the dialogue a user is presented with only those functions valid at that point. These are presented in the form of labelled, touch-sensitive areas. To operate, the user simply has to touch the required function.

Command languages

Command language dialogue is a user-initiated type of dialogue which minimizes system overheads and response time. It is generally preferred by experienced users (e.g. designers and programmers) because it is more like computer programming than the other dialogue modes. However, this dialogue mode also provides least assistance to the user in acquiring a 'mental model' of the system and a knowledge of the functions and syntax of the language; syntax can often be complex and involved.

Except for very simple command languages, this dialogue style requires a trained user, with the amount of training varying directly with the complexity of the language. It should be aimed at sophisticated users and more or less dedicated users who use the system frequently enough to maintain familiarity with the commands, command procedures, and syntax. Inexperienced users can be catered for by embodying the command structure in some other dialogue type (e.g. menu selection); they can progress to direct command input after gaining the requisite experience.

Graphic interaction

Interactive graphics is not actually a separate dialogue, but can incorporate or supplement the more conventional dialogue types dependent upon the design emphasis. Graphic displays allow certain classes of information to be conveyed to the user at very high rates, because of the sophisticated visual information processing capability of humans. Complex interrelationships can be explicitly conveyed using graphics by presenting them as spatial relationships. This exploits humans' highly developed spatial awareness and ability to extract meaning from spatial patterns.

Natural language

The ultimate goal of natural language is to allow a user to converse with an electronic system in the same way as with another person. It requires a high level of machine intelligence and is still a highly experimental area whose requirements transcend the requirements of conventional user–system dialogues. Human interpretation of natural language is sensitive to context and prone to ambiguities; conventional machines have insufficient powers of inference to interpret similar imprecision in the user's input. Constrained 'natural language' offers a partial solution, but is essentially just a sophisticated command language with similar advantages and disadvantages.

Hybrid dialogues

During a period of interaction with an electronic system a user may undertake a variety of transactions (e.g. enter data, retrieve documents, etc.), each of which presents the user with a different set of requirements. In this case, it does not make sense to constrain the dialogue to one style only, but to present a suitable style specific to each transaction, or set of common transactions. This mix of dialogue styles forms a hybrid dialogue. The current trend towards integrated systems with distributed resources has increased the complexity of user–system interactions and will allow a greater diversity of tasks to be performed at a workstation. Users will be more familiar with some system facilities than others and so initiative in the interactions will have to change between the user and the system accordingly. Also, different tasks will require different dialogue styles to facilitate interaction in accordance with their particular features.

Parallel dialogues

As users gain experience with a system their requirements change. They will want to exercise greater control over the interaction, use the system in a more flexible manner, and employ some of the more sophisticated facilities available; this embodies a change in initiative from the system to the user. To cater for this requirement, two, or several, dialogue styles are necessary, available in parallel. Users can then progress from one to another in accordance with their level of sophistication. A classic example of parallel dialogues is menu selection in conjunction with command language input. Naive users express their requirements by choosing options from successive command menus. As they, and their requirements, become more sophisticated, they can take the initiative and enter commands directly, thus using the system in a more flexible and powerful manner.

Dialogue choice matrix

The important factors which influence the choice of dialogue style are summarized in Figure 10.2. Two asterisks denote that a particular dialogue is recommended with respect to the respective user types or task related considerations; a single asterisk denotes that a dialogue style can be considered but with care; no asterisks means the type of dialogue is not recommended for the users or tasks concerned.

GUIDELINES FOR INPUT TO THE SYSTEM

The following guidelines focus on four of the more widely used types of dialogues:

- form filling
- menus
- function keys
- command languages.

	Q&A	FF	QL	MS	FK	CL	GI	NL‡	HD	PD
USERS										
Naive users	**	**		*	**		**	*	*	
Casual experts		**	*	**	**	*	**	*	**	**
Associative experts		**	*	**	**	*	**		**	**
Experienced pro's		*	**		**	**	*		**	**
TASKS										
Procedural	**	**	**	**	**	*	**	*	**	**
Non-Procedural			*	*	**	**	**	*	**	**
Information input		**		*	*	*	*	*	*	**
Information retrieval		**	**	**	*	*	*	*	*	**
Processing				*	*	**	*	*	*	**
Information creation		*		*	**	*	*	*	*	**

NB: ‡ With explicit constraints to decrease ambiguity etc.

Q&A = Q&A
FF = Form Filling
QL = Query Languages
MS = Menu Selection
FK = Function Keys

CL = Command Language
GI = Graphic Interaction
NL = Natural Language
HD = Hybrid Dialogues
PD = Parallel Dialogues

Figure 10.2: Selecting an appropriate type of dialogue. (Reproduced by permission of ITT Industries Ltd.)

Form filling

- On initial appearance of the display, ensure the cursor is placed automatically at the first character position of the first input field.
- Minimize user actions required to move the cursor from one parameter entry field to the next (e.g. using a tab key).
- Allow the user to move to the next field easily when an input does not occupy the total field.
- Do not require users to overwrite lines of text with their responses.
- Use default values at least for the frequently specified inputs.
- Provide full exit, back-up (i.e. move to previous input), and 'Help' facilities in any field.
- Accomplish entry of logically related items by a single explicit action at the end rather than by separate entry of each item.
- Protect field labels and make them transparent to keyboard control so that the cursor skips over them when spacing or tabbing.
- Delineate each entry field (e.g. using underscore) and achieve input by replacement of delineating characters.
- It is best not to validate input until the end of a transaction (e.g. specifying print parameters), especially where parameters are interrelated, unless the nature of the user's task demands it (e.g. high precision required). ('Validation' means checking by the system to see if the input matches a type that has been designated as 'legal', e.g. that it is not too long or that it does not contain letters when the system is expecting only numbers.)
- Provide easy editing for input correction after validation; i.e. allow the user to move easily through the parameter fields to alter incorrect inputs.
- Make areas of the display not needed for input inaccessible to the user, under computer control, to allow the cursor automatically to jump input fields.
- User action confirming entry of multiple inputs has to be valid irrespective of cursor position on the display.
- Allow the user to RESTART, CANCEL, or BACKUP and change any item before taking the final ENTER action.

Menus

- One input selection only per menu is optimum.
- Options must be displayed at time of selection.
- Display only the options appropriate at a particular step in the dialogue.
- Structure menus in a way meaningful to the users.
- Position the most frequently selected options at the top.
- A fast system response time is needed to allow smooth transition from one stage to the next.

- Allow input stacking.
- Serial presentation of successive menus in the same area of the screen is preferable to simultaneous presentation in different areas of the same screen.
- To resolve errors or confusion, allow the user to exit from the current menu sequence.
- Ensure the wording of menu items reflects the current concerns and likely questions of the user at that step in the dialogue.
- If the user population is variable, provide various menus with different levels of detail.
- Provide multiple paths to accommodate both experienced and inexperienced users (e.g. experienced users should be able to bypass levels by entering a page number or identification code).
- Sequence menu frames in an order that is natural for the user. This may mean holding choices in computer memory until they are needed by the system.
- An initial menu of control options must always be available to serve as a consistent starting point at the beginning of a transaction sequence.
- Minimize the number of stages of selection.
- Permit users immediate access to critical or frequently selected options.
- Try to avoid more than 3 or 4 levels of selection in a hierarchical structure.
- Indicate current choice position in the menu structure.
- Allow easy backtracking, e.g. by a single key action.

Function keys

- Distinguish different function keys by label, position, colour, highlighting, shape, size, or other means.
- Label informatively to designate the function concerned.
- Avoid user confusion by distinguishing between closely related functions.
- A single function key requires only one label.
- An assigned function should remain consistent in various tasks and transactions.
- Function keys which are not valid options for current input should be temporarily deactivated.
- Do not use mechanical overlays manipulated by the user to de-activate function keys.
- The current subset of active keys should be conveyed to the user (e.g. by illumination).
- Accomplish a function with a single key press.
- Avoid using multi-function keys—i.e. keys which can control more than one function—unless the constraints of the keyboard demand it,

or unless there are too many functions to represent with single function keys.

- In the case of multi-function keys, the user should always have a convenient means of knowing which function is currently available.
- Provide an easy means to return to the base-level function.
- Avoid changing function by repeated activation.
- Place frequently used function keys in convenient locations.
- Locate potentially disruptive keys (e.g. 'escape', 'erase', 'delete', etc.) to avoid accidental operation.
- Group function keys meaningfully in distinctive locations on the keyboard, according to:

    — function
    — priority
    — frequency of use.

- Locate emergency functions in prominent positions distinguishing them by distinctive coding (e.g. size and/or colour).

Command languages

- Provide a command entry area in a consistent location on every display, preferably at the bottom.
- Use commands which reflect the users' task concepts, not system concepts.
- Do not use commands longer then 5 to 7 letters long.
- Impose a clean structure on the command language by:

    — making commands meaningful and distinctive
    — using commands consistently throughout the dialogue
    — avoiding ambiguity
    — relating commands explicitly by meaningful categorization
    — emphasizing significant differences in function.

- All words in a command language must be consistently used and standardised in meaning.
- Provide flexibility:

    — permit the user to assign personal names to files
    — allow meaningful and unambiguous abbreviation
    — allow entry of a sequence of commands in anticipation of prompts in future dialogue states (i.e. input 'stacking')
    — allow macro formation (i.e. frequently used command sequences called up by one command).

- Avoid commands which sound similar (e.g. store and restore).

- Free commands from system constraints (e.g. file size).
- Minimize punctuation.
- Use simple delimiters (e.g. slash).
- Use delimiters consistently.
- Ensure the system tolerates single or multiple blanks in command entry.
- Allow users to review their input of commands which are subject to misinterpretation or have potentially disruptive consequences.
- If a command entry is not recognized, the system should not reject it outright but initiate a clarification dialogue.
- Use keyword argument formats in which both the argument and its value are specified. Positional argument formats, in which argument values must be specified in a given order, impose a greater memory load and result in more errors.
- Use argument menus to construct commands which have many, often-used arguments.

GUIDELINES FOR INPUT TO THE USER

The following guidelines focus on four key aspects of designing visual displays for outputting information from the system to the user.

- content of the display
- format
- coding
- use of colour

Content of the display

- Display information on a screen bearing in mind user requirements:

  - decide what context information, instructions and prompts are necessary
  - indicate the context of the display (e.g. title), and any instructions or prompts necessary to process the information, at the top of the screen
  - instructions or prompts concerning the disposition of a completed screen should be at the bottom
  - impose a meaningful structure on the displayed information
  - provide an obvious starting point in the upper-left corner of the screen.

- Provide only information that is essential to making a decision or performing an action. Do not flood the user with information.

- Provide all data or information related to one task on a single display if possible; it is difficult to remember information from one screen to the next.
- If several screens are used for information related to one task, link them in an explicit manner.
- Present information in a usable form. Do not require reference to documentation.

Format

A well formatted display:

- is developed within the physical constraints imposed by the terminal
- fully utilizes the capabilities of its software
- if used for data entry, is developed within the constraints imposed by related source materials, such as worksheets, forms or manuals
- is consistent within itself and with related displays.

In particular, it is often helpful to:

- Group items according to some logical principle, for example:

    - by sequence of use (i.e. where spatial or temporal order corresponds to the usual order of encounter)
    - by function (i.e. items with a common purpose)
    - by importance (i.e. most significant information, or that requiring immediate response is grouped at the top)
    - by frequency of use (i.e. most frequently used items at the top of the display).

- Arrange grouped data in the display with consistent placement of items, so that users can detect similarities, differences, trends and relationships.
- When there is no appropriate logic for grouping data, some other principle should be adopted, e.g. alphabetic or chronological ordering.
- Reserve specific areas of the display for certain kinds of information, such as commands, status messages and input fields, and maintain these areas consistently on all displays.
- Provide cohesive groupings of display elements by using blank spaces, surrounding lines, different intensity:

    - blank spaces: leave 3 to 5 rows or columns between groups
    - do not break the display into too many small windows.

- Provide symmetrical balance by:

  - centering titles and illustrations on the vertical axis
  - placing like elements on both sides of the areas
  - placing lighter elements further from the vertical axis and larger elements closer to it.

- Minimize the number of items simultaneously displayed, with the following considerations:

  - present all information relevant to a user's task if possible
  - remove interim data from the display when it is no longer needed
  - allow users to eliminate irrelevant items from the display.

- The sequence in which information is presented should be logical both in terms of a user's task and/or other information sources being used.
- Avoid clutter on a display; it usually increases search time, causes items to be overlooked, and can cause misreading.
- Avoid displaying redundant and irrelevant information; it can mislead the user.
- Consistent formats will stimulate users to develop expectations concerning the location and form of information or data. This will ensure that:

  - displays are easier to use and interpret
  - unfamiliar or new information is more readily and accurately interpreted.

- Change established display formats only as necessary to distinguish one task or activity from another.
- Display the appropriate level and quantity of information in the simplest way (e.g. tables, graphs) in keeping with users' requirements.
- Use partitioning to present information items coherently; do not partition the display into too many small windows.
- Begin each partition with a title or a header, reflecting contents or purpose, separated by one blank line from the body of the partition.
- Reserve the last several lines at the bottom of every partition for status and error messages, prompts and command entry.
- Ensure that partitioned areas are clearly delimited.
- Use the partitioning of various display features as consistently as possible for all displays.

Coding

The coding of information is especially useful to help discriminate among different classes of information simultaneously present in a display.

- Alphanumeric coding:

  - abbreviations of words can be used but they must be meaningful
  - numeric codes are not very meaningful, but can be used to convey sequence information
  - joining abbreviations with numeric codes gives greater flexibility.

- Spatial coding: allows particular classes or types of information or the relationships between items of information to be emphasized by their respective positions.
- Brightness coding: two or three levels of brightness can be easily distinguished on a VDU. Selective brightening of parts of the screen clearly indicates critical or interesting items of information.
- Flashing: urgent or important information can be distinguished from other parts of the display if the appropriate lines or characters are made to flash (typically 2–4Hz).
- Shape coding: shaped symbols are particularly useful, in addition to alphanumeric characters, if they can be associated easily with the objects they represent.
- Size coding: the size of characters can be varied to indicate relative importance or different status (e.g. headings).
- Colour coding: see below.
- Others:

  - depth
  - line type (solid, dashed, etc.)
  - line width
  - motion
  - focus or distortion.

Use of colour

Colour is selected for special attention here as it is rapidly becoming much more widespread than before and is often misused. It warrants being considered under several headings, as follows:

Colour discrimination:

The eye's ability to resolve fine detail (visual acuity) does not depend on a difference in hue between foreground and background but on brightness difference ('contrast').

- The brightness of colours, from highest to lowest is:

  - white
  - yellow
  - green
  - blue
  - red.

- Very bright colours produce glare, and dull colours not enough contrast. Green provides good general visibility over a broad range of brightness levels.
- Do not use more than 8 colours, and no more than 4 per display.
- For normal discrimination, select colours that are widely spaced along the visual spectrum:

    — the most generally useful are red, green, yellow and blue
    — other acceptable colours include orange, yellow-green, blue-green and violet.

- The most legible colours on a dark background are green, yellow, cyan and white.
- Various national and international standards, e.g. DIN standards, are likely to become important during the next few years in governing the use of colour in electronic displays.

Colour as a formatting aid:

- Use colour to:

    — relate or tie data/information fields into meaningful groupings
    — differentiate groupings from one another
    — relate data/information fields that are spatially separated
    — partition the screen where the amount of data/information precludes effective spatial grouping
    — assist reading of long lines of text by changing the background colour of every alternate group of three lines (NB. Use two colours of similar brightness).

- Represent only one category of data/information with each colour.
- Apply colour coding as an additional aid to the user on displays that have already been formatted as effectively as possible in a single colour.

Colour as a visual code:

- A colour coding scheme must be:

    — relevant to the performed task and meaningful to the user
    — known to the user
    — a redundant feature of meaningful display items (e.g. symbols, number, text) which conveys context rather than specific application data/information
    — consistent with known conventions; e.g. red for danger or debit amounts.

- Use a consistent 'base' colour for standard text.
- Nothing is gained by using more than 3 or 4 different colours for text.
- Use colour to identify:

  — categories of data/information: enables user to identify a category without reading the contents (e.g. age over 35 years)
  — source of data/information (e.g. department)
  — status of data/information (e.g. processed vs. unprocessed; important vs. unimportant)
  — order of operations (e.g. prompting and guidance through a complex transaction).

- Use bright colours to emphasize data/information and colours lacking brightness to de-emphasize data/information.
- For emphasis and separation, use contrasting colours such as:

  — red and green
  — blue and yellow
  — red, green and blue
  — red, green, blue and white.

- To convey similarity, use similar colours such as:

  — orange and yellow
  — blue and violet.

Use of colour in graphics:

- Large areas of luminous colours (e.g. white or yellow) should be avoided because of glare and a potential flickering effect.
- Adjoining areas of colour will appear to merge if they are too close in luminosity or spectral hue.
- The appearance of a colour will be different dependent on the colour surrounding it.
- It is best not to use red and blue on the same screen, and never next to each other.
- Use blue as a background or block colour only, not for critical data/information.
- Warm colours (red and yellow) usually appear larger than cooler colours (green and blue).

Some important precautions with colour:

- Use colour consistently and judiciously.
- The high attention-getting quality of colour may be distracting, causing a person to:

- notice differences in colour, regardless of whether they have a task-related meaning
- visually group items of the same colour in a way unrelated to the task, or in conflict with another task-related group of items.

- Indiscriminate use of colour on one display may interfere with its attention-getting power on another display.
- Do not use too many colours on one display; i.e. normally a maximum of 8, but probably safer to use no more than 4 or 5. As the number of colours in a display increases, the following will also increase:

  - the time required to find, interpret, and respond to a specific colour
  - the probability of confusion among colours
  - the demands on hardware for reliably reproducing each colour.

- Do not use colour in an attempt to compensate for a poor display format; redesign the display instead.
- Remember that a small proportion of the population (approximately 8% of the male population) is colour blind.

GUIDELINES FOR FACILITATING USER–SYSTEM 'MESHING'

Whilst it is convenient to consider the inputs from the user to the system and from the system to the user separately in order to focus on some aspects of dialogue design, the two do not occur separately in the course of a dialogue. The user and the system 'hold a conversation' and, like conversations between two humans, this requires the two parties to 'mesh in' with one another's behaviour and not act independently. In the case of user–system interaction, this means especially that the user must be able to interpret the behaviour of the system and must be able to exert some control over the course of the 'conversation'. In particular, it is important in designing the dialogue to:

- provide appropriate feedback from the system
- allow users to interact with the system in line with their 'real time' expectations—the question of response times
- provide users with an acceptable means of exerting control over their interaction with the system
- control errors; that is, design the system to avoid mistakes or to allow easy recovery from them.

Space precludes doing more than just touching on some of the key points relating to these aspects of dialogue design.

Providing feedback to the user

System feedback is necessary for users to ascertain system status and to interpret the response of the system to their actions; i.e. to establish a context for their interaction. It is important that system context is explicitly conveyed to users to ensure that only valid dialogue options are chosen, or to enable them to anticipate the outcome of inputs whose function varies with system state. At any stage in the dialogue the user must be able to answer the following questions:

— Where am I? (i.e. what dialogue state?).
— What can I do? (i.e. what options are open?).
— How did I get here? (i.e. what sequence of actions have brought me to this state?).
— Where can I go? (i.e. to what other dialogue states can I progress?).
— How do I get there? (i.e. what control options are necessary to take me to the desired dialogue state?).

It is also important that the system response to each user action is explicitly conveyed, not only to check the validity of the input, but also to ascertain its function.

Feedback is a prerequisite for learning how a system operates. Moreover, without appropriate feedback it is difficult to exercise any control over the dialogue.

Special cases of feedback include:

— feedback about the current status of the system and the major tasks it is performing
— feedback about the meaning of commands and system actions
— feedback about the validity of inputs from the user to the system, i.e. whether the system is able to accept them
— prompts and help messages
— error messages.

Response times

The control of hardware, software, and system-generated delays in response to user activities has a significant impact on task performance, acceptance of the system, and work-related stress. System response time is a parameter which must be controlled during the design and development of a dialogue. The provision of response times appropriate for various human activities can satisfy the interactive requirements of users whilst providing cost benefits in terms of system productivity.

Response time is the time from the last key depression, lightpen selection, or other input action, until the system responds to the user with an output signal or action.

Figure 10.3 presents some key categories of dialogue events requiring different 'maximum' response times, based on figures provided by Miller

(1968) and Engel and Granda (1975). Optimum values are probably less than those stated. As far as possible, system response times should be held constant; variation in response times can be even more unacceptable than relatively long response times.

Category	Examples
1. Very long (up to a minute)	loading a programme
2. Long (up to 10 seconds)	response to complex inquiry in graphics form; updates requiring access to a host file
3. Medium (up to 2 seconds)	response to user identification; response to simple status inquiry; response to graphics manipulation, e.g. change of scale; selection of a function; error message in response to an input error
4. Short (up to 0.5 seconds)	to initiate scrolling of text; to display a new menu
5. Very short (up to 0.1 seconds)	drawing with a light pen; a 'click' from a key depression

Figure 10.3: Recommended 'maximum' response times. (Reproduced by permission of ITT Industries Ltd.)

User control of the dialogue

The user must be given the means with which to exert direct control over the course of the interaction with the system, subject to the constraints of task objectives, level of skill and the dialogue style. It must be possible to:

— initiate an interaction (i.e. log on)
— control the sequence of interaction
— interrupt the interaction at various levels
— use system default values.

Log on. Whatever the type of dialogue it will be necessary for the user to log on to the system. There is one simple rule, which is almost always violated:

• Only one simple action should be necessary to initiate a log on. Difficult or cumbersome log on procedures can discourage system use before its benefits have a chance to be demonstrated.

Obviously, this requires careful consideration because of security implications. One possible solution might be to incorporate an identity card reader into the system, or to use voice recognition.

Sequence control. Having achieved successful log on, user inputs to the system must be accepted and system outputs to the user presented in a sequence of user–system transactions which correspond to the task requirements and dialogue type.

- Provide a flexible means of sequence control and allow the user to obtain guidance as required.
- Simplify control inputs as much as possible.
- Ensure the means of sequence control is compatible with desired ends; frequent or urgent control actions have to be easy to make, whereas potentially destructive control actions must be designed to require close attention and minimize accidental triggering.
- Allow users to take the control initiative in accordance with their evolving level of skill.
- Do not allow the system to pace the user; pace has to be determined by users in accordance with their needs, attention span, and time available.
- Do not allow delays in system response to interfere with user control inputs.
- Design displays to reflect features relevant to sequence control in a distinctive position and/or format.
- In a time-shared system, inputs by one user must not interfere with those from another.
- Use consistent terminology to refer to control inputs in online messages and instructional material.
- Allow the user to make at least some sequence control inputs directly at any step in the transaction sequence (i.e. from any display frame) without having to return to a general options display (i.e. menu dialogue) or basic interaction mode; an example of this requirement is the use of edit functions.
- Treat control options which are generally available at any step in the transaction sequence and are frequently used as implicit options and implement them as function keys.

Interrupt. A specific means of dialogue control is to allow the user to interrupt the transaction sequence. Interrupts must be used judiciously in line with defined user needs.

- Flexibility in dialogue control is provided by permitting the user to interrupt, defer or abort a current dialogue sequence in a consistent manner.
- Different levels of interrupt require differently named control options to avoid confusion.

- Provide a CANCEL option to erase inputs and regenerate the display in its original form without processing the interim changes.
- A BACKUP will allow the user to return to the display entered in the previous dialogue state.
- A RESTART option will return the user to the first display in a defined sequence and permit a review of the sequence of entries and any necessary changes.
- An ABORT option will consistently cancel all entries in a defined sequence; its consequences are such that the user must be required to confirm.
- An END option will have the consistent effect of concluding a repetitive transaction.

Defaults. Default values are a means of reducing user workload and promoting a natural and concise dialogue.

- Design sequence control software to carry forward a representation of the user's knowledge base and current activities wherever possible.
- Employ a profile of the user to set up defaults for program processing to match the language the user generally employs.
- Explicitly convey a predefined default value and allow the user to indicate its acceptance or otherwise.
- Allow the user to select default values for the system if they cannot be predefined.
- Use defaults between commands to supply missing commands, missing arguments, and missing commands when arguments are given.
- Consider the following points when using defaults with form filling dialogues:

  — display currently defined default values automatically in their appropriate data fields with the initiation of a data entry transaction
  — allow default values to be accepted by a single confirming keystroke or by tabbing past the default field
  — allow replacement of any default value during a particular transaction without changing the current default definition.

Control of errors

Users will inevitably make errors. It is essential to protect the system and its data from the errors without antagonizing the user. It is also essential to protect the user from the consequences of the errors.

The following general guidelines should be considered:

- Minimize the information processing load on users.

- Minimize dialogue state changes, mode changes and display changes. Display information relevant to a particular dialogue state concurrently if possible.
- Provide adequate 'help' facilities.
- Design the system to adapt to the users' changing skill levels.
- Tolerate common errors, e.g. misspellings.
- Incorporate redundant information for critical items to aid interpretation.
- Permit commonly used alternatives.
- Use a consistent and explicit ENTER action to indicate completion of input, unless implicit activation is preferable.
- Use defaults.
- Require explicit confirmation of defaults.
- Require explicit confirmation of control inputs which will cause extensive change in stored data, procedures and/or system operation, especially if the operation cannot be reversed easily.
- Use an explicit command input or labelled function key to CONFIRM a control input which is different from the ENTER key.
- Word the prompt for CONFIRM action in a way which clearly states the consequences of the action.
- When a user logs off, sequence control software must check pending transactions and, if data loss seems probable, display an advisory message requesting confirmation.

Validating input. With respect to input, it is important to allow users to achieve psychological closure, that is the completion of a logical unit of input in a transaction sequence, or a whole transaction sequence, without serious disruption in their train of thought. Therefore, validation of input must be compatible with these requirements. Designers must seek to:

- Validate input immediately unless to do so will disrupt a user's train of thought; i.e. where a significant proportion of input must be reentered.

In addition to these general guidelines, care should be taken over the design of those aspects of the dialogue relating to input validation and error recovery procedures. Space precludes a detailed treatment of these here.

BIBLIOGRAPHY

It has only been possible in this chapter to present an introduction to the guidelines that can be drawn from the human factors literature on dialogue design, focusing on the more general guidelines and a few more specific areas, e.g. use of colour. The following short bibliography is provided for readers who wish to go further. In the next chapter we go beyond a consideration of specific dialogue design guidelines to consider the role of mental models in influencing product usability.

Bergman, H., Brinkman, A., and Koelega, H. A. (1981) System response time and problem solving behaviour. *Proceedings of the Human Factors Society 25th Annual Meeting.*

Clark, I. A. (1980) *How to Help 'HELP' Help.* IBM Technical Report MP. 198, IBM UK Labs. Ltd., Hursley Park, Winchester, England.

Eason, K. D. (1976) A task-tool analysis of manager-computer interaction. Paper presented at NATO Advanced Study Institute on Man–Computer Interaction, Mati, Greece, September 1976 (Reprinted by the Department of Human Sciences, University of Technology, Loughborough, Leicestershire, England).

Engel, S. E., and Granda, R. E. (1975) *Guidelines for Man/Display Interfaces.* IBM Technical Report TR 00.2720, Poughkeepsie Laboratory, USA.

Flohrer, W. (1982) Man–machine dialogue prerequisites for user employment. Presented at 1982 International Zurich Seminar on Digital Communications: Man–machine Interaction.

Foley, J. D. (1981) Tutorial: *How to Design User-Computer Interfaces.* ACM.

Galitz, W. O. (1980) *Human Factors in Office Automation.* Life Office Management Association.

Galitz, W. O. (1981) *Handbook of Screen Format Design.* QED Information Sciences.

Gallaway, G. R. (1981) *Response Times to User Activities in Interactive Man/Machine Computer Systems.* Report HFP 81-25: Revision B, published by NCR Corporation (Corporate Human Factors).

Kidd, A. L. (1982) *Man–Machine Dialogue Design.* British Telecom Research Laboratories, research study number 1. Martlesham Consultancy Services, British Telecom Research Laboratories, Martlesham Heath, Ipswich.

Lochovsky, F. H., and Tsichritzis, D. C. (1981) *Interactive Query Languages for External Data Bases.* Telidon Behavioural Research 5. Department of Communications, Ottawa, Canada.

Martin, J. (1973) *Design of Man–Computer Dialogues.* Englewood Cliffs, New Jersey, Prentice-Hall.

Miller, R. B. (1968) Response time in man–computer conversational transactions. *Proceedings of Fall Joint Computer Conference,* 33 (Part 1).

Ramsey, H. R., and Atwood, M. E. (1979) *Human Factors in Computer Systems: A Review of the Literature.* Technical Report SAI-79-III-DEN, from Science Applications, Inc., 7935 E. Prentice Ave., Englewoods, Co 80111.

Shackel, B. (1981) The concept of usability. Paper presented at the ITT Human Factors Symposium, ITT Shelton, Connecticut, USA, 5th October.

Shneiderman, B. (1980) *Software Psychology: Human Factors in Computer Information Systems.* Winthrop Publishers, Inc. Cambridge, Mass.

Smith, S. L. (1982) *User–System Interface Design for Computer-Based Information Systems.* Reference ESD-TR-82-132, The Mitre Corporation, Bedford, Massachusetts.

Stewart, T. F. M. (1976a) Displays and the software interface. *Applied Ergonomics,* 7, 137–146.

Stewart, T. F. M. (1976b) The specialist user. Paper presented at NATO Advanced Study Institute on Man-Computer Interaction, Mati, Greece, September 1976

(Reprinted by the Department of Human Sciences, University of Technology, Loughborough, Leicestershire, England).

Sutherland, S. (1980) *Prestel and the User.* Commissioned by the Central Office of Information, from the Centre of Research on Perception and Cognition, Universty of Sussex, Brighton.

Williges, H. B., and Williges, R. C. (1982) *User Considerations in Computer-Based Information Systems.* Technical Report, Computer Science, Industrial Engineering and Operations Research, Virginia Polytechnic Institute and State University, Blacksburg, Virginia 24061.

Wilson, P. *et al.* (1980) *Designing Systems for People.* NCC Publications.

Chapter 11

Beyond Dialogue Design Guidelines: The Role of Mental Models

MARK LANSDALE

INTRODUCTION

Dialogue design guidelines can be very helpful to the systems designer and much research effort needs to be directed towards their further development. However, in this chapter we argue that we need to go beyond such guidelines—not discarding them, but going further to consider not just the catalogue of recommendations represented by the guidelines but an overall mental model of the USI within which the specific guidelines can be applied.

However, before discussing mental models and what we mean by the term, it is useful to take a general look at the way human beings process information. One way of doing this is in terms of the constraints placed upon human information processing. These constraints fall roughly into two categories: basic limitations on the psychological processes of memory and attention, and the problems created by the fuzziness and redundancy of much of the real-world information encountered.

Human memory constraints

Memory constraints differ in the short and long term. In the short term, information seems to be retained accurately (although rarely a literal copy of the original) only if it is consciously attended to. Unattended information is lost quickly. Also, a limit on short term retention is placed by the number of 'chunks' of information humans can hold concurrently. Controversy exists over how many 'chunks' this is, but a rough generalization gives this as being somewhere between three and ten, depending on the circumstances.

A consequence of this is that human capabilities in logical processes such as mental arithmetic are severely constrained. Except in the simplest of operations, the memory load in remembering the initial values, the intermediate values (e.g. carry units in subtraction) and the results is simply too great for reliability, given the concurrent processing load.

Long term storage is less prone to loss of information, although items can become temporarily unavailable (such as forgetting a particular command word, for example). This form of memory is more characterized by the difficulty of retaining details, the information stored tending to be of a general and less specific nature (i.e. more related to context).

Attentional constraints

Apart from the fact that human sensory input and output is necessarily limited in terms of parallel processing (e.g. we can only look in one direction or say one thing at a time), central processing capacity is also poor when dealing with more than one channel of information at a time. This is particularly true when the characteristics of several channels are similar. For example, listening to two conversations at the same time is particularly difficult because of the extra processing effort involved in distinguishing the two sources of information.

Some parallel processing is possible, especially in highly skilled situations such as driving a car and holding a conversation. However, the limitations on processing capacity mean that increases in the difficulty of one task affect the other. For example conversations become more difficult to sustain as more processing resources are allocated to the primary function of driving in difficult conditions.

Fuzziness of information to be processed

The world we live in is predictable, but only in terms of *how likely* something is. This means that the occurrence of an event is often not accompanied by a unique or defining set of information items, i.e. the circumstances can vary widely. The information is therefore 'fuzzy'. Because of this, information is often open to more than one interpretation.

Redundancy of information

Another clear feature of information in the 'real world' is the vast quantity of irrelevant or redundant information from which humans must extract the relevant items.

Human information processing systems overcome these limitations by behaving as selective processors: rather than attempting to process all information as a computer would do, attention is focused on particular aspects to the exclusion of others. The information processed depends upon what is perceived by a person to be relevant, on the basis of available context information. This process is therefore quite different from that in a computer, which operates in a pre-programmed way on all data impartially. Unlike a computer, human information processing is not entirely data-driven but is also a function of what the human is thinking about.

One way of looking at this strategy is that the human mind is subconciously making assumptions about its environment based on previous experience, and is processing information accordingly. These assumptions, based largely on incoming data, serve to 'de-fuzz' the incoming information by providing a frame of reference or context for it. An example of these assumptions working can be seen in the following sentence:

Finger at the ready, Hank prepared to shoot Big Jake as he came
out of the door, ... he liked taking suprise pictures.

Most people, seeing the words 'Hank', 'Big Jake' and 'Shoot' assume a context of a Western shoot-out. The possibility of processing the same information in terms of photography is not normally considered until the reinterpretation is suggested at the end of the sentence.

Reading is a good illustration of selective processing. Much evidence exists to show that little of the text may actually be processed (in the sense that the visual symbols are decoded). Consider this sentence where all the short words have been replaced:

When xxx sun xx shining, girls xxx boys come xxx xx play.

The fact that it is intelligible shows that the replaced words are probably little used in reading. Given the redundancy of the text, the strategy is one of taking as few key words or grammatical structures as possible to be in a position to determine the gist of the text. Hence, proofreading for spelling errors is difficult and tiresome: we are not accustomed to reading all the text. Similarly James Bond books are much easier to read than Tolstoy because much less of the text need be processed to understand what the author intends.

Selective processing serves to overcome a number of problems in processing by reducing the quantity of information processed. It does this by providing a framework, in the form of 'assumptions' within which incoming information is assessed and assimilated. A critical component in this process is the organization of information in human memory. In the example given, the association of words like 'Shoot', 'Hank' and 'Big Jake', lies in memories of countless westerns which provide a context or mental image in which terms the sentence was initially interpeted.

Selective processing is, therefore, exploiting the ability to recall from memory a vast amount of background information which is compatible with, or associated with, the incoming and already processed data. This capability, which is essentially a thesaurus of knowledge about the world, is often referred to as 'semantic memory'. Its availability makes selective processing a useful strategy because the context it creates around incoming information, based on prior experience, is usually a good predictor of what the information actually means.

Semantic memory provides a general context whereby information can be interpreted into familiar patterns. However, particularly with technological developments, human processing problems have become more critical, as for example in the control of a computer system. In these cases, the information is too detailed and specialized to be interpreted purely by reference to previous general experience; we need specialist knowledge of how the system works in order to carry out tasks with the system reliably.

Mental models

Such knowledge is said to be represented in human memory by a 'mental model'. But what do we mean by this, and how useful a concept is it? Consider the following quotations taken from recent literature on the subject:

> An operator's internal model is defined as the internal or conceptual representation of how the operator thinks the system should or will work. (Eberts, 1982.)

> The mental model can be conceived of as a higher level control mechanism that functions to determine the various ways in which an initial situation can be changed in order to reach a goal state. The internal representation is a plan for skilled performance. (Hale, 1982.)

> The user's conceptual model is the set of concepts a person gradually acquires to explain the behaviour of the system. It is a model developed in the mind of the user that enables that person to understand and interact with the system. (Smith *et al.*, 1982.)

These quotations are taken from recent papers concerned with user–system interaction. They are intended as definitions of the mental models (to settle on one term) of interactive systems which users have in their minds to guide their use of those systems. It seems a generally accepted axiom of cognitive psychology that such models exist in some form or another (see below), and for every definition supplied, tens of authors will emphasize the significance of 'models' or 'users' representations' without feeling the need to specify what they mean by these terms.

However, when we begin to look at models in detail, as in the above quotations, we begin to feel uneasy: Does Eberts really extend our understanding of 'internal models' by referring to them as 'internal or conceptual representations of how the operators think the systems will

work? What does Hale mean by a 'higher level control mechanism'? Like-
wise for Smith *et al.*'s definition to be helpful we would want to know more
about what the 'concepts a person gradually acquires' are.

In short, most definitions of mental models are vague and use imprecise
terms to describe other imprecise terms. There is some commonality of
meaning, but this is less in the area of what a mental model is, as to what it
achieves. As Young (1981) puts it:

> The notion of the user's conceptual model is a rather hazy one,
> but central to it is the assumption that the user will adopt some
> more or less definite representation or metaphor which guides his
> actions and helps him interpret the device's behaviour.

The desire to understand more about mental models is not simply of
academic interest. Cognitive psychology is also in the business of advising
on the design of interactive systems. Clearly the understanding of models
and how their effectiveness is modified by system design is central to this.
Definitions such as those given above are therefore simply inadequate. To
quote Allport (1980):

> To my understanding, the only way that psychologists' theories
> have succeeded in doing justice to the richness and complexity of
> human cognition is simply by being so general and underspecified,
> in process terms, as to mean almost anything.

The first objective to be identified in this area, therefore, is an attempt at
closing the gap that exists between the state-of-the-art psychological understanding of mental models and how this translates into guidelines for system
design.

Now consider the following quotations:

> To design the interface is to design the user's model. (Moran,
> 1980.)

> The first task for a system designer is to decide what model is
> preferable for users of the system. (Smith *et al.*, 1982.)

This implies a quite different understanding of the term 'model'. Vague
as they might be, the previous comments seemed to be in consensus that
the user's model was something to be learned and developed by the user to
help him or her use the system. On the other hand, these quotations imply
that the user's model is something to be imposed on the user by the system
designer as a *modus operandi* for using the system. The two views are not
mutually exclusive, of course, but they do place different emphases on the
role of the psychologist in the design process. The former view focuses upon

the learning process by which the active user learns how to use the system. The latter view presupposes the notion that designer-originated models can be generated and the 'best' selected as the user model. In this case the psychologist is less concerned with a developmental learning process than with the qualities of a model which make it a good one to provide.

Thus a second problem that quickly emerges in the discussion of 'mental models' is that the term is not merely underspecified, but is used to refer to different concepts; sometimes in the same articles (e.g. Smith *et al.*, 1982). The problem is exacerbated by the indiscriminate use of different terms such as 'user's model', 'mental model', 'internal representation', 'cognitive model' and so forth, where the way in which they are related or differ is rarely made clear.

Although other authors consider the options (e.g. Young, 1983), there are three meanings of the word 'model' with which this chapter is concerned. These are:

Conceptual representation model. This terms refers to the type of model in which the user is given, or explicitly generates, an analogy, metaphor or notional machine which acts as a descriptor for the system. Examples of such models would be to describe a word processor as 'like a typewriter', or to build systems around a 'desktop' metaphor such as in the Lisa Personal Computer or the Xerox Star workstation.

Implied dialogue model. This concept of 'model' refers to the fact that features of the user–system dialogue make implications to the user as to the function of the system. Thus, for example, to use the word 'archive' as a command term carries with it a number of implications about the function associated with it. The implied dialogue model may well suggest a conceptual representation model to the user in the absence of one being given explicitly, but the two are not synonymous: the conceptual representation (we may drop the 'model') provides an *overall* model of the system, whereas the implied dialogue model refers to the recruitment of past experience to make inferences perhaps only about small aspects of the system.

Cognitive model. The previous two uses of the term model are based around the idea that the user's prior knowledge can be recruited either to provide an overall system model, or to describe particular aspects of the system. The notion of a cognitive model is that the user learns with experience particular aspects of the system which cannot be described purely in terms of recruited information. This information may consist of a representation of how the system differs from a given model; perhaps may include a more abstract or unarticulated level of understanding: such a model would incorporate a capture of what the system *cannot* do as well as what it can. In short, the cognitive model is the (as yet) undefined essence of what makes an individual skilled as opposed to merely informed.

Together, these definitions of models provide a framework for the concept of a mental model. However, it should be emphasized that these categorizations are essentially for clarity of discussion rather than representing necessarily important theoretical assertions. For one thing, these concepts are in practice intimately interwoven. For example, the conceptual representation and the cognitive model can be distinguished in the sense that the representation is something given to the user whereas the cognitive model is acquired with experience; but this does not say how this learning process is affected by the model itself, or in what form, if any, the conceptual representation model is maintained in the cognitive model.

The distinctions nevertheless do point to different areas of interest. Both the conceptual representation and the dialogue model are concerned with the role of past experience of system use. Conceptual representations involve the provision of coherent 'stories' or descriptions provided for the user whilst, on the other hand, implied dialogue models are the upshot of the user's making his or her own inferences on the basis of what (s)he sees of the system and how it relates to past experience. In the former case the model is provided, and in the latter it is generated.

Cognitive models do not so much involve the use of prior knowledge as the modification of it and the acquisition of new knowledge. Here the interest is in how the user's understanding of systems in general and one in particular is modified by experience beyond the limitations of prior knowledge. Thus the cognitive model covers the representation of information which is necessarily not represented in the conceptual representation or the dialogue model, such as the way in which they need to be supplemented or modified to match themselves to the system function.

Mental models and USI design

The USI is not merely a communications channel between user and machine; it also plays a major role in developing and sustaining the user's expertise. If we accept this assumption, then the quotes of Moran and of Smith *et al.* earlier are put into perspective: the USI is a basic datum from which the user induces a cognitive model. Hence the structure of the USI and the dialogue in particular will strongly constrain the development of that model. The designer can therefore strongly influence the development of the user's expertise by the design of the USI.

As has already been seen, a common approach in USI design is for the designer to decide what model to impose upon the user. At present, however, no guidelines exist to indicate what model should be chosen, and the designer is obliged to intuit a model. This methodology is flawed in a number of ways. Firstly, if the designer chooses to implement what (s)he thinks is his or her own model of the interface, (s)he is unlikely to be able to articulate it comprehensively in a form which can be transcribed to the USI.

For one thing, we do not have intuitive access to all our own knowledge (e.g. Allport, 1980)—much information seems to be stored and processed in a preconscious form. Thus any specification the designer may come up with is unlikely to be a faithful capture of the true state of his or her knowledge. In particular, the designer has a problem in circumscribing his or her model as an entity, since it will relate to similar systems in the areas where there is functional correspondance. These relations may not exist in the user's mind, and what the designer thinks of as given or common knowledge is not. Thus commonly USI designers make unreasonable assumptions about the user's knowledge.

The alternative method of interface design has been simply to implement the USI piecemeal without an overall model. This situation is worse because the designer(s) (there may be more than one) will still use their own intuitions, but are not constrained in any way to relate these implementations to a common model. Given that, by definition, a designer is an expert, his or her knowledge is more developed than that of the prospective user, and may be based around concepts with which the user will not be familiar. It is these concepts that form the common model that the designer(s) are implicitly implementing, and thus the interface is incomprehensible to the new user.

If we accept this analysis, then clearly what is required is a set of guidelines which enable the designer to generate a USI which both elegantly supports the functions of the system, but also communicates well with the user. Such guidelines would complement more conventional guidelines of the sort presented in Chapter 10. The more conventional guidelines are useful as far as they go but they do not address the issue of mental models. They concentrate upon a number of rules of thumb which ameliorate a number of problems in dialogue design, but do not give guidelines as to the relationship of these rules to each other in the overall structure of the dialogue. This results in a number of outstanding problems that can arise in the use of these guidelines: some may be contradictory (e.g. Maguire, 1982); others may have more than one possible implementation. There is little guidance as to what should be considered a priority when compromises are required; and finally there is no guarantee that when followed there will not be residual difficulties of use. The contention here is that judicious use of models in the design process will considerably improve this position.

The remainder of this chapter is structured around the three meanings of 'model' described in the introduction above. First, we consider conceptual representations of systems. One approach is to look at the so-called 'machines within machines'—notional machines or surrogate models which can be used to describe system function. This includes a review of the differing approaches to the use of metaphor and analogy in conceptual representation.

We then consider issues of dialogue design and implied dialogue models. In particular we concentrate on the structure and use of command languages, which has been the focus of most of the research work.

Thirdly, we look at cognitive models from a more general point of view. This includes a short review of principles of information processing, and a discussion of some of the theoretical distinctions used to describe cognitive models.

Finally, we attempt to summarize the preceding sections and to highlight issues for further research.

CONCEPTUAL REPRESENTATIONS

Models of calculators

One of the common views of the user's model of an interactive system or device is that the user is provided with or conceptualizes a machine within the system: a notional or mechanistic representation which can be used to simulate or 'run' the workings of the system. Such a model could be used to predict the behaviour of the system, monitor its current state, and to reason about its workings.

Young (1981), referring to such models as surrogates, set out to study the utility of such models in different types of calculator: a Reverse Polish Notation calculator (RPN), a four-function calculator (FF), and an algorithmic calculator (ALG). Of these, the RPN calculator explicitly uses a stack register model for evaluation of an expression (see Figure 11.1).

The FF calculator can be described in terms of a derived register model (see Figure 11.2) while the ALG calculator has no comparable simple surrogate. While such a model is possible, it is not necessarily desirable. As Young (1983) points out,

> If we were to pursue that path, it would be appropriate after a while to remind ourselves that the purpose of the exercise was to develop a conceptual model for the person operating the calculator, not an implementation design for the engineer building it.

Young's exercise in looking at the properties of such surrogate models reveals a number of difficulties with them as cognitive models, i.e. descriptions of the user's expertise. Firstly, as noted above, it becomes clear that register models (models which attempt to capture the internal state of the system in terms of the contents of operations upon registers) do not necessarily represent conceptually simple devices. Indeed, in the case of the ALG calculator, a register model will be fantastically complex compared with Young's alternative model, which is as follows:

> You type in an expression. The machine examines it, analyses it, and calculates the answer according to the rules of arithmetic.

This represents, in behavioural terms, a far more useful representation of the ALG calculator than a surrogate register model, since it mobilizes in the user's mind the relevant information concerning the rules of arithmetic.

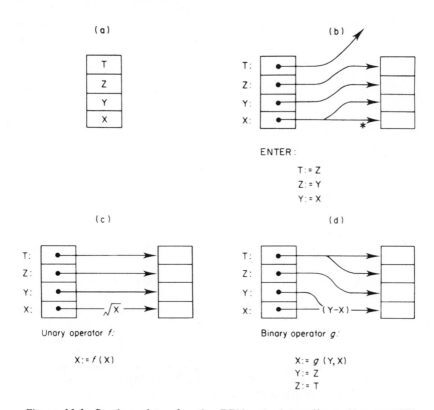

Figure 11.1: Stack register for the RPN calculator (from Young, 1981, p. 54). (With permission from International Journal of Man Machine Studies, 15, 51–85. Copyright by Academic Press Inc. (London) Ltd.)

Given that the ALG calculator works according to these self-same rules, nothing could be more appropriate. On the other hand, in the FF and RPN calculators, the rules of arithmetic are not applied in the same way, but are implemented in a modified or novel form. In this case the rules of arithmetic are inappropriate. It is therefore conceivable that surrogate register models which possibly embody the procedures and rules to be used in the calculations are more useful models in this case.

Further consideration of the FF and RPN models reveals that although they may appear to satisfy the criterion of simplicity and correspondence with good design principles, their utility either as an aid to the user or as a conceptualization of what is going on in the user's mind is not assured. For one thing, the models themselves do not indicate their ease of use. The FF calculator is, in advanced use, rather difficult to understand, but the preferred model can give no indication as to why this should be. Nor is the function of this model inherent in its structure: to 'run' such a model requires further background knowledge as to the interrelationships between the components than is evidenced by the surrogate alone.

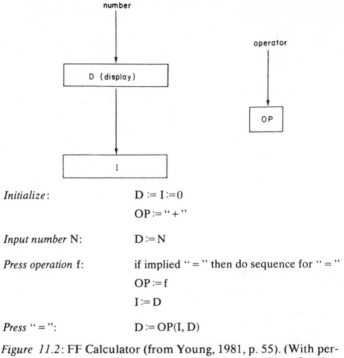

Initialize: $D := I := 0$

                         $OP := " + "$

Input number N: $D := N$

Press operation f: if implied " = " then do sequence for " = "

                         $OP := f$

                         $I := D$

Press " = ": $D := OP(I, D)$

Figure 11.2: FF Calculator (from Young, 1981, p. 55). (With permission from International Journal of Man Machine Studies, 15, 51–85. Copyright by Academic Press Inc. (London) Ltd.)

A conclusion of Young's analysis is therefore that surrogate models are unlikely to be *sufficient* models of the system; they may represent conceptualizations of features of the system or can be used as methods of communicating aspects of the system's functions, but do not in themselves represent cognitive structures which can be 'run' like simulations. To use these models requires additional information which negates their initial theoretical simplicity since it admits the activation of cognitive structures not present in the notional machine.

Young's analysis is entirely theoretical and considers surrogate models in terms of their ability to model all functions of the calculator. In fact, as Norman (1983a) points out, users of calculators do not generally exploit all the facilities that calculators offer, particularly those involving complex manipulation of partial results in registers or constant-value operations. Thus for many users the range of functions their mental models need to encompass is a subset of those Young was considering. The surrogate models give no indication of why this should be. Indeed, the question arises as to whether these users are in fact exploiting less than all the facilities of the device precisely because the models they are using can only support or 'run' a subset of them. This further lays doubt upon the utility of surrogates which may be logically sufficient but can offer no explanation for this.

Further, Norman's informal description of calculator usage reveals aspects of mental models which do not correspond with the device's function. An example of this is the overuse of ENTER or CLEAR keys which is superstitious in nature—they are overused apparently to ensure avoidance of the problems created when not used correctly. Also, users carry out clumsy or ancillary actions in calculations, such as storing intermediate results on notepads or carrying out more calculations than strictly necessary to avoid having to keep intermediate results. Such additional aspects of mental models seem to represent strategies designed to avoid or diminish problems previously encountered. Again, the implication seems to be that the user's mental models contain more information than is necessarily required by a logical machine-like representation of the calculator.

Work by Mayer and Bayman (1981) also relates to Young's work. They were concerned with modelling user's level of skill in using calculators in terms of their understanding of the internal workings of the device. Naive and expert users were asked to describe the internal state of the device at particular stages in a calculation. For example, the subjects might be asked to describe what state the calculator display would be in following the input: '2 + 3' or '2 + 3 ='. Different models predict different answers: users who believe that calculators carry out simple binary operations as soon as the second character is entered predict the answer '5' to the first example, whereas a model based on evaluations being carried out only when the '=' key is pressed predict '3'.

In summary, Mayer and Bayman's study shows that people differ widely in their assumptions as to how devices function, particularly among novices. In this study, these differences were described in terms of different sets of production rules, but as is pointed out, this probably underestimates the degree of divergence: different users may behave in accordance with the same production rules, but their conceptual representation of the system might be quite different. Such differences would only be apparent when the conceptual representations ceased to correspond when predicting the system's function.

The ability to extrapolate from calculators to more complex devices is a little uncertain, and is not purely a matter of scale. Complex interactive systems are not as passive as calculators; they can instigate and control transactions, they handle more types of data in a more complex way, and they tend to serve a wide number of applications. Nevertheless, the study of users' models of calculators has a number of things to say about mental models in general:

Surrogate models. Young's analysis of surrogate models indicates that surrogate models are limited in value. The value is greatest in those cases where the system has been built around a simple notional model such as the stack model (the RPN calculator). Similar models for other calculators may be difficult to use or even impossible to conceptualize. Thus surrogates may sometimes represent a method of modelling, but are not necessarily sufficient on their own to support a user's model.

Compatibility with prior experience. An important determinant of the usability of a device and the utility of a model is the relationship between the manner in which the device carries out its tasks and the way the user has previously carried them out. For the RPN calculator, a surrogate model is feasible and compensates (after a suitable learning period) for the lack of correspondance between the device and the traditional rules of arithmetic. In the ALG calculator, a surrogate model would seem to be an irrelevance, since the calculator's functional correspondance with the rules of arithmetic means that these rules provide an adequate model in their own right. Somewhere in between, the FF calculator fails to support all the rules of arithmetic and does not have a simple surrogate to represent the operational rules and those which are novel to the system. One way of interpreting the sub-optimal use of FF calculators may well be that users restrict their use of them to those rules of arithmetic which do apply. In this case, the user's mental model may be a modification of the model employed by users of the ALG calculator containing added information delimiting the range of arithmetic operations which apply. This speculation apart, it is clear that the utility of surrogate models is determined by a subtle interaction between device function, prior experience of the tasks involved, and the surrogate model.

Models are fragmentary and complex. The observations of Norman and of Mayer and Bayman, in conjunction with Young's analysis, bring home the point that mental models, particularly for novices, are representationally complex, and contain diverse, fragmentary or even incoherent knowledge. A similar conclusion is arrived at by a number of other authors (e.g. Hammond and Barnard, 1982).

Analogy and metaphor

Another popular view of how humans understand, reason, and learn about unfamiliar situations (such as a new computer system) is that they invoke existing knowledge structures (schemas or schemata) within which they interpret the new information. Metaphor and analogy are seen as principal mechanisms for this. Learning is a process of assimilation of new information into existing schemas followed by some process of restructuring or accommodation in which new schemas are created by modification of the old in the light of newly acquired information (e.g. Rumelhart and Norman, 1981).

Such views are based on a number of converging lines of evidence. One approach arises from experiments such as those carried out by Bransford and Johnson (1972) in which subjects were given the following passage:

The procedure is actually quite simple. First you arrange items into different groups. Of course one pile may be sufficient depending on how much there is to do. If you have to go somewhere else due to lack of facilities that is the next step; otherwise you are pretty well set. It is important not to overdo things. This is, it is better to do too few things at once than too many. In the short run this may not seem important but complications can arise. A mistake can be expensive as well. At first, the whole procedure will seem complicated. Soon, however, it will become just another facet of life. It is difficult to forsee any end to the necessity for this task in the immediate future, but then, one can never tell. After the procedure is completed one arranges the materials into different groups again. Then they can be put into their appropriate places. Eventually they will be used once more and the whole cycle will have to be repeated. However, this is part of life.

Subjects were divided into three groups. The first (No Topic) were given no added information as to what the passage was about. The second (Topic After) was given the additional information that it was about 'washing clothes' after reading the passage; and the third (Topic Before) was given this topic before reading the passage. The results showed clearly that subjects who read the passage without first knowing what it was about rated it low in comprehensibility and remembered little compared with the Topic Before group. This illustrates how the learning process can be facilitated by a context and prior experience activated at the time of receiving the information. In schema theory, the appropriate 'washing clothes' schema can facilitate assimilation of the information in the passage.

Another source of evidence implying a significant role of prior knowledge in thought is to be found in problem solving, in which a consensus of evidence has accumulated to suggest that abstract problems are more difficult to solve than identical problems couched in a concrete context. An example is performance on the Wason Selection Task. In an experiment by D'Andrade (cited in Rumelhart and Norman, 1981), subjects were shown four cards. In the abstract case, subjects were essentially asked to turn over the minimum number of cards to verify the statement: 'All cards with a vowel on one side have an odd number on the back.' In the 'realistic' case, subjects were asked to look at four receipts and verify that receipts over $30 had been signed on the back. Despite the fact that the two problems are formally identical, only 13 per cent of the subjects in the abstract case did the task correctly whereas 70 per cent did so in the realistic case. The imposition of a realistic framework, although strictly speaking irrelevant to the logic of the reasoning, was somehow facilitatory.

A third area which strongly links the use of prior information to learning and reasoning is the observation of consistent use of analogy and metaphor.

This is seen as ubiquitous in language, is noted with respect to scientific reasoning (Roediger, 1980; Gentner and Grudin, 1983), and is seen as common in reasoning about computer systems (Halasz and Moran, 1982).

This type of evidence leads to the generalization that an important feature of human thought is that it be contextualized; that is, thinking is not carried out in the abstract but is organized in terms of pre-existing concepts. The evidence is such that the nature of the context applied can materially affect the course of reasoning. Rumelhart and Norman, for example, comment that in the learning of the mathematical concept of fractions, use of the pie analogy provides an excellent model for the addition and subtraction of fractions, but obscures the concepts of multiplication and division. More recently, Gentner and Gentner (1983) experimentally compared reasoning about electrical systems in the context of an hydraulic analogy compared with that of a 'milling-crowd' analogy. The results indicated that whereas the hydraulic analogy was superior in predicting the behaviour of batteries in different configurations, the 'milling-crowd' model was better for reasoning about resistors. In other words, in the absence of deeper understanding, the choice of analogy constrains the way in which one can reason.

This chapter does not consider exactly how the use of analogy or metaphor works in terms of psychological mechanisms. The interest here is more upon what the effect of analogy or metaphor is on learning and whether they can be exploited in any way in interface design. Before discussing in detail the approaches taken to metaphor and analogy, two general points are worth emphasizing, as follows:

Analogy and metaphor are powerful. We would not be discussing the role of analogy and metaphor if it were not for the fact that they are extremely powerful ways of communicating the functionality of systems. Whatever their disadvantages (discussed below), they can transform complete bewilderment to a reasonable level of competance and confidence. The problem is, as with strong medicines, that they can be used detrimentally as well as beneficially.

Analogy and metaphor are important for novices. By definition, experts understand how systems work. On the assumption that analogies or metaphors are open-ended and may predict certain system aspects incorrectly or fail to describe others, experts have therefore developed mental models which have moved beyond the level of the basic analogy or metaphor. There is little that the basic metaphor or analogy can do for experts beyond helping them communicate their understanding. The interest here therefore is mainly in the use of analogy and metaphor by non-experts in learning about a system and helping them to reason about it. This is not to say that experts do not use analogy or metaphor, but simply that this has less effect on their use of systems than is true for naive users, who are still learning how to use the system.

A number of different approaches can be taken to the use of analogy and metaphor at the USI, as follows.

Advance organizers. Advance organizers are short expositions about systems which lay out the general ideas and concepts before the user begins to use them. As such, they have their theoretical foundation in the principle of 'contextualization'. Recently Mayer (1981) has used concrete models of computer systems as advance organizers for subjects learning programming skills. Results typically show that for reasoning problems involving some creative reasoning from information learned (as opposed to solving problems identical to those encountered) performance can be improved considerably by the use of such models.

Mayer's research shows the advantages of advance organizers, but gives no theoretical basis for them. Consequently the guidelines he provides (Mayer, 1979) are descriptive rather than prescriptive:

— The organizer should allow the reader to generalize all or some of the logical relations in the text.
— The organizer should provide a means of relating the information in the text to existing knowledge.
— The organizer should be familiar to the reader.
— The organizer should encourage the learner to use prerequisite knowledge that the learner would not normally have used.

Under the burden of trying to comply with such unhelpful guidelines, it is not suprising that advance organizers should sometimes fall flat and produce insignificant effects on learning.

Complex metaphors. One of the common complaints about the use of metaphor, which applies also to advance organizers, is that the model provided cannot explain all aspects of the system. As the user's experience develops (s)he therefore comes across the problem of accommodating those features which do not have corresponding representations in the existing metaphor or analogy. One approach to this problem is to elaborate an existing metaphor or analogy with further metaphors or analogy.

This is the concept of 'complex metaphor'. Halasz and Moran (1982) strongly argue against this notion, using filing as an example. They point out that as the user's understanding increases, a concrete model of filing (such as a filing cabinet analogy) needs 'fixing-up' with added means of explaining concepts such as passwords, directories and multiple copies. Ultimately, the metaphor becomes more complex than the system it was designed to describe and may actually hinder, rather than aid, learning.

Halasz and Moran's view is a common but not universal opinion. Carroll and Thomas (1980), for example, in considering the problem of leading users through a sequence of metaphors towards a more detailed understanding of a system, argue for an elaborative method within a single domain:

> ... if one began by comparing a computer storage handling system to a warehouse, one would be better off introducing new jobs, objects, and places in the warehouse to describe additional aspects of the storage handling routines than in changing the metaphor to how the body deals with its energy demands.

The problem is how one chooses the right basic metaphor in the first place, an area where much further research is needed.

Literary metaphors. An alternative use of metaphor proposed by Halasz and Moran and also by Rumelhart and Norman is the so-called 'literary metaphor'. This refers to the traditional view of metaphor in which the metaphor does not embrace whole systems or functions, but is designed to highlight a specific feature. To say of someone that he was a vulture is not to imply that he has wings, talons and a sharp beak, but to emphasize aspects of his character. In describing features of a text editor, Rumelhart and Norman give three such metaphors: a 'tape recorder' model to emphasize how once in 'append' mode the editor records all input slavishly; a 'secretary' model which describes how text input is related to command input by illustrating the difference between requests for dictation, etc, and is dictated; and a 'card index' model to provide a description of how lines of text are stored in the editor (see also Norman, 1983b).

In Halasz and Moran's analysis, the problem with analogy is that ultimately it can misinform the user about the system in those cases where the analogy ascribes functions non-existent in the actual system. In their view, literary metaphor avoids this by using metaphors whose scope is evidently limited in applicability. They therefore see the use of them as being to convey single concepts in an overall abstract model, hoping in this way to avoid the abduction of misinformation while at the same time enjoying the benefit to learning.

The common theme of the above positions is that analogy or metaphor imparts knowledge structures or schemas which allow the user to reason about and use the system. In this view, the user is seen as a recipient of provided information. This is equally true if the metaphor or analogy is provided by the system or by the user, in that one is assuming in both cases that the choice of metaphor causes a schema to be unpacked from memory to be applied directly to reasoning about the system.

There is little doubt, however, that users are not passive recipients of information; faced with a new computer, all but the most careful users will attempt to log-on and use it before consulting manuals. All the indications are that, in learning, humans interpret, process and restructure information as it is received, and that it is the effort involved in this process which enhances learning. This is the 'levels-of-processing' view (Craik and Lockhart, 1972).

This notion leads Carroll and Mack (1982) to consider a possible active role that metaphor and analogy might play in learning. The suggestion is that the utility of the devices may lie not in the information they automatically make available, but the active exploration process that must take place to map the chosen metaphor or analogy onto the target domain. In Carroll and Mack's terms:

> Metaphors are not 'right' or 'wrong' descriptions , as models are: rather they are 'stimulating' (or unstimulating) invitations to see a target domain in a new light.

Thus to say 'an expert system is like the legal system' is inviting the reader to explore the numerous ways in which the legal system works, some of which will be appropriate and some of which will not. In this view, a sharp distinction is drawn between metaphors and models. Models provide a complete description and present a passive learning situation. Metaphors are on the other hand active, invite a deeper level of understanding at the cost of some miscomprehensions on the way.

The principle is a very inviting way in which to see a number of the approaches to metaphor and analogy described above. However, a lack of theoretical understanding still persists. If one were to create guidelines that suggest that open ended metaphors be used, which ones should we choose? Are we sure that active learning will in the long run be beneficial? Is the pay-off of sacrificing commonality and precision of models for deeper learning worth it?

IMPLIED DIALOGUE MODELS—COGNITIVE FACTORS IN COMMAND LANGUAGES

Command names

The major focus of attention in research on command languages has been on the choice of vocabulary. This follows the obvious criticism that command languages often incorporate jargonistic or incomprehensible names (Long *et al.*, 1982; Norman, 1981). In this research the theme has been to attempt to identify different parameters of command words which improve their learnability and memorability. The identification of such parameters would provide a set of prescriptive guidelines which would help system designers make an optimal choice of command words.

The first problem in this is that preferences in the use of words seem to vary widely both between individuals and between user types. Agreement between experts is generally higher (e.g. Rosenberg, 1982), but never unanimous. Landauer (1982) also found this trend, but more significantly, found a low level of agreement between designers and users, which he put at 15% to 20%. These results therefore highlight the danger of a lack of compatibility between designers' choices of words and what the user would prefer. They also show that for any single function, no particular word is likely to be the 'best' word for more than a low proportion of users.

One response to the lack of compatibility between users and designers has been to attempt to use command names generated by the users themselves on the grounds that these might in some way be more acceptable. Given that users do not agree on command names, it is not surprising that this does not produce a tangible improvement. Black and Moran (1982), for example, found no differences in the memorability of command sets generated either by designers or by users. In this, however, they did identify systematic trends in the types of words these different groups generated: designers tended to use highly specific, relatively uncommon words (e.g. INSERT, DELETE, REPLACE and MOVE) whereas users tended to use less specific, more common names (e.g. ADD, OMIT, CHANGE and PUT). The relevant factors may not necessarily be who chooses the names, therefore, as much as the semantic properties of the words chosen.

Subsequent experiments have shown that there are, indeed, significant differences to be found in the learnability and memorability of low frequency, specific words (the reader should note that these attributes are intimately linked) as compared with high frequency, unspecific words (Black and Moran, 1982). The differences, however, are not generally great and manifest themselves in subtle ways. Barnard et al. (1982) found virtually no differences in the memorability or speed of learning of the two types of command sets in an editing task. However, in terms of the subjects resorting to a HELP facility, they found that the general command sets caused users to access HELP more frequently and more quickly than with the specific command sets. When users of the specific command set did access HELP, it tended to be only after a longer period of deciding to do so.

Model of learning command names

These results are fairly easily interpreted in terms of a simple model. Any command label, be it a random string of letters or an English word, will have a number of meaning associations. For words such as *table* this will be fairly high, for nonsense strings such as *dsw* it is low. For specific words such as *invert*, the domain of related association will be relatively narrow, whereas for more abstract words such as *change* it will be wide.

As the user understands a system function, the command name is needed to become a mnemonic handle for it. The more the meaning(s) of that word overlap with its functions, the more easily it will be remembered. Thus RUBOUT is a more potent recall aid to the workings of a deletion function than OURTUB. Hence English words related to the function are always going to be the most potent prompts for recall of what the function does.

However, a second process in the learning of a command name is to learn in detail what the function does. This requires isolating those features of the meaning of the word exactly apppropriate to the function and excluding those that are not. Further, it requires adding to the meaning those aspects of the function not represented in the word's original meaning. Thus the learning of a command name, assuming it to be a common word, involves three elements:

— excluding the meanings not appropriate to the function
— identifying those meanings appropriate to the function
— incorporating new meanings to the label needed to represent the function.

These all involve a psychological cost. The incorporation of new information takes time and effort and is subject to forgetting. The excluding of information (technically known as unlearning) is also a time-consuming process and also prone to forgetting. In this case the effect of forgetting is the re-inclusion of information that has been excluded. That is, the command word begins to revert from its purely technical meaning to its natural, fuzzy, meaning.

From this model we can attempt to make a number of generalizations, as follows.

Non-words. Non-words as command labels are theoretically blank slates to which the user must make the association of the function. They have the advantage, therefore, that information does not have to be excluded in the learning process. On the other hand, more information has to be learned and they are intrinsically poor cues for recall of function. If, however, a function were highly abstract and novel, then few words would exist that had much overlap in meaning with it. In this case the added learning required might be outweighed by the amount of unlearning saved. The prediction is therefore that non-words are most suitable (although how useful cannot be quantified) as command names when the functions to be labelled are most unusual and ill-suited to description by single English words.

Specific words. Specific words appear to be advantageous because their discrepancy with the required meaning is not great. Thus the amount of learning effort, (i.e. gaining and shedding of meanings) is reduced. This, of course, assumes that the user understands the specifics of the term in use.

General words. General words are easy to learn but have the disadvantage that a great deal of meaning has to be shed to specify their exact meaning in a command label. In behavioural terms, this means that non-expert users have only a vague understanding of what the command means and are likely to overgeneralize its functions. General words may nevertheless be more useful when a general meeting is required, as in, for example, a set of menu options describing generic ranges of commands available.

If this model is useful in terms of explanation, it is also useful in that it highlights the deficiencies of our understanding in this area. If we wished to provide powerful guidelines as to the choice of command names, it must involve some quantifiable comparison of different words to indicate which is 'best'. This in turn would require a much more detailed specification of the meaning of the attributes of words such that the amount of learning required to associate a label to a function may be more precisely defined. We are nowhere near such precision in our understanding of the psychology of words and naming behaviour. (See Rosenberg, 1982, for an example of this kind of work.) At the moment, on the use of command words, we can only deal in the realm of generalities.

Command syntax

A second aspect of command languages is their syntax and structure. In general, most systems are syntactically complex because this is the simplest and most consistent way of representing a wide range of functions and representing the relationship between them. This is best illustrated by an example: suppose in a general suite of editing routines one program modifies only the source file whereas another leaves the source file intact and copies the modifications to a temporary file. There are three possible ways in which this could be implemented in a command language:

  (i) Separate command names: One solution is to use two separate command names such as EDIT and MODIFY. This keeps the syntax of the commands simple but has a number of disadvantages. First, the relationship between the functions is not truly represented by the names. (MODIFY is a more general word than EDIT, implying a hierarchy of meaning.) Second, the increased number of command names places a greater pressure on the user's memory. Third, the need for more command names puts more pressure on the designer to find appropriate names.

  (ii) Compound labels: Another possible method of labelling is to use complex names which are hybrids of two or more words, such as SEDIT and MEDIT. These words preserve the relationship between the functions, but also have disadvantages. They are not necessarily easy to learn and remember, and can create ambiguities or pseudo-relationships between unrelated command names (e.g. Carroll, 1982).

(iii) Syntactical expressions: The most common solution to the problem of complex command languages presently in use is to use syntactically complex commands such as EDIT COPY FILE or EDIT SOURCE FILE, which have the syntactic structures such as |COMMAND^ |QUALIFIER^ |FILENAME^. Theoretically, the advantage of this method is twofold. First, if the user learns the syntactic structure of the language, then for a relatively small command set (s)he can generate a large number of different commands. Secondly, such a method implicitly makes for a direct representation of the relationship between similar commands such as those above. On the debit side, the user is required to learn the syntactic structure of the language, which may be complex, or even inconsistent. In the latter case the system effectively regresses in its usefulness to (i) or (ii) above, since the user has to learn particular constructions as well as exceptions. Work by Carroll (1982) has shown that users designing their own command languages spontaneously introduce such syntactic methods.

The effect of consistency in command structures has been studied in detail by Barnard *et al.* (1981, 1982). In the first of these experiments, the efficiency of different methods of ordering arguments was studied. The commands to be tested were all of the form: COMMAND [arg 1] [arg 2], in which one of the arguments was recurrent and the other not. One method of structuring is simply to apply a consistent position to the recurrent argument (first or second). An alternative is to consider the relationship of the string to a natural language representation. For example, given the command:

DELETE [message Id] [string]

This could be 'naturalized' by the sentence 'Delete from file [message Id] the string [string]'. The reverse argument ordering gives 'Delete [string] from file [message Id]'. The latter, which gives the direct object first, is a more natural and encodeable sequence in natural language terms.

In general, linguistic compatibility is achieved by the ordering of [COMMAND] [DIR OBJ] [IND OBJ]. The cost of this, however, is that the consistency of the location of the recurrent argument cannot be maintained. This can be seen in the two commands:

DELETE [string] [message id]

STORE [message id] [file location]

Thus the notion of natural language ordering and consistency can lead to contradictory guidelines in argument ordering. Barnard's experiment indicates that consistency can be a vastly more potent aid to learnability than natural language ordering, although to a large degree this difference was invested in the advantages of having the recurrent argument consistently first (but see below for a discussion of this). Notably, in a comparison of command argument sequences in which the direct object was placed first (compatible with natural langue) with the reverse (incompatible with natural language), both were equally poor. This would imply that natural language orderings carry little significance in the context of command sequences. Certainly, any effects that might exist are dwarfed by the user's sensitivity to consistency in command languages.

This does not mean, however, that consistency is the only relevant attribute in command syntax and structure. Research by Carroll (1982), for example, has isolated a property of command vocabularies he calls 'congruence'. The notion here is that many functionally related commands should have this relationship apparent in the structure of the command language.

Thus for example, the command 'Up' to move up lines in an editing task is most suitably complemented by the command 'Down' to move in the opposite direction. 'Lower' or 'Descend' or 'On' are not congruent to 'Up' because they differ in particular ways: they may be different parts of speech; they may imply a subtle change in meaning (e.g. Down vs Descend); they may actually refer to metaphysically different functions (e.g. 'On' is more congruent with a forward–backward model than an up–down model).

Carroll's research (1982) has shown this property to be an extremely potent variable. Congruent command paradigms seem easier to learn and remember, and create fewer errors. In a complicated language in a simulated control situation (controlling the movements of a robot), the effectiveness of congruence in the language was tested by independently varying congruence and hierarchy in the syntax to produce four command paradigms: Congruent, hierarchical syntax (e.g. Move Robot Forward); Congruent, non-hierarchical (e.g. Advance); Non-congruent, hierarchical (e.g. Change Robot Backward); Non-congruent, non-hierarchical (e.g. Back).

Subsequent analysis showed that the congruent paradigms performed better in all tests, whereas in a simple sense hierarchicalness had no effect. Interestingly, however, there was a significant interaction between non-congruence and non-hierarchicalness, which seemed to be particularly difficult.

Methodological warnings

Thus far, much of this work has been presented uncritically. In fact the area of linguistic factors in command languages is a researcher's minefield, and all results should be treated with some circumspection. There are a number of complicating aspects in this area which make it such a difficult area to study:

Frame of reference. In other sciences, experimental tests of alternative methods are set in a theoretical frame of reference which provides an overall structure within which comparisons can be made. For example, in a chemical engineering process there are rigid disciplines—chemical theory, economic practice and engineering technology—which effectively provide guidance as to what the relevant variables in any particular situation are. From this one can experiment in a systematic way by varying the value of these and observing the results. In psychology such theoretical backing is much more flimsy. There is no simple way, for example, of defining congruence or degrees of congruence in a command language independently of the

designer's intuition. Defining congruence by an empirical assay of usability—
i.e. the more useable, the more congruent—is circular. Secondly, when one
identifies features of languages or words such as specificity/generality, frequency of use, hierarchy or congruence, there is no guarantee that these
words capture the essence of the potent variable, or that they are actually
overlapping concepts with unseen but more significant variables. As a result,
any experiment which purports to compare the utility of different properties of command sets (e.g. specific vs general) has no metric with which to
ascertain the degree to which those properties have been achieved. Nor can
similar experiments be compared in anything but a general sense since there
is no way of specifying exactly how alike they are.

Empirical assessment of usability. As has already been argued, the value of a
command paradigm vocabulary cannot be assessed simply: it depends upon
a complex range of criteria. Adopting only one or some of these, such as
memorability or preference, cannot give the whole picture. It is not even
clear that it is known what range of measures would define the whole picture.
(See Chapter 9 for a discussion of this point.)

Generality of results. Most of the research in command languages has been
carried out as controlled experiments with users who were aware of their
role as experimental subjects. Controlled experiments (as opposed to simulation) basically run into the problems: a) that the task induces task-specific
strategies which affect the data in systematic but unpredictable ways, and b)
that subjects are sensitive to task demands and will adjust their behaviours
in response to minor changes in task. Given that many experiments tend to
be artificial situations, there exists the very real danger that the experiments
do not provide representative results.
    A very good example of this is to be found in Barnard and Hammond
(1982), in which they examine the curious effect that consistency in command argument order is really only of benefit when the recurrent argument comes first (Barnard *et al.*, 1981). Subsequent experiments show that
this effect disappears when the experiment changes: when the subjects have
to select the command names, as opposed to having them presented as in
Barnard *et al.* (1981), which is a far less common event in command use,
the advantage of placing recurrent arguments first is lost. In other words,
the original effect seems to be specific to the original experiment and does
not generalize to more common situations. This certainly casts into doubt
the wisdom of translating experimental results literally into guidelines. For
example, Black and Sebrechts (1981) offer:

> The best argument order for commands is a consistent one that
> places first the arguments that are present in the greatest number
> of commands (and hence likely to be the given rather than the
> new information of the command).

This probably succeeds only in overloading design engineers with spuriously over-specific guidelines.

Barnard and Hammond (1982) see the significance of this danger to be that experiments on command languages should incorporate a deeper analysis of behaviour:

> In sum, a major implication of this research is that our analyses of human–computer dialogue should not solely focus on the formal properties of dialogues. Rather it should take the form of an integrated analysis of the cognitive skills underlying user performance and the different sources of information and knowledge which these skills draw upon. Guidelines or recommendations which avoid these issues are unlikely to prove fruitful.

One can have sympathy with the sense of what is being said. It is, after all, the object of this chapter. However, 'the integrated analysis of the cognitive skills underlying user performance' is potentially a bottomless pit into which the analyst is drawn, endlessly chasing his or her own tail as new experiments reveal new and sophisticated subject strategies. There is a case for such research, but only when the experimental stimuli and design allow for a close specification of users' strategies. It is not clear that this is possible in this case. Alternatively, therefore, one might take Barnard and Hammond's demonstration of the lack of generality of results to raise a serious question-mark over the use of the experimental method in this area.

COGNITIVE MODELS

An early distinction was made between cognitive models (the user's internal knowledge about a system) and conceptual representations (knowledge structures provided explicitly or implied by the user–system interface). In truth, the distinction is a matter of convenience rather than theoretical necessity: the conceptual representation is a contributory element in the development of the cognitive model. It may provide a surrogate model in early system use (as in analogy); and may provide chronic obstacles to usability (as in poorly designed command languages), but ultimately expert users develop their cognitive models beyond conceptual representation through learning by experience.

Although we can specify the relationship between cognitive models and conceptual relationships in this vague way, little is known or understood theoretically about the nature of this interaction. We are as yet unable to predict the usability of an interface from its formal specification. Attempts at specifying formally the psychological properties of interfaces have begun (e.g. Moran, 1981; Reisner, 1982), but it is not yet clear how useful these approaches will become.

Because of this gap in our knowledge, empirical research has fallen roughly into two types. The most common form is the operational or objective research. Examples of this are the utility of different ways of presenting analogies, or the usability of different command language paradigms. This method is characterized by the empirical comparison of different conditions which are created by the experimenter on part-theoretical and part-intuitive grounds. The other method can be described as structural or theoretical research. Examples of this are Gentner and Gentner's approach to analogy, Rumelhart and Norman's views on learning by analogy, and Bransford's work on contextualization in thought. This research is often characterized by less empirical rigour and a stronger desire to understand the psychological mechanics of the processes of interest. These two methods correspond roughly to the objective study of conceptual representations and the investigation of cognitive models respectively, and therefore these labels can be said to denote areas of methodology as much as anything else.

Categorizing the empirical research in this way may be a useful device in the absence of any other organizing factor, but at the stage of summarizing this research and attempting to extract the important points it is less useful. Young's (1981) work on surrogate models of calculators began with the aim of studying conceptual representations, such as the RPN stack model. However in the discussion of models for the 4-function and Algorithmic calculators, and other research on how people actually use calculators (e.g. Mayer and Bayman, 1981), the issues became more centred around users' cognitive models. Equally, the research on command language paradigms is ostensibly a matter of conceptual representation, but when one considers aspects such as Carroll's principle of congruence, cognitive factors come into play. It is clear from this that definitions of mental models which concentrate exclusively on the cognitive aspects or upon conceptual representation are misleadingly simplistic.

Another simplistic notion which can therefore be dispelled easily is the idea that a cognitive model is a mental entity, distinct from other forms of mental activity, whose specific properties can be investigated and potentially exploited. Much of what user's know about systems seems fragmentary, apocryphal or even superstitious (e.g. Hammond and Barnard, 1982; Norman, 1983; Mayer and Bayman, 1981), and can hardly be said to reflect a coherent mental structure. It would not be sensible to reject outright the idea that users can use analogies or surrogate models to assist with aspects of system use. However, a reasonable working assumption would seem to be that cognitive models of interactive systems are no different from the cognitive mechanisms we use to interact with the world on a day-to-day basis, other than that they are based upon systematic entities which allow for systematic representation. In the study of cognitive models therefore, there is no tractable sub-issue which can be investigated while the problems of mainstream cognitive psychology look after themselves.

This discussion of mental models has therefore turned full circle. Starting from a discussion of the general properties of human information processes, we find that detailed examination of what we know about 'mental models' leads us to recognize that we know little more about them just because we consider them in a particular context.

Cognitive economy

A common thread underlying much of the work reviewed here which is useful immediately and points to a failure in research, is the concept of 'cognitive economy': one interpretation which can be placed on many aspects of system use is that users operate according to a principle of cognitive economy.

By this it is meant that users in learning and using a system do so in such a way that certain types of mental effort are minimized. By this it is not necessarily meant that the efficiency of users is maximized. A trivial but illustrative example lies in the use of manuals. Users perceive manuals as a tiresome and difficult way of learning how to use a system. Commonly they will begin to use the system by learning as they go along, making mistakes and attempting to correct them. In the long run this probably means that the users take longer to reach (or never achieve) necessary levels of expertise.

Cognitive economy in system use is manifest in a number of converging lines of evidence:

The need for consistency. The role of consistency in interface design is paramount. As yet, no other principle of design has shown such potent effects on usability and no benefits seem to be gained in creating inconsistency in order to satisfy some other principle. In terms of information processing, the more consistent an interface is, the more it can be described in terms of a small number of rules which capture and create the consistency. In short, consistency is directly related to the effort involved in understanding the interface. This learning of consistency is not a passive process. The indications are that users go to some effort to determine the nature of these rules by actively applying 'working hypotheses'. Hence, many errors of system use can be traced to the inappropriate commissioning of responses generated by false hypotheses or over-generalizations. A component of cognitive models is therefore a representation of the redundancies in the system.

Pay-offs between complexity and economy of functions. Studies of the use of calculators show an almost universal trend of users using inefficient or superstitious methods to ensure correct system functions. Two examples of this are the excessive use of CLEAR or ENTER keys and the reluctance to use complicated functions, preferring instead sub-optimal methods, including the manual noting of intermediate results (Norman, 1983). There may well be a number of factors here. The users may simply not bother to learn more complicated methods, perceiving the cost of learning to outweigh the benefits of the more effective methods. Alternatively, even though the users understand these methods, they may well be seen as undesirable because of the greater mental load placed on the users (such as remembering which intermediate results are stored in which locations), or the greater perceived likelihood of error.

Economy of effort in decision making and problem solving. In research on decision making a consensus of opinion has arrived at the conclusion that strategies of problem solving and decision making are evolved to reduce 'cognitive effort' (Rouse, 1982; Keen, 1979.) The well-known example of this is Simon's (1982a) concept of 'satisficing': that a decision maker will seek a solution which is adequate for his or her needs, but will not necessarily seek the optimal solution (See Chapter 8).

Information management. A number of recent studies of behaviour in real office situations (e.g. Malone, 1983; Cole, 1982) have illustrated sub-optimal performance due to economy of effort. In his analysis of the problems of storage and retrieval, for example, Malone points out that office workers are faced with a difficult decision at the point of storage: where to file and under what descriptions. Rather than actually make this decision, people defer by placing items in piles. This short term expedient becomes inefficient when the piles become large and unmanageable.

Numerous examples exist also in the literature of theoretical psychology whereby subject's reduce their cognitive load by means of an expedient strategy. A good example of this is an experiment by Bower (1961). This essentially looked at memory for paired-associates in which the response was the digit 1 or 2 and the stimulus was a consonant pair, e.g.: RT-1; SW-2; KP-2; etc. It was assumed in this experiment that subjects would learn the stimulus items and then associate them with the correct response. In fact, many subjects were cleverer than that. They adopted the strategy of learning only those stimulus items associated with one or other of the responses. Thus at recall (the test being the presentation of the stimulus item and requiring the suitable response) the subjects could respond with the chosen response if they recognized the stimulus and the other if they did not. In this way their learning load was at least halved. The frequency with which this strategy was used stands as a testimony to the ingenuity of the subjects and their sensitivity to methods of reducing effort.

What is being economized in such behaviour? Eason (1981) returns to the idea that information processing relies preferentially on using existing strategies and schemas rather than indulging in productive or creative thought. In other words, in thought, we are creatures of habit rather than analysis. A similar principle underlies Rouse's (1982) view of decision-making. In this view, the decision-maker where possible attempts to interpret or mould decisions in terms of familiar patterns.

Clearly at this level of analysis we are some way from fully understanding mental models or explaining behaviour in terms of them. However we are beginning to focus upon issues that will ultimately yield useful research in this area.

CONCLUSIONS

Mental models of the USI provide a framework within which the more specific kinds of dialogue design guidelines considered in the previous chapter can be applied. Many different terms relating to mental models are often used rather vaguely in the literature. We have attempted to provide some degree of clarification in this chapter by distinguishing between three major types of USI mental model, and considering some of their key properties. The three types are: conceptual representation models; implied dialogue models; and cognitive models. Ultimately, it is the user's cognitive model that is most important as it is this that can be considered to guide the user's behaviour—the other two are really means to this end. A key factor in helping the user to learn an optimal cognitive model is consistency in the way the USI is designed.

This chapter concludes our consideration of designing electronic systems to be usable. In the next two chapters we consider what is involved in introducing new systems into organizations.

Part Four
Introducing Systems into Organizations

Chapter 12

Introducing Office Systems: Guidelines and Impacts

BRUCE CHRISTIE

INTRODUCTION

The focus of this book has been on the design of the hardware and software that the equipment manufacturers make available in the market place. It is up to user organizations to choose whether and how to use these systems, and how to introduce them into their organizations. This chapter presents some broad guidelines for the introduction of office systems and considers briefly their possible organizational and broader societal impacts. Within this general context, the next chapter then considers in more detail the need for a new approach to identifying what sorts of systems an organization really needs.

GUIDELINES

The following main steps are suggested for the introduction of office systems and will be described briefly. Further discussion can be found in Christie (1981), Birchall and Hammond (1981), Kling (1983), and Buchanan and Boddy (1983).

1. Be sure there is a need
2. Ensure commitment at the highest level in the organization at the very start
3. Ensure adequate financial and administrative support
4. Involve all key parties at the earliest possible stage
5. Establish goals

273

6. Maximize awareness
7. Analyse the nature of the problem
8. Anticipate the spread of effect
9. Identify the effects on job security, payment systems and career opportunities
10. Assess the time scale of the changes realistically
11. Take account of laws and regulations and consider technology agreements
12. Plan for the changes
13. Prepare for the changes
14. Select technically appropriate equipment
15. Attend to 'global' USI factors
16. Attend to specific USI features
17. Attend to the needs of different user groups
18. Implement the changes
19. Monitor and evaluate the changes
20. Modify the programme for introducing the systems as necessary.

This is not the only possible list of steps that one could devise. Its purpose is largely to structure discussion rather than to suggest a rigid, sequential scheme. Many of the elements identified in the list will occur in parallel or at several stages during the introduction of new systems. The results from some stages will feed 'back' to modify 'earlier' elements and feed forward to provide a framework for later stages.

1. Be sure there is a need

There are many different reasons why an organization might want to consider introducing new systems, e.g.:

— to maintain competitiveness
— to improve efficiency
— to improve working conditions
— to provide for diversification into new areas
— to reduce costs
— to promote a 'leading edge' image
— to try out new technology in preparation for the future.

It is vital to be clear from the outset what the reasons are because they are not necessarily always compatible. Trying out new technology, for example, or promoting a 'leading edge' image may require a significant financial investment and may not always be compatible with saving money, at least in the short term.

Technology driven change is not always necessarily bad but some products do appear to be solutions looking for problems. It is important to relate proposed systems to potential real applications, either current or potential.

Systems are rarely if ever installed once and for all. They need to be able to change as the organization changes, so likely future needs should be taken into account as well as current needs, and perhaps more so. For example, to maintain competitiveness an organization must not simply plan to match what its competitors are doing now, because by the time that it is achieved the competitors will be perhaps one to five years further on and the organization will still be uncompetitive; it is necessary to aim to match a future state, not a current one—much as when intercepting a speeding car one does not plan to arrive at where the car is seen to be right now but where it is expected it will be. It is important to compare future situations, not a future situation with a current one.

2. Ensure commitment

It is essential to ensure commitment at the highest level in the organization at the very start. This is important for gaining the more specific kinds of support required, including financial and administrative support. Electronic office systems have an impact on all areas of an organization's activities to some degree, and so are an appropriate subject for the attention of the most senior people.

It may be especially appropriate at this level in the organization to assess the benefits of new sytems in terms of where the organization is going rather than where it is at the moment. This point has been explained above. It is especially relevant for senior management who need to be concerned with general organizational strategy.

Commitment from senior management and from the highest levels of other relevant bodies (e.g. trades unions) also helps to establish a strong 'normative belief' that significant people expect the various people involved to support the introduction of the new systems. It often helps if the most senior people can be the first to have access to the systems and be seen to be using them. This supports the image of the systems as being important and 'high status' rather than being associated with more lowly activities.

3. Ensure adequate financial and administrative support

This is essential if the new systems are not to founder before they have a chance to 'get off the ground'. Users will very quickly become disenchanted and withdraw their support if the introduction of the systems does not go smoothly and effectively. They will soon resort to conventional procedures if they begin to believe that, for example:

— 'It's easier to ask my secretary.'
— 'It's easier to go and see the person.'
— 'I cannot be sure the information on the system is complete and anyway it's often not up to date.'

— 'The terminal is always being used by someone else when I need to use it—except when it's not working.'

This kind of situation is certain to reduce interest in using the new systems and the people involved are likely to become more entrenched in their traditional ways of doing things, making it even more difficult to introduce new ideas in the future. It is particularly unwise to confuse what is a true pilot trial with what is actually simply a poorly funded, poorly supported system. Sufficient funds and administrative support must be made available to allow a service to be provided that is better (even if only in limited, 'pilot' areas) than that provided by the systems already installed. This may involve recruiting people with special skills that are required, or using specialist consultancy services. It is counter-productive, an expensive way of 'saving money', to attempt to 'muddle through' until the system 'takes off' and the greater investment is 'justified'—with such a policy it may never be justified, because the system may never 'take off'. It is not simply the amount of funding. The appropriateness of the resources made available is also a key to the success or failure of the programme. For example, it is vital to have staff with the skills needed to negotiate contracts with vendors, as well as being able to design and implement systems, including the human aspects.

4. Involve all key parties

A number of writers (e.g. Conroy and Ewbank, 1980) have commented on the importance of involving all key parties in the introduction of new office systems, and practical experience (e.g. Bo-Linn, 1980) suggests this may be an essential element in achieving successful implementation.

Ways should be found of having the various parties work together so they feel part of a single team, with common norms. They should share information so they can have a common appreciation of what options are available and what the likely outcomes of the various options are.

An advocate for the new systems should be identified for each rank within the organization, and each part of the organization. The advocates will set norms and provide information that will be important in influencing the success of the systems.

5. Establish goals

The different various parties involved will differ in the emphasis they put on different potential benefits of the new systems. For example, management may emphasize the benefits of improved productivity, improved efficiency, and so forth. The trades unions may emphasize the benefits of improved working conditions, new career possibilities, and other potential benefits for their members.

Birchall and Hammond (1981) propose a method of systematically interpreting the interests of the various parties in terms of forces for change and resistances to change. This is illustrated by the hypothetical example in Figure 12.1. This form of 'mapping' can assist in identifying the goals of the various parties, areas where common interests exist and pressures resisting change.

Sufficient points of common interest have to be found to establish an agreed set of goals and to commence the planning process.

6. Maximize awareness

It is important to go beyond involving representatives of the interested parties. A programme of structured discussions, demonstrations, and training sessions—as well as other means—should be used to maximize awareness throughout the organization of: the general system concept; the particular applications for which it is being used; and how the system is being implemented.

7. Analyse the nature of the problem

Throughout the programme, and especially in connection with the identification of applications and the subsequent implementation of the new systems in relation to these, it will be necessary to move from a general appreciation of the objectives of introducing the systems to a much more detailed analysis of the problems they are intended to address. In the course of this analysis it may be helpful to

- examine existing solutions so that previous mistakes can be avoided and worthwhile features of previous approaches incorporated in the new systems.
- examine particular situations where problems are thought to exist and compare these with the 'ideal' to identify what kinds of improvements are necessary
- ensure that the views of different parties are taken into account
- follow preliminary diagnoses with more detailed work based on attitude surveys, interviews, and other methods as appropriate (much of this can often be undertaken by working parties)
- ensure that the problem analysis covers analysis of all key aspects, especially those relating to: organization, tasks, people, and technology.

8. Anticipate the spread of effect

The effects of introducing new systems can spread quickly through the organization and can have an impact on aspects of the organization's functioning not directly planned for. Information can be manipulated, copied and transmitted faster. This can improve efficiency in some areas but may also provide an environment where the results of errors or faults

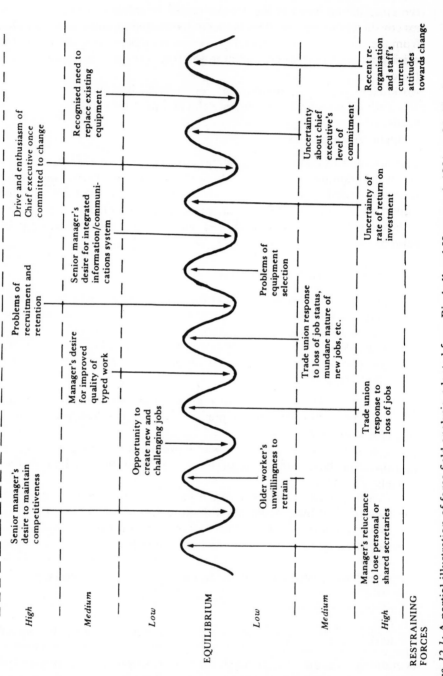

Figure 12.1: A partial illustration of force field analysis, adapted from Birchall and Hammond, 1981, p. 95. (Business Books Ltd.)

can escalate faster. More people than those directly involved in operating the systems may need to know about them. They may need to understand exactly how the information they use is generated, and how that differs from the previous systems. The easier access to information and greater processing power provided may shift the centres of power and control within the organization—between departments, and between the management and workforce.

9. Identify the effects on job security, payment systems and career opportunities

The notion of 'resistance to change' as an abstract, diffuse kind of force is not very helpful. Any resistance there may be will reflect more fundamental concerns. Concerns about job security, payment systems and career opportunities are often among the most important of these.

When introducing new systems, it is important to take account of these sorts of concerns and especially to

— review policies that impact job security, redeployment and retraining needs and opportunities
— anticipate effects in these and other areas of concern to those involved and then negotiate, consult and organize to maximize the opportunities and minimize any disadvantages
— avoid producing changes that damage career opportunities but build on opportunities that allow for skill enrichment and the development of new kinds of career paths.

10. Assess the time scale of the changes realistically

It is very easy to underestimate the total elapsed time required to introduce new systems. What may at first thought seem possible to achieve in a few months may end up taking several years.

Delays are common in regard to

— selecting suitable systems and sorting out the technical aspects of installing them
— working through the organizational aspects with all the interested parties
— building up awareness of the new systems and what they can do
— training users or operators, where this is necessary.

11. Take account of laws and regulations and consider technology agreements

A number of countries have been active in developing laws and regulations governing the introduction of new systems. It is essential that the user organization concerned takes full account of these. Some key examples are presented in Figure 12.2, taken from Evans (1983, p. 155) in a publication from the INSIS Programme of the Commission of the European Communities. This also gives an indication of the level of activity in developing national, sectoral and local agreements.

12. Plan for the changes

Whatever courses of action are decided upon, they must be

- technically feasible
- socially and politically acceptable
- economically viable
- administratively convenient
- organizationally feasible.

Kling (1983) recommends the following elements in planning for the changes:

- develop a social and functional plan
- examine institutional as well as functional feasibility and appropriateness
- involve a diverse and representative group in planning, design and implementation reviews
- continue design and implementation reviews through the system's life cycle
- evaluate the potential social impacts of different solutions
- maintain the adaptability of the system through incremental design and implementation.

Some of the elements represent a more detailed planning of some of the broader principles reviewed above. The results of the more detailed planning should include detailed proposals setting out modified procedures and practices, the necessary agreements, and detailed plans for implementing the changes. An action plan should result which should include sub-goals, target completion dates, resource requirements and specifications of standards to be achieved.

13. Prepare for the changes

Once planning has reached a suitable stage, preparatory work will need to be undertaken. The scale of this will depend on the scale of the changes involved. The preparatory work can involve, for example:

- finding and acquiring new buildings
- structural alterations or rearrangement of existing offices

Collective Agreements

Country	Laws and Regulations	National Agreements	Sectoral level	Company or Plant Level
West Germany	Works Constitution Act 1972 Works Safety Act 1973 plus VDT regulations 1981	None	Job protection agreements in metalworking, textiles, footwear, leather, paper processing, printing	Over 100 agreements concluded
UK	Health and Safety at Work Act 1975	None	Parts of public sector	Over 200 agreements concluded
Norway	Working Environment Act 1977, plus VDT Regulations, 1982	1975 Employer/Union Agreement on computer-based systems	Banking	Most of industry and services covered by local agreements
Sweden	Working Environment Act 1978, plus VDT Regulations 1981 Codetermination Act 1977	1976 Employer/Union Work Environment Agreement	Technology agreement in printing. Codetermination agreements in public government & private industry.	Use of legislative rights
Various	Health and Safety laws (France, Italy and elsewhere) Codetermination laws (e.g. in Austria) Statute of Workers Rights 1970 (Italy)	1981 agreement for private sector in Denmark	Printing sector in Netherlands, Belgium, Austria and Greece; metal working in Italy	In USA, 1979 Ford agreement on procedures for introducing technology. General Motors quality of working life programme. In Japan, company unions consensus in return for job security and income sector.

Figure 12.2: Examples of the regulation of technological change (from Evans 1983, p. 155) (Reproduced by permission of The Commission of The European Communities.)

— recruitment, selection and training of staff
— redeployment
— briefing of managers and specialists
— training
— organization of work.

Very often the development of the necessary software will require most time and resources, but the preparation of operators and end-users (managers and others) is also time consuming if done properly and is of most interest here, in the context of human factors.

Awareness of what is being done and when users can expect what sorts of services needs to be heightened during this phase. This may involve a combination of, for example:

— introducing 'updating items' into the agendas of peripherally relevant meetings as well, of course, as meetings of committees and other groups directly involved
— special workshops and briefing sessions for key groups at key stages of the preparation
— a special newsletter to inform users of progress, demonstrate that real progress is being made, and help to encourage realistic rather than unrealistic expectations of the new systems
— relatively informal interviews with key 'opinion leaders' in the organization to make sure that the inevitable rumours are as far as possible based on fact rather than misunderstandings, and to foster positive attitudes.

Other methods may also be tried but it is important not to 'go overboard'. Important as the changes will be, people can tire of hearing too much about them too often. The emphasis should be on quelling fears due to misunderstandings, fostering a positive attitude, and cultivating a realistic appreciation of what the new systems will and will not be capable of. As far as possible, this should include showing how the various interested parties (e.g. trades unions) are involved, and providing a further vehicle for people to feel they have some influence over what is being done and are being treated as intelligent participants in the overall programme rather than having changes imposed on them arbitrarily.

Training programmes may need to be set up for both the operators and end users of the new systems. These may concentrate on training users in an appreciation of the systems' capabilities and the need for detailed training of how to perform specific operations will be minimized if appropriate attention has been given to systems psychology in the design of the hardware and software.

Alternative forms of work organization should be considered, and the most appropriate selected. This means, for example, that wherever possible

— design jobs with variety, meaning, feedback, discretion, control, and development and use of knowledge skills
— design the physical layout to proper human factors standards and ensure, for example, that people can communicate with one another and see the results of the operations they are performing
— organize work and layout to develop skills and knowledge through work experience—skill enrichment rather than deskilling
— provide a means for manual back-up during the introduction of the new systems and subsequently when there are system failures
— reconsider the boundaries between operating and managing jobs and consider new forms of organization that may be more beneficial.

It is necessary at this stage to develop as full as possible an appreciation of what an appropriate information environment is for the particular organization concerned. This will probably mean

— interviewing and observing the people whose work will be affected in order to develop an understanding of their 'work worlds'
— developing a range of scenarios concerning the use and evolution of the system as it is likely to be experienced by different groups
— developing prototypes in key areas, integrated into the real operational environment, to get early feedback on the usability of the various components of the systems proposed.

14. Select technically appropriate equipment

It is vital to select equipment that meets all the relevant technical criteria. It should be capable in technical terms of doing the job, it should be reliable, it should be available, it should be backed up by good after sales service, and so forth. These are vital but they are not part of the human factors of systems and will not be considered here. What is important here is to appreciate that this is just one of many elements involved in introducing new systems. It is essential but (as many organizations in the past have learned at great cost) it is by no means sufficient.

15. Attend to 'global' USI factors

Following the distinction made elsewhere (Christie, 1981) between 'global' USI factors and specific system features, the global factors are the broad dimensions of judgement—the broad criteria—which one can apply to the user-system interface (USI). They relate to such notions as

— 'user friendliness'
— 'humanness'
— 'social presence'

— ease of use
— ease of learning
— how stressful the system is
— and other factors.

Some of these have been discussed in previous chapters and will not be dwelt on here. What is important here is to appreciate that they are just as important as 'technical' factors in influencing the success of the system, and it is most important that user organizations insist on equipment (both hardware and software) being satisfactory in these terms as well as technically.

16. Attend to specific USI features

These are the particular features of the USI that determine the global factors referred to above. It is most important that user organizations satisfy themselves that the equipment they are considering using meets the criteria discussed in the various chapters on usability, especially

— hardware ergonomic standards and guidelines
— dialogue design guidelines.

17. Attend to the needs of different user groups

There is no such animal as 'the average user', any more than there is the average human—people vary, their environments vary, the demands on them vary, and their needs vary (both from person to person and from time to time). Somehow, systems have to be designed and implemented to address this problem adequately. All too often in the past, electronic systems have assumed that all users are alike—more specifically, like the people who designed the system.

There are two main aspects to the problem. First, some groups may simply show a lower level of acceptance of the new systems than other groups. Secondly, assuming some minimal level of acceptance, groups may differ in the kinds of USIs they need.

The first aspect of the problem has been recognized for some time, and is illustrated by an early example of introducing an online bibliographic retrieval system into an academic environment (Borman and Mittman, 1972). There was a marked difference between the research workers and the students. The researchers typically showed little interest in using the system—even after careful preparation and help—and preferred to continue with their more traditional methods. The students, in contrast, were enthusiastic. It appeared that the researchers—who had already established ways of working—would have needed continuing 'marketing' and 'hand holding'. The students—who came to the experience without the same set patterns—saw the system in a different light.

In addressing the second aspect of the problem, it is important to recognize that for many users the USI has to be both a link (with required facilities) and a barrier (protecting the user from other aspects of the system). Acceptance of a complex system will reflect an interaction between the type of user and the kind of thing the user is trying to do. An illustration comes from an early study by Eason (1976). The managers in the study apparently experienced a sharp increase in difficulty with operating procedures as the complexity of the task they were asked to do increased. Specialists also experienced an increase in difficulty, but more gradually. There was also a difference between levels of seniority. The more senior people were especially likely to stop using the system or use only a few, familiar routines—or to introduce a human 'buffer' (someone skilled in using the system who could use it on their behalf).

It is important to identify who the key user groups are, and what special needs they have, before embarking on the implementation phase proper.

18. Implement the changes

Once the planning and preparation for the new systems have been completed, it is possible to begin the process of implementation. The time required for this will depend on the scale of the changes involved, from perhaps a few months to several years. Where large systems are being introduced, it is often best to proceed in stages. The first sub-system or part of the organization to be completed can act as a 'show piece' to foster user interest and enthusiasm and help overcome any remaining barriers to change. It is especially helpful if the systems can be introduced during these initial phases in such a way as to create an image of them as high-status rather than low-status. Staged implementation also gives an opportunity to assess impacts directly on a relatively contained basis, and to test back-up procedures.

It will often be necessary to operate dual systems during implementation, in which both the old systems and the new systems are operated in parallel so there is no risk of interfering with the flow of the organization's business.

Managers will need to be especially careful during implementation to protect the operators of the systems from excessive demands of users who may not fully appreciate the learning period which operators must go through or who may have unrealistic expectations of what sort of service can be expected with the new systems.

19. Monitor and evaluate the changes

Criteria for evaluating the success of the new systems will have been established at an early stage—see (5) above. The evaluation itself may be carried out by management directly, through external consultants, or by a working group made up of representatives of all the main interested parties. As with the other stages in the change process, the monitoring and evaluation stage has to be planned in advance. Appropriate and sufficient resources have to be available, sufficient time allowed, access to information agreed, and deadlines established.

The groups who need feedback from the evaluation must be identified and, in addition to written documentation of the findings, presentations to these key groups should be organized.

Some of the key human factors criteria for evaluating the impacts of new systems are discussed below.

20. Modify the change programme as necessary

In the early days of computer systems it was not unusual to conceptualize the life history of a system as a linear sequence of successive stages, such as: requirements specification; design; coding; testing; training; documentation; and maintenance. These elements are still important, but it is acknowledged much more today that they should be thought of as components of a spiralling, iterative process where feedback and feedforward are important.

Most systems are not installed on a complete, once and for all basis. They spread, and they evolve. To be kept on course in terms of the original human, technical and economic goals, systematic design reviews are needed throughout the total life of the system, not just at the start when it is first declared to be 'operational'.

IMPACTS

The main dimensions

The new systems need to be evaluated in terms of the goals established for them. These should include human factors goals. In addition, however, the systems are likely to have effects that were not specifically planned for. We are concerned with the human impacts.

Kling (1983), in a publication from the INSIS Programme of the Commission of the European Communities, proposes thirteen main dimensions in terms of which the human impacts of office systems can be evaluated at a broad organizational and social level. These include impacts on:

— the quality of life
— the 'social fabric' of group life—e.g. home-based work and use of electronic communications could mean in some cases that users would see less of one another
— the distribution of power and control within the organization
— dependency on specialised resources and suppliers—e.g. increased dependency on the technical resources required to maintain and modify the systems
— intelligibility of social and technical arrangements—the systems should improve users' understanding of their work world, not make it more mysterious

- employment patterns—altering the mix of jobs and skills required
- privacy
- the fit between computing arrangements and social/organizational culture—e.g. a misfit could occur if one were to try to impose a highly analytical computing method in an area where intuitive judgements and extended intergroup negotiations had been used very successfully in the past
- infrastructural demands—e.g. on the electrical power, telecommunications lines, and physical environment required
- equity in receiving benefits and bearing costs—the 'symbiotic relationship' between the user and the organization referred to in Chapter 2
- ideology—well integrated systems of beliefs about what is important in the world and how people and social groups should act
- social accountability—which needs to go beyond legal contracts, professional standards, and administrative authority
- social carrying costs—such as accidents and environmental pollution, which could be markedly reduced through the impacts of electronic systems.

The impacts of office systems in terms of these thirteen broad dimensions remain to be researched but they suggest at the very least a check list of major areas of impact and potential benefits that user organizations should consider when embarking on a programme for introducing new office systems.

Impacts on managers and professionals

Managers and professionals account for about 60 to 75 per cent of office labour costs, and are the key consumers of information in the organization.

It is possible to paint a fairly gloomy picture of the impacts on managers and professionals—gloomy from the point of view of the individuals concerned. Wynne (1983) presents such a view when he writes, in a publication from the INSIS Programme of the Commission of the European Communities,

> Many of the 'middle' management decisions could become dependent on computer-based processing, with a very small degree of freedom allowed for an individual's interpretation of the computerized results compared to a much higher degree of autonomy for similar middle management functions in the past ....Many previous managerial and supervisory roles could be almost completely automated, being turned into primarily clerical or machine-operation tasks. (Wynne, 1983, p. 139.)

He goes on to identify six main threats to management effectiveness, as follows:

- loss of management autonomy if systems are introduced that force managers to comply with a simplistic, 'rational' view of what is required to run the organization
- organizational inflexibility if the systems assume that objectives can always be defined clearly, evaluated unambiguously, and that information is only ever needed in order to influence decision-making
- deskilling and increased stress and alienation if the systems reduce the priority and freedom given to the processes of informal persuasion, bargaining and team building
- avoidance of risk taking if the systems overly formalize and impersonalize information flows
- emphasizing information quantity rather than quality if the systems are designed to provide managers with more information, and this could overload the manager and create deleterious effects
- increased organizational conflict because the manager will have to deal with an extra dimension of complexity.

Although it is possible to paint such a picture, it is very difficult to accept it. It is a 'worst case scenario' in which the systems suppliers provide the wrong systems and the user organization introduces them into the organization incorrectly. Also, it focuses almost completely on the manager and fails to consider other professionals. Wynne himself acknowledges this implicitly in his brief discussion of how the threats can be turned into opportunities.

Whilst it is important to bear potential threats in mind, it is probably more realistic—and certainly more helpful in facilitating progress—to focus on the benefits that managers and professionals might wish to gain from new systems. This is also more in correspondence with the way managers see the situation. Jarrett (1982, p. 25) cites an Urwick/Computing Magazine attitude survey of UK managers which found that 90 per cent of them expected office automation to improve management effectiveness, not make it worse.

The key benefits which US managers in a Booz, Allen and Hamilton survey saw (Jarrett, 1982, p. 26) included the following:

- enhanced decision-making
- improved managerial and professional productivity
- improved clerical productivity
- and other benefits.

Enhanced decision-making. This is a key benefit that managers appear to seek from new systems. How the systems can provide that benefit has been discussed in Chapter 8.

Improved productivity. This benefit can be provided by using electronic systems to reduce the proportion of time that is wasted in each main activity category. However, it must be said that the major sources of productivity improvement may lie elsewhere. A Fortune survey (Jarrett, 1982, p. 27), for example, suggests that about a third of a US manager's time is wasted because the work done could in principle be delegated. One reason why delegation does not take place in practice to the extent it could theoretically may be that there are often not enough people available to whom to delegate work. This suggests managers will not benefit fully from electronic systems until either: a) systems are sufficiently intelligent to be able to accept the kind of work that 'can be delegated', or b) systems help staff to save enough time that the staff can then accept more work from the managers, or c) a mixture of those two.

It is important to appreciate that managers and professionals do much more than just make decisions; indeed once one attempts to identify exactly what the decisions are that are made it becomes very difficult to do so without imposing very arbitrary definitions of what constitutes a decision as opposed to something else. Much of the time potentially saved by electronic systems will be used not necessarily to make decisions but will be spread across a diverse range of activities, such as:

— browsing to stimulate creative thinking
— thinking
— discussing
— persuading
— building teams
— improving the quality of work done, e.g. in preparing reports.

It is particularly important in connection with the last example to recognize that fewer and fewer jobs are concerned with producing goods, and more and more are concerned with producing information (consultancy reports, journals, contracts, policies, catalogues, manuals, software, videos, proposals, and so forth). The quality of these products is dependent upon, amongst other things: sufficient time for thinking, adequate access to appropriate information, and adequate production facilities (e.g. printing facilities).

Changes in the content of work. The capability of electronic systems to support the manager and professional through enhanced decision-making, improved productivity, and in other ways, will facilitate a shift in the content of work. The work of the manager and professional can be thought of in terms of three main levels:

— Orientation activities: these enable an adjustment to changes in the environment; they are often novel, unstructured, complex, and involve information about new possibilities and new constraints (new markets, new technology, new products, new competitors, new legislation, new risks, and so forth).

— Planning activities: these take the findings and strategies or policies based on orientation activities and implement them by setting up new projects, new departments, new companies, or in some other way establishing programmed activities.

— Programmed activities: these are routine, repetitive and standardized; they involve information in the form of routine reports, standard letters, contracts, invoices, and so forth.

The support provided by electronic systems will facilitate a shift in the work of the manager and professional away from programmed activities to a greater emphasis on orientation and planning activities. This is likely to mean more time spent:

— meeting people
— scanning information sources
— discussing ideas
— thinking
— problem-solving
— and related activities.

Motivation and temperament. These changes suggest that more than ever the manager and professional in the modern organization will need to:

— be highly motivated and interested in investing energy now for returns in the future
— like working with people and be capable of building new teams and working relationships
— be flexible in viewpoint and able to get to the root of a problem quickly
— be concerned to maintain an appropriate social image
— be bold in dealing with people and not too easily threatened
— be willing to experiment with new ideas and ways of doing things
— be capable of dealing with criticism and failure realistically without it wounding his or her ego too much—to be capable of absorbing the knocks of everyday life without crumpling.

These aspects of the modern manager and professional are discussed more fully elsewhere (Christie, 1981) where they are related to more precise psychological definitions and methods of assessment.

Ability to cope with stress. Electronic systems, if designed and introduced into the organization properly, may in themselves help to reduce the stress on users. Nevertheless, the general trend towards a faster rate of change in organizations and their environments suggests that this may be more than offset by increases in stress from broader organizational factors. In addition, the adaptation to new systems is itself likely to be somewhat stressful during the adaptation period. The personality characteristics mentioned above will help managers and professionals to cope with stress; but not everyone has such characteristics, and even if they do there are likely to be an increasing number of people who could benefit from special training and counselling if they are to avoid the 'executive burnout' of which we are beginning to hear more and more. This is also discussed more fully elsewhere (Christie, 1981).

Impacts on secretaries and clerks

Many managers and professionals would like to have more secretarial support than their organization can afford to provide. Gone are the days when the typical situation was for every manager to have his or her own secretary. Sharing is now the norm.

Intelligent electronic systems will help to restore the situation by providing managers and professionals with intelligent electronic support, where human support cannot be provided. It may not be as versatile as a human but it will be better than not having such support at all (the current situation), and in some ways (e.g. ability to remember things) the electronic systems may prove better than a human.

This will be of direct benefit to the managers and professionals concerned, who will find they can devote more of their time to the higher order aspects of their work.

The availability of such systems will also have an impact on the work of secretaries and clerks. The human secretary will have more status in the organization and will assume more of the functions of a personal assistant. What this means in practice will depend upon the work of the particular manager or professional involved, but it could mean taking some of the burden of the more routine administration which requires meeting and talking to people in the organization. It will probably also mean supervising and controlling whatever 'macro' aspects of information processing the electronic systems are not capable of handling, and these will become fewer as time goes on. In the short to medium term, it might involve, for example, filtering information to help sort the relevant from the irrelevant, and perhaps taking responsibility for the final polishing up of routine documents.

Typing and other skills. The way in which typing skills are used is likely to evolve along the following lines:

- secretaries spend a large proportion of their time typing
- this proportion is reduced as a new type of job emerges: the (centralized) word processing operator

- managers and professionals begin to do some of their own typing in order to take full advantage of what word processing has to offer— there is a shift from 'batch mode' to 'interactive mode'
- word recognition technology evolves to a state where managers and professionals can dictate directly to a machine, and the need for typing is reduced in favour of a need for more general keyboard skills.

Clerical work will make very much greater use of 'personal computer' type skills.

Psychological dimensions. Bjorn-Anderson (1983) discusses the changes in secretarial and clerical work in terms of six key dimensions, as follows:

- Formalization and specialization: organizations must retain their ability to adapt to changes in their environment—changes that occur much more now than in the past—and this means, amongst other things, that too much emphasis on formalizing office work and specializing the roles of office workers is to be avoided.
- Rule-oriented decision making: electronic systems are capable of applying very large numbers of rules to particular situations to determine a solution; this can be a powerful aid to decision-making, but it is essential that the system is adaptive, that the rules can be changed, and that the ultimate control lies with the human (the superior 'computer') and not with the electronic system (an inferior computer).
- Autonomy and control: there is no intrinsic reason why electronic systems should be used to reduce autonomy and control; and there are reasons why this should be avoided, partly because humans need autonomy and control for job satisfaction and partly because it does not make any kind of sense for the superior machine (the human) to serve the inferior machine (the electronic office system)—it must be the other way round.
- Work performance control: monitoring of work can be useful to the humans controlling their own and the machine's work, but it becomes counter-productive if carried to excess.
- Stress: as discussed elsewhere in this book, it is important to avoid creating undue levels of stress.
- Social contact and alienation: social contact amongst staff is important at all levels and it is important to introduce systems in such a way as to provide adequate opportunities for social contact.

A further dimension which Bjorn-Anderson does not discuss so explicity but which is also important is that of

- Skill enrichment: intelligent electronic systems are capable in principle of working with users having minimal skills gradually to enrich the skills of the user whilst actually doing real work (not 'training' in the conventional sense).

Impacts on patterns of working

Telecommunication is for practical purposes distance independent. This means that work can be done or delivered at one location by someone working at a different location. This has led to much speculation over the years about the impacts the new systems might be expected to have on patterns of working. We shall focus here on just four possibilities:

— working from home
— neighbourhood work centres
— flexibility in time spent working
— the 'headquarters city'.

Working from home. Electronic systems may allow more work to be done from home, but the impact is likely to be relatively small—especially for managers and professionals. The reason is clear when we consider the following factors. About 30 to 40 per cent is a fairly good 'rule of thumb' for the amount of time managers spend on paper-oriented activities. This proportion varies from one organization to another but it is not too bad as a rule of thumb. Even if all of this could be done at home, which does not seem entirely plausible in the foreseeable future, it amounts to just two days a week. The proportion of time involved is greater for some other categories, especially secretaries and clerical workers, but one needs to take account of the social isolation, reduced security, and other factors that mitigate against such work being done at home on a large scale. If we assume that some meetings—say 40 per cent (see Short, Williams and Christie, 1976)—could be conducted using systems based on the ordinary telephone network, then this increases the estimate of the amount of work that a manager or professional could do from home—by perhaps an extra day, making three days a week in total. This is probably a rather generous estimate since it assumes that all the activities that require one's presence in the office can be scheduled to fit exactly into the two or three days available; in practice, of course, it often happens that one needs to be in the office to attend a meeting, or for some other activity, that only takes up a small part of the day. And that sort of situation arises most days.

Neighbourhood work centres. The alternative to working from home does not have to be to commute into the centre of the city or to the organization's main site, wherever that may be. It would be possible in principle for the organization to disperse its operations or set up a number of satellite offices, establishing a series of small units over a broad area. In fact, the number of such units would have to be quite large—say, of the order of 20 to 50 in the case of an organization around London—for this is to offer any commuting advantage. With only a small number of units, people would still have to commute relatively long distances, and this could be even more arduous than the traditional scheme since it would be necessary to travel across instead of radially into and out of the city (which is usually the pattern on which public transport systems are based). It would also be expensive for the organization to equip all of the units with everything needed to provide a full service. In the light of these considerations, the idea has begun to emerge of neighbourhood work centres. Different variants have been proposed. One possibility is that different organizations 'club together' somehow. Alternatively, the centres could be set up as independent businesses, offering their services to whoever wished to use them. These would be similar in some ways to hotels today (offering conference rooms, communications services, and other facilities), 'temp' agencies—and the companies that offer a business address, answering service, offices to rent on an hourly or daily basis, and various secretarial services. The neighbourhood work centre would be larger, with a wider range of facilities, and with greater emphasis on electronic information and communications systems.

Flexibility in time spent working. Whilst the possibilities for working from home may be somewhat limited, and fully-fledged neighbourhood work centres may be some way off, the greater flexibility afforded by electronic systems does pave the way for a less rigid approach to the working day, emphasizing an already existing trend in the work of many professionals today. There may be greater emphasis on working to deadlines with less emphasis on working '9 to 5'. As today, but more so, some people may choose to work early and/or late with perhaps a break of several hours in the middle of the day. Such a staggering of activities across hours of the day and days of the week could result in more efficient use being made of facilities and greater opportunities for some people (e.g. parents) who may need to integrate their work activities into an already complex schedule.

The headquarters city. Even with the kinds of changes outlined above, the central city is unlikely to disappear. What may happen is that it may evolve towards something that Goldmark called the 'headquarters city' (Goldmark, 1973), which shares many of the characteristics of the 'ideal' city or 'essence of a city' described by Mead (1965). Mead saw the city as a 'point of confrontation', where strangers could meet and do business or engage in other, e.g. cultural, activities. According to this view, the city, in essence, is not 'miles and miles of factories or people working in factories which have no relationship to the city or to any of its benefits' (Mead, 1965, p. 19); it is a place where strangers belong, where groups separated by ethnic, political, or organizational lines can interact to learn from one another, build teams, and explore possibilities. The headquarters city would be a place top level management and others would use to conduct high level negotiations and decision-making. As well as the necessary hotels, restaurants, cultural and other facilities, such a city would require 'new and imaginative architecture, a 20th century administration and mass transporation' (Goldmark, 1973, p. 25).

A near-term scenario

Geoffrey Baker, Director of the Energy Divison of Twentieth Century Consultants Ltd., awakens with a bad head after having snatched a few hours sleep following the dinner engagement of the evening before. After struggling through the transition to something passing for normal waking life he switches on his home workstation. A single push of the button connects him to his electronic office. A mumbled 'Geoffrey Baker' results in a synthesized voice coming back with, 'I'm sorry, would you mind repeating.' It sounds genuinely apologetic. 'Geoffrey Baker' he repeats, more carefully this time. And with a friendly, 'Good morning, Dr. Baker,' the system comes alive, 'I have urgent messages for you from Dr. Phillips and Ms. Hammond.' Yes, through still bleary eyes he can see the priority light glowing. He will attend to them in a moment. First, check the day ahead.

His electronic diary reminds him of his engagement with a client in London in the morning and indicates a teleconference has been arranged with the Manchester Regional office in the early afternoon—a private teleconferencing room has been booked at the Sheroda Hotel. He notes the suggestion that, since he will be in London, he could think about a present for his wife's birthday which is only two weeks away. There is a note from Marie, his personal assistant, that she finished off the quarterly report last night and it can be distributed as soon as he agrees it. She will not be in the office until the afternoon. He transfers the report to his 'executive traveller', along with a unique code that will tell the system to go ahead and distribute the document.

He checks on the latest advice from the Energy Expert—a specially designed expert system his company developed for use in their consultancy operations. As he half expected, the general picture remains the same, despite the new inputs. Reluctantly, he transfers the output to his executive traveller. He will have to think very hard about how he can put it tactfully to the client. He wishes he had more case material but Energy Search—his intelligent information seeking system—has only been able to find three relevant cases in the UK. It is a new kind of problem that simply did not exist even five years ago.

The 'urgent' message from Dr. Phillips will have to wait until he has discussed the point with the client later in the morning, but he can deal with Ms. Hammond's query now and get that out of the way before breakfast. He looks through his files for the document he was looking through last week. A quick voice message attached to the document and Ms. Hammond's query is dealt with very well. He sends the electronic package to her and turns his attention to breakfast.

On the train down to London he uses his chord keyboard to type a couple of quick memos into his executive traveller. The chord keyboard allows him to use just one hand (pressing keys in combination to produce different characters). Some of his colleagues prefer to use a conventional keyboard but he finds the chord keyboard easier and faster to use on a bumpy train journey, and it only took him a few hours' practise to become proficient in using it. And it is certainly a lot more private than dictating, and less disturbing to other passengers. Then he settles back to look through the quarterly report.

The passenger telephones on the train are out of order and so on his arrival in London he stops at a telephone booth. As soon as he hears the connection signal he presses the device from his executive traveller against the mouthpiece of the telephone and presses a button which tells his office it is his executive traveller. Taking the telephone handset again, he speaks his name. 'Geoffrey Baker', so the system can check his voice pattern and check he is the owner of the executive traveller. 'I hope your journey went well, Dr. Baker.' He presses the device against the mouthpiece and sends the code he stored earlier, telling his office to distribute the quarterly report. Marie, as always, did a good job. Finally, he selects the two memos he wrote on the train and presses a general 'send' button on his traveller to tell the office to distribute the memos. Now he has safely dealt with those items, he can go and meet the client. But first he must buy some aspirin.

CONCLUSIONS

The incorporation of human factors technology into the hardware and software of office systems is necessary for the success of such systems, but it is not sufficient. The way in which new systems are introduced into an organization is also a critical element in their success or failure. This chapter has presented some key guidelines for the introduction of new systems and reviewed the impacts they are likely to have.

Twenty major elements in the introduction of new systems have been identified. Failure of any of them is likely to reduce the degree of success achieved. Organizations considering introducing electronic office systems need to appreciate that they will have far ranging impacts, and need to take an imaginative and creative approach to maximizing the benefits in terms of managerial, professional, secretarial, and clerical work, and in terms of overall patterns of working.

Only when the systems are seen as new tools to be used creatively and imaginatively, rather than machines to be followed slavishly, will the full benefits of electronic office systems be realized in practice.

In the next chapter, taking our discussion of guidelines and impacts as general context, we focus on the question of deciding just what type of systems a particular organization might need, and we consider the need to move towards a new methodology for deciding this.

Chapter 13

Identifying Future Office Systems: a Framework for a New Methodology

JACK FIELD

INTRODUCTION: BRAKES ON PROGRESS

The breakthrough in production technology once had to wait for developments in communication and information technology to catch up. Today, conceptions of the automation of office functions often seem to be simplistic and to leave the social–political milieu of the office to drag along behind. There are three reasons for that:

— One reason is that manufacturers—and even government—have been forced to take the initiative and convince clients that productivity gains in the office can be as worthwhile as productivity gains in production: the hard-sell reason.
— Secondly, the conceptual framework brought to bear in the analysis of the office has been dictated by a sectional or overly narrow view of the office: the self-fulfilling prophecy.
— Thirdly, the prevailing methodology has forced office work into the straightjacket of currently available technology: the straightjacket problem.

These three factors act as brakes on the progress that can be made in developing and implementing office systems, for the reasons explained as follows.

The hard-sell

A common approach to office technology is to survey office activities or functions, in order to show the amounts of time spent on each function.

297

Panko (1982) and Doswell (1983) list 16 such surveys between them. Such surveys are used to demonstrate that, because a particular function can be identified, is performed and is time consuming, it is therefore desirable to transfer the performance of that function to technology. Such a demonstration rests on four assumptions:

— that out of the whole variety of activity that goes on in an office—from time-wasting to overtime, from flirtation to harassment and from productivity to paper-shuffling—one can pick out the essential functions relatively easily. For example surveys agree that 'communicating' is what most time is spent on. What such surveys have not yet investigated is the nature of such communications: are they 'noise' in the system or are they ceremonies that sustain trust, responsibility, loyalty? Are the replicating, monitoring, feedback or parallel communication channels required by minor inadequacies of current systems or are they integral to the essence of the task? Are the interests the communication serves organizational, sectional or personal? Surveys do not investigate under what conditions—which could then be supported by technology—communication is effective or what different kinds of effectiveness there are
— that whatever functions are performed actually need to be performed; and its corollary—that functions unperformed or performed elsewhere should be left that way
— that a reduction in the time necessary to perform a function in itself ensures that such time will be more productively employed elsewhere or saved (to contribute to the cost of transferring the function to technology) and not to cause queuing, space or storage problems or new problems of updating and retrieval
— that technology will perform the function to a criterion satisfactory to all the interested parties: management, and staff, customers of various kinds, external regulatory bodies; and the corollary—that it is possible to arrive at such a criterion.

These assumptions can be seen to have operated in the case of the promotion and selling of the first steps in the move from typewriter to networked microprocessor, namely word-processing equipment. Text processing is obviously essential, it has definite locations and time saved could avoid recruiting and providing space for extra typists. Performance criteria are uncontroversial. While the speed and quality of text processing has improved immeasurably, the background factors in its milieu have scarcely changed. The input to text processing may actually have deteriorated. Originators know they can now afford to take less trouble in dictating or handing text over, because inserting a comma or changing the position of a paragraph is almost equally troublefree. Operators soon learn that the finished state is, with the exception of the penultimate one, several print-outs away and anyway can be run against a spelling-check. The cost of the paper consumed by extra drafts gets lost in general overheads, but unfortunately spelling programmes cannot distinguish 'right' from 'wrong' when both are correctly spelled, nor detect a missing or surplus 'not' or 'no'—or zero. The easier it is to process text, the greater the temptation to produce it and to respond promptly, so that the receive–respond cycle frequency increases. As the hard-sell assumptions spill-over from processing to output they encompass intelligent copiers too. There the whole image of a page is reproduced in one pass, including graphics, more quickly than by a line printer—as well as collated, distributed, digitalized and filed electronically so as to be reproducable in a variety of formats.

The fundamental question may not surface: does the communication need to take place at all, or can the source and destination become either integrated or entirely separate; does there have to be an explicit communication or could it not have been expressed otherwise (e.g. an unspoken understanding; as part of another transaction; via another medium with different electronic or technological parameters)?

The self-fulfilling prophecy

Along with logging the time consumption of functions, system analysts traditionally have proceeded to survey the connection between activities and identify interdependent functions linked together into familiar procedures and sub-routines: dealing with correspondence and messages received, processing orders and invoices, updating internal records and accounts. This is no accident of course, because the analyst will have been briefed as to what services the office was intended to provide. If the work is so arranged that the survey is designed according to a brief agreed with, for example, the sales director alone, then the contribution a newly automated office can make to the organization as a whole will be minimal. For example, it will probably have relatively little impact on the traditional gap between sales and production—even though, given the potential of even currently available technology, the sales office could give 'real time' information on fluctuation in orders, and could bring production capacity and customer satisfaction into closer harmony (by trading off quality, delivery time, and price). A technological solution is only as good as its brief. If an organization is itself out of phase, then technology will lock it into that disphasia rather than release it.

The straightjacket

The current approach of defining technological requirements is squeezed between the demands of the client's ongoing operations and the technology that is available or which may be assembled off the manufacturer's shelf. As the Butler-Cox (1982) survey of office systems puts it:

User: 'You don't have the system I need.'
Supplier: 'Tell me what you need and I'll see if I can supply it.'
User: 'How do I know what I need? All I know is you haven't
           got it.'

In other words, certain functions need to be abstracted from total office
activities and linked up to comply with agreed procedures. Realistic market
forces pressurize the approach into expressing the functions and linkages in
terms of the manufacturers' now-technology.

This level of surveying the existing formal functions of an office arrives at
a model of an office that can be represented inside technology, in terms of
available input and output switching, queuing, counting, storing, matching,
retrieving, reproducing and transmitting devices. This is the methodology
that has given us electronic mail and conferencing, local and international
networks, word and graphics processing, intelligent copiers, and optical character readers—with expert systems, animation, and voice recognition and
generation beginning to form part of the package.

Nevertheless, as a conceptual approach this methodology is rather like
observing a crowd at a fire and coming up with the recommendation to
pass buckets along a straight line between water and the blaze. The design
of a modern firefighting device as part of a fire-prevention system awaits a
wider-ranging survey in a next-generation methodology.

A FRAMEWORK FOR A NEW METHODOLOGY

To prepare the next-generation technology for the office of the future one
needs to drop the twin constraints of working to a specific brief and of
translating findings into immediately available technological representations.
While this may at first sight seem a flight from reality, the new approach will
reveal layers of reality from which the next generation of office technology
is most likely to emerge.

Working without a specific brief would be foolhardy if we were considering an office pre-installation survey. What is under consideration here,
however, is an exploration of the nature of an office wide enough to cover
offices in general and yet precise enough to provide the ground-rules or
parameters for a new technology feasibility study for a particular organization. Such an exploration is in the nature of a scientific enquiry which
must have a certain amount of discretion to follow where its data take it. It
must penetrate what is directly observable, reportable, obvious and publicly
acknowledged—to what in private is contentious and problematic. Management objectives, clear and inspiring in the boardroom, when translated into
office procedures, turn into cynical remarks in the workshop and retail out-
let.

In laying out the framework—the mix of principles and action that makes
a methodology—three stages are laid down:

— stage 1, multiple definitions
— stage 2, content of methodology
— stage 3, putative conclusions, expressed as guidelines.

Stage 1: Multiple definitions

Organizations are neither isolated nor homogeneous. Hence any particular office will be perceived as serving multiple demands depending on the location of the perceiver. Stage 1 of the new approach searches out the demands that others have of the office functions about to be surveyed. There are two locations and several tiers within each:

Internal	*External*
Management	Regulatory bodies
Supervisors	Associated companies
Operatives	Suppliers
	Customers

One needs to maximize the differences between the demands of each tier and minimize the differences within each tier. Thus extra tiers might be required. Management might split into local and central; operatives might split according to skills or unionization, similarly for the tiers in external locations. Regulatory demands of office procedures may be complementary or in conflict with what suppliers and customers want. For example, for large organizations Value Added Tax returns are made to the authorities on magnetic tape, more promptly and precisely than many suppliers and customers would desire.

As a more detailed example, a Sales Director might want an office to tell him or her the value of orders unfulfilled because of the inability of production to meet certain specifications (e.g. one-off finishes, prepackaging, delivery dates). The Production Director, particularly if he or she finds out about the Sales Director's interest, will want to get together the figures of machine setting-up costs, the frequency of short and interrupted runs, the percentage of time at full capacity. The information an office provides may be objective, accurate, valid and up-to-date, even neutral. What gives it coherence, however, is its mobilization behind a case or an argument on behalf of or in response to a sectional interest inside or outside the organization. This is what makes information and its technology useful and worth paying for.

The new approach concentrates on the cases which identifiable interests— the tiers—need to collate and present in their efforts to sustain or perhaps even contain or redirect the contribution the organization makes. So far the tendency has been for the top management of organizations to control the quality and quantity of information and ration it out, internally and externally, to each according to their station. The situation is changing, however, although there will always be need for privacy and confidentiality to safe-guard individuals and commercial security. Employees and regulatory bodies (taxation authorities, social insurance, government contract surveyors) are gaining greater rights of access to organizations.

It is anyway a by-product of information technology to have information spillage. Where once the accountant prepared the balance sheet in his own fair hand before bringing it personally to the Chairman, now the running totals are likely to be on a dozen screens throughout the organization.

Units of analysis are now closer to real activities and real time. With the unfolding of organizational information the new approach can afford to pursue the information needs of what might appear to be sectional or vested interests. Organizations can benefit from diverse interests confronting each other with alternative selections and interpretations of information within an arena of objectives and timescales devised by top management. To create a methodology that demands data that interlock even if this sometimes leads to conflict may help the organization to face complexity productively. A methodology that aggregates or simplifies information so that the action called for is always obvious, is not doing the organization any favours. It is the clash of different opinions—given they reflect real constituencies of interest (the tiers) and good quality information—that avoids stagnation and provides adaptation, development and growth.

How to isolate and differentiate the individual interest groups will depend on circumstances (e.g. the professional background of the personnel apply-ing the new methodology, the time and resources available, the sensitivities of the groups involved).

The public labels people carry (management, supervisors, operatives, cus-tomers, suppliers, etc) are only hypothetical. It is quite possible that man-agers identify more with their own department and their own operatives, than with managers in other departments, and similarly for other categories of people. Whether such identifications are in the interests of the organiza-tion or not will either be settled as survey data accumulates or not.

If not, then (as far as resources allow) allowances should be made for the distinctiveness of that department or section, with the possibility that it will split off from the main organization. Equally allowance should be made for bringing the department or section closer into the fold of the whole organization. In other words technological links into the rest of the organization should be less permanent, and compatible with other systems with which the division or sector may ultimately be aligned.

Stage 2: Content of the methodology

Here, indicative rather than prescriptive, are three lines of approach. The new methodology, by a process of triangulation (do two lines attached to a third, enclose a coherent figure?) seeks to define the outline of the new technology in terms of these three. They are:

- to formulate organizational issues which may be regarded as openers to a discussion, and to consider the relationships between groups within the organization in terms of these issues
- to consider the relationships between tasks, technology and people
- to consider the individuals involved in terms of their particular organizational needs and the possible benefits/impacts of new systems.

The *first* line of approach is to formulate organizational issues which may be regarded as openers to a discussion. The discussion could be with individual informants taken one at a time. In that case each informant must be given time for reflection, either by forewarning, by call-back, by being given a schedule to be completed subsequently, by being interviewed in the security of their home away from the pressures of work and the curiosity of colleagues.

Instead of interviewing informants individually, informants could be included in a group. This would allow the questions to be recast to suit the spectrum of situations the whole group encounters and responses could be qualified and amended to include the experiences of the group in its interaction at work. As mentioned, the isolation of groups is problematic, as each interest group will see the total organization from a perspective bounded by its own interest.

Those using the new methodology will not of course be entirely naive, and a balance of individual interviews and of groups of the same individuals constituted in different formats (e.g. horizontally—all of similar grades; vertically—those in the same up-and-down line of command; self-selected; by randomized invitation) will allow a rounded picture of the organization to emerge. Nor is it necessary or advisable for an interviewer always to be present in a group. The group could be given the task of discussing the listed issues. It will be found that such a group can often very perceptively reformulate the issues and come up with agreed answers—or possibly a minority report as well.

Not only should the interviewer not always be present, but particularly for those issues that have been difficult to resolve by individuals or groups, detailed observations or logging of the activities involved in such issues, and experimentation, should be constructed. By experimentation is meant the rearrangement of those aspects of the work situation that information obtained earlier has fastened on as being ambiguous or contentious.

Experimentation, and indeed more discussion, is likely in itself to change perceptions, so that an extra dimension must be allowed into the methodology—time. The Rand Corporation's Delphi technical forecasting technique can be adapted to this purpose (see Helmer, 1966, for a statistical treatment of these kinds of data, sample working forms, etc). While Rand limits itself to technical experts, it is the time-bound aspect of the technique that has most to contribute here. Essentially the members of the organization are informed of shifts of organization-wide perceptions over time (e.g.

monthly), so that individuals can take account of these and either jump on the bandwagon or dig their heels in and sharpen up their arguments, which would then be included in the next mailing of perceptions. The expression of perception is decoupled from individuals, so that the effect of status is removed from influence.

The eleven issues that form the first of the three lines of approach of the new methodology's triangulation exercise are shown immediately below. These are presented here simply as a list of issues, not in any particular questionnaire or other format.

Issues

1. Whose work/decisions do *you need* to do your work?
2. Who *needs your* work/decisions for them to do their work?

For these various pieces of work/decisions:

3. How frequent/how regular is the need?
4. How critical is the timing/duration of the need?
5. In what formats/modalities would the need be best fulfilled from various points of view (e.g. task performance; task motivation; wider activity, ruboff and side effects)?
6. Is the need for something specific, standard, variable, or complex?
7. Should the need be provided for alongside (shared with others cooperatively or consecutively) or integral to performance, or as a deliberate interruption or rest pause?
8. How well is the need met currently?

If these needs were met *(perfectly/to some extent)*:

9. In what would that show itself (more time/energy spent on core rather than peripheral activity; less waste; fit in better with other organizational activities; improve morale; could take on extra responsibilities; would get bored in longer run; other signs and effects)?

Contextual factors

10. In planning work or evaluating it retrospectively, in discussing work, in comparing yourself or the section you work in to others, what indications of objectives, job descriptions or roles, measures of performance or satisfaction would you be helped to have, to have updated, and which are lacking?
11. In organization disputes (e.g. resource allocation, centralization and devolution, redefining responsibilities, etc.), how would information, its exchange and modification have to be arranged to prevent the misframing or avoidance of issues, the smoothing over of differences that could generate growth or resolution at a higher level?

Commentary

It is important not to restrict the methodology to information, instructions or messages. Just as information has a tangible expression (memos, sound-waves) so actual products or the way value is added to them (checking that something has been done, fitting a part so that the whole finished product matches a technical drawing) have a potential technological component. Hence the issue is work broken down into tasks and sub-tasks as well as built-up into wholes, such as the aims and objectives of tasks (see also Issue 10).

Issues 3–7 have implications for the media balance (sound, vision, touch), intensity (amplitude, frequency, pacing), connectivity (network, switching), complexity (signals, free text, speech), etc.

Issues 8–10 have to do with cost—economic and personal—and growth. The new technology should alert organizations to changing circumstances and increase the flexibility of responding, rather than do what is already being done better and quicker, at the price of flexibility.

Issues 10 and 11 have to do with the organizational fit of work/decisions. As this information is obtained from individuals and groups, from interviews, discussions, observations and experimentation, accumulates and falls into place, a specification for a new technology will emerge that allows information to creep up the organizational hierarchy even as it modifies and adapts it to changing circumstances, allowing for growth while containing the amount of administration.

The *second* line of approach to the triangulation goes beyond the relationship of individuals to the organization, to the relationship of the tasks they perform. The approach is twofold and in tandem:

— What is it people are inherently at an advantage at, relative to technology (and vise versa)?
— Where in relation to that advantage (disadvantage) stand the tasks people have to perform?

While we cannot here break down the tasks performed in particular organizations, a broad general categorization derived from particular organizations has been adapted from Hackman and Lawler (1971), immediately below.

Autonomy: how frequently supervised (e.g. surveillance, hourly, daily, weekly, monthly, etc.)?

Identity: how clearly defined, self-contained, separated-off or distinct, as opposed to grouped, integral with other tasks; is there a single-constant or a multiple-variable input?

Variety: how much variation is there in the input to the task, how many and how severe the interruptions?

Feedback: does the performance of the task involve the input being modified by the output?

Immediately below, we list the advantages and disadvantages people have over technology in performing tasks. Matching up the list above with the list below, since they have been packed with examples from a particular organization, will point to crucial new technology areas.

The advantages of people over technology

1. Detecting a wider range of energies.
2. Sensitivity to a wider variety of stimuli.
3. Wider ability to perceive patterns and generalize from them to others.
4. Detect signals (including patterns) in high noise environments (the cocktail party phenomenon: recognizing own name being spoken in a crowded room).
5. Storing large amounts of information for long periods and retrieving relevant facts in appropriate but unexpected circumstances.
6. Judgement.
7. Improvisation and flexibility.
8. Handling low probability alternatives.
9. Arriving at new and completely different solutions to problems.
10. Ability to profit from experience.
11. Ability to track a wide variety of situations.
12. Performing a wide variety of fine manipulations.
13. Putting on an extra good performance when required—and when overloaded.
14. Reasoning inductively and intuitively.
15. Ability to lead, imitate, exercise responsibility.

The advantages of technology over people

1. Monitoring performance of people or machines.
2. Performance of routine, repetitive, precise tasks.
3. Responding quickly to control signals.
4. Exerting large amounts of force smoothly and precisely.
5. Storing and recalling large amounts of precise data for short periods of time. ·
6. A certain kind of computing ability.
7. Sensitivity to low, intermittent, or very regular stimuli.
8. Handling of complex operations (i.e. doing things simultaneously).
9. Reasoning rigorously.
10. Insensitive to extraneous distraction.

The *third* line of approach to the triangulation goes beyond considering the individual's relationship to a future technology in organizational terms. As has just been shown, the attributes of people are so unique that they remain essential to organizations, even if their skills are displaced and have to be restructured and redeployed. People, however, move in and out of commitments all the time—in the organization and outside it. In the introduction of new technology the organization has to compete for the individual's commitment, and must avoid undermining the feeling that individuals have of their importance to the organization. Indeed a correct specification for a new technology should make the individual feel even more committed to the organization's goals. The items in the list below are draft items for a questionnaire to individuals at various levels, to be fitted into a wording conditioned by the environment of particular organizations. The responses will show where improved technology would heighten individual commitment and forms the third strand in the triangulation exercise for a new technology.

1. Do you know what the result of your work will be?
2. How adequate are the solutions you have adopted to tackling the problems in your work?
3. How well are you doing relative to what is expected of you?
4. Do you receive the necessary information and suggestions on how to improve your work?
5. Do you spend enough time in the central core of your job?
6. Can you control the pace of your work; do you know what will be required of you far enough in advance?
7. Are you personally held responsible for what is your own responsibility—or for more or less than what you are responsible for?
8. Can you take time to plan and train to improve on your performance?
9. Do you feel your work is as strategic as you can make it; do you feel your work is sufficiently integrated with that of others—or too integrated?
10. Were you employed to do the work you do now, and how suited to do it do you feel?

Stage 3: Guidelines

Whether the output of the methodology is based on large enough numbers to allow statistical analysis or whether it is impressionistic or even journalistic is less important than whether it takes account of the weaknesses and strengths of the technology and the social milieu in which the office is embedded. Here are thirteen conclusions set against the background of the socio-technical literature of organizational life that implementers of the new methodology will want to display to themselves on that internal screen we all have implanted on the inside of our foreheads. They are expressed as guidelines forming part of the new methodology additional to and to be applied within the general framework described in Chapter 12.

Compatibility. Ensure the compatibility or at least state the date, source and method of obtaining underlying data units, so that needless friction about interpretation is minimized and conflict is converted into enquiry.

Decision level. Provide data aggregated at a level appropriate to the responsibilities of the decision-makers so as to contain the potential for the resolution of difficulties (e.g. in the above case of Sales and Production seeking to resolve their conflict, the investment cost of extra production machinery and the cost of subcontracting should be revealed as relevant considerations).

Meeting of minds. Make arrangements for meeting not necessarily face-to-face but certainly mind-to-mind, to build up trust and facilitate ultimate agreement (electronic mail, diary and conferencing; the use of a computer projection and simultaneous distribution system) that will allow different sides to call up data, examine the origins and question the nature of the data, extrapolate and simulate, seek briefing and comment from others not able to participate in the whole meeting, etc.

Extra space. It has been said that technology creates space and time but destroys the working group. In 'selling-in' the new methodology, it must point to how some of the extra space and time generated by technology can be used to increase the desire to collaborate, to concern oneself with areas adjacent to one's primary office responsibilities, to search for the multiple meanings of a fact, to reduce any fear of failure that is prevalent, or any desire for unilateral control.

Communication. Trace communications, their quality (style, ambience), that reconcile differences in purpose into constructive decisions. In processing orders for example, have sales and production priorities ever been reconciled? Was that due to speed of information, the status of the people involved, some promise of benefit? Either express those qualities in the technology (e.g. ease of personal contact) or preserve them in parallel with the technology.

Boundaries. Check how far office boundaries—the rules for access—were drawn according to then-technology or other dated circumstances. With networks and portable input and output devices office work can be distributed widely. On the other hand, while dispersal may favour the flow of information, it may disrupt its meaning and credibility and the flow of work.

Degree of integration. Explore opportunities for integrating some activities vertically (along the dimension running between retailing/service and resource acquisition/investment); and horizontally (associated companies, suppliers including media, markets, banks, regulatory and trade associations, etc.). For example, sales figures can be obtained directly from the cash registers rather than input in the office. The central office, however, could relate the figures to local background factors that in a single locality are too unstable to systematize (e.g. unemployment figures, local competition). On the other hand, with improved information flow factoring, devolution and delegation become equally viable options. The new technology gives greater weight to physical proximity—who needs to be near to whom and what.

Alternatives. When the new methodology cannot decide between alternative ways of doing things, include the alternative ways of handling the technology in some form (as a kind of control-group or baseline). Even the new methodology is not going to provide permanent solutions to changing situations.

Flexibility. If alternatives cannot be provided or are incompatible, err on the side of flexibility if complexity is not increased proportionally (e.g. a customer ordering system with Cash on Delivery or bill-later options should perhaps allow for ordering by credit card number or by videotex), even if the current customer base is low on bank accounts. The facility itself has a persuasive image potential which is in some circumstances capable of changing what might be regarded as givens. The advantages flowing from the introduction of new technology are often seen as improving quality and output. Flexibility, however, would be its most valuable consequence (Wild, 1974).

Algorithms. Recognize that activities are not fully represented by algorithms (activities do not proceed linearly in one direction from one choice point to the next—indeed even laboratory rats only behave like that in mazes too narrow for U-turns). Typically, activities consist of false starts, explorations around imagined, unreal as well as real options, the trial-and-error of several options, the retracing of steps and boundary conditions that change out of recognition. Therefore search out the elements of a support system which can compensate for the rigidities of algorithms and either preserve those elements or simulate their ability to cope with exceptions as when operators combine, rotate and interchange to overcome breakdowns and bottlenecks.

Menials. Production that is concentrated rather than dispersed, particularly if productivity is increased, creates increased demand for materials (paper, inks, tapes) and maintenance (cleaning, calibrating). While the technical solution is to move to purer information (e.g. electronic rather than consumable media) in the meantime a decision must be made as to whether operators, specialists or contract staff would be more suitable for the performance of more menial tasks. The problem is common enough in the switch to new technology, but lacks the glamour to capture the attention it deserves. Poor morale among staff stuck with menial tasks in the vicinity of high technology is dangerous.

Balance. If time allows, the new methodology should become a stimulus to organizational thought and open-ended enough not to close with a single recommendation at the end of it. It must provide alternative recommendations such that the various interests involved would be faced with alternative costed ways of running office-work. The outcome may introduce new technology piecemeal while dispatching the rest of the work done in the office elsewhere (e.g. to a time-shared system at an associated organization or at a trade association; or close to production so that the information never needs to reach the office, or less often than before). In that way manufacturers can respond to situations they have had a hand in developing, rather than to ones backdated to the original methodology survey.

User-friendly. Manufacturers put a premium on making their systems 'user friendly'. Under this heading software has been designed with such ghastly bonhomie as to greet this writer's log-on with 'Hi Jack—great to have you aboard'. Friendliness, if that is the usage, has to fit the mood induced by the opaque nature of modern technology with commonly only a screen to indicate what impression one has made on the machine. In the absence of anything to the contrary, the screen acts like a window into the workings of the machine. Creating confusion, erasure or hesitation on the screen intuitively feels like having created confusion, erasure and time-wasting inside the machine. Worse, the speed with which commands are executed, unquestioned, unflustered and often uncomprehending of the operator's true intention, together with the machine's use of direct second-person language is powerful evidence consistent with an omniscient slave sitting on the other side of the screen, deliberately misunderstanding instructions whenever possible, despite contextual clues obvious to the operator. The actual mechanism is invisible and anyway too complex to retain in one's mind. Worse still can follow. As operators overcome their resistance and master the technology—although they will often still speak to the machine as if to the person behind the screen—they are distanced from the human qualities of the data they now smoothly manipulate. Generally speaking, software designers pride themselves on concealing the subjective components of data: the variety of circumstances under which it was obtained; who obtained it, how carefully, with what intentions; who checked and updated it. Data on a display or printout carries an extra measure of authority that too easily beguiles. The specification for a new technology may very well take some emphasis away from speed and apparent accuracy of data presentation, and place it on taking operators over the steps between the origins of data and its final presentation. (See also Chapters 9, 10, and 11.)

Vignette

Here is a little vignette of the outcome of a piece of new methodology, shorn of data and applicable to certain office locations only, as one of many possible simple illustrations.

The office telephone is frequently used to convey time-bound information, yet established methodologies repeatedly demonstrate its deficiencies—mainly that the calls do not get through to the right person often enough. What established methodologies neglect is the participants' point of view. The recipient's view of the telephone is ambivalent—it is intrusive and simultaneously (when answered) compelling in its involvement. The more involving, the more calls to the recipient do not get through (because the line is still engaged); the more involving—and time-consuming—the more likely the recipient will seek to avoid further calls (unless the recipient lacks other responsibilities), which telephone technology allows him or her to do without any embarrassment as the caller is unaware the call is being refused deliberately. Meanwhile callers waste their time as it costs nothing to let the number ring, although the line is kept needlessly in use. Callers could meanwhile occupy themselves by keying in their name, function or number, the degree of urgency that they attach to the call and perhaps the topic they wish to discuss.

Linking a screen to a telephone could allow recipients to glance at incoming messages and impose their own order of priority on them without any commitment. The telephone on its own, on the other hand, gives no indication of degree of urgency, of who the caller is, how long the conversation will last—until the call-recipients commit themselves by answering. A telephone answering machine is somewhat like electronic mail on the receiving side, except that disk and vision on the VDU give quicker selectivity than tape and hearing, although tape captures more of the event at the point of input (e.g. intonation, personality). Even the perfect electronic mail system suffers the defect that it either gives too much discretion to the recipient as to what or whether to respond to a received message or becomes as intrusive as the telephone. The missing specification of a negotiating procedure to bring the receptive states of people who have to communicate into conjunction is what the new methodology points to, in this instance by bringing the 'screen' and the 'telephone' closer together.

CONCLUSIONS

In this chapter we have seen that it is important for a user organization to go beyond the general guidelines for introducing new systems discussed in Chapter 12, but working within that general framework, to consider carefully what office systems are actually all about—what the organization hopes to achieve by introducing them—and how to avoid the triple traps associated with

— the hard-sell
— the self-fulfilling prophecy
— the straightjacket.

Equipment manufacturers can also benefit by coming to a better understanding of what types of systems really need to be designed to serve organizations better.

This is not something that can be done simply. It requires an understanding of and commitment to the reality that different groups in an organization have different frames of reference—there is no single objectively true picture of the way any particular organization works or what it needs in order to work well. In coping with this actual complexity (rather than some simplistic theory of what the organization ought to be like), it is necessary to take a complex approach. A threefold approach has been discussed which includes

- formulation and discussion of key organizational issues with the various groups involved
- consideration of the interrelationships between technology, tasks and people
- involvement of individuals in a consideration of the benefits of new systems for them.

Some practical guidelines are presented which reflect the general approach advocated and which supplement the more general guidelines presented in Chapter 12.

Part Five
Conclusions

Chapter 14

Conclusions

BRUCE CHRISTIE AMD JOHN MCEWAN

INTRODUCTION

As a species we have evolved a capability to communicate and process information that is superior to any other life form on this planet. Using this capability we have been able to build on each other's knowledge and develop the civilizations we see around us today. We have stood on each other's shoulders to climb out of the primeval slime and enter a world of high technology, a world in which we can reach out to the stars—both physically and with our voices. We have extended our biological communications and information processing capabilities by building electronic tools that would have passed the understanding of our cave dwelling ancestors—electronic systems that have become such an intrinsic part of our world that the very security of our species depends upon them. If and when the missiles are launched, they will be under computer control and will have been launched because the telephone and other communications links between the superpowers of our time ultimately proved inadequate for the most important human communication of all.

To avoid the human species being an experiment that ultimately went horribly wrong, we must develop our communications and information processing capabilities to the next level—where we have knowledge instead of just information, and where we can be much more confident we will make the right choice in times of crisis.

There are small but discernable beginnings. We are crossing a threshold, leaving our 'industrial society' behind us and entering our 'information

315

society'. Viewed at the detailed level of everyday business and government this represents enormous opportunities and challenges, and out of these are emerging the hard-nosed commercial pressures to improve our communications and information technology. The European ESPRIT programme and the UK's Alvey programme are two key examples.

Office systems are at the leading edge of these developments. It is becoming essential to develop systems that can help ordinary human beings to function effectively in an infinitely complex world of information. The technology has been developing so rapidly that it threatens to outstrip our capability to harness it, to use it effectively, to convert its potential benefits into actual benefits. It has become critical to turn serious attention to the design of the user–system interface.

A psychological perspective

In developing office systems, one is doomed to failure unless one starts with a psychological perspective—that is, a perspective that recognizes that office systems are communications and information processing tools to be used by humans in a particular kind of context called 'the office'.

Go into any organization and you will see people. Organizations are, in fact, primarily about people working together. They may or may not use various kinds of electronic and other machines to help them, but primarily they are people working together. It is important to consider what is involved in this in order to understand what it is that electronic systems need to be designed to support. Without such an understanding, such systems are solutions in search of a problem.

What office systems need to support is human behaviour in an office environment. This behaviour does not take place in a vacuum, and nor does the human's interaction with the electronic system. It takes place within an organizational context, and within the context of the office system user's life as a human being.

The organizational context can be described in terms of broad organizational objectives to which everything else about the organization can be regarded as subservient. These objectives determine the general style of the organization, what is important and what is not. The organizational objectives are served by setting up the necessary organizational functions (e.g. marketing, accounting), and these in turn determine the roles that users of office systems (and other tools) need to fill.

Within the context thus defined, information is processed by users to fulfil their organizational roles, so that the organizational functions can be carried out and the organization can achieve its objectives. Both the organization as a whole, and individual users can be regarded as information processors.

User–system interaction also needs to be recognized as something that is not an isolated part of a human's life, but is something that takes place within the context of the user's life as a human being. Personal telephone calls, office parties, and sexual liaisons, are just some of the more obvious pieces of evidence for this. The user is not just an information processor in a simple sense but a human being with physical, physiological, motivational and other needs—needs that mean the user cannot sensibly be treated simply as another type of computer. It is vital to the success of new technology to understand the psychological basis of marriage between people and machines.

The successful design of the user–system interface requires that the interface be designed to serve the needs of the intended users in both the organizational and life contexts. Anything less is avoiding reality.

A historical perspective

To ignore the fact, when designing office systems, that the intended users are human beings with human characteristics would be a flight from reality. It would be equally unrealistic to assume that the needs which office systems are intended to address are all new phenomena, that did not exist and were not tackled before the advent of electronic systems. Offices have been around for a long time—at least since the time of the Pharoahs—and to understand what it is that current developments in office systems need to address it is helpful to take cognizance of the historical context. Neither is it the case that it is only very recently that management has considered the potential for electronic systems. The new technology has been under the scrutiny of management literature for a quarter of a century. The recent upsurge of interest in office systems needs to be understood against this backcloth.

From an historical perspective, the office can be seen as separate from production activities *per se*. Its role has emerged from a need to aggregate representations of production activities, and to interpret and negotiate their value. Electronic office systems, then, need to support the processes of representation, interpretation, and negotiation. Two rather different processes need to be supported in fulfilling this role, and these are: on the one hand, the rational or functional; and on the other hand, the expressive or ceremonial. Electronic systems will fail to the extent they disregard either one of these.

In performing its role within organizations, the office has gone through one 'revolution', as a response to the Industrial Revolution, and is going through a second 'revolution' now—one that started in the 1960s and can be seen as a response to the Post-Industrial Revolution.

Out of the second, current revolution, are emerging not a single 'office of the future' but many different types of 'offices of the future', running between the two extremes of the largely ceremonial and the mainly functional. The new 'electronic' offices facilitate a better relationship between 'office' activities and 'production' activities—one that is less dependent on physical proximity. The new kind of relationship depends upon and reflects the possibility of better communications.

The new, electronic office systems will be successful only to the extent they can handle adequately both aspects of the office to improve communications by facilitating shared understandings.

PRODUCT TRENDS

By the close of 1983, a wide range of office products in addition to the conventional voice telephone were widely available to meet two broad categories of user needs, for:

- — Type A communication — person to person interactive communication in real time
- — Type B communication — person to 'paper'/'paper' to person.

There was also a growing awareness of the possibility of a third kind of communication:

- — Type C communication — person to 'intelligent' machine, including interaction with electronic 'knowledge bases'

but there was next to nothing commercially available by way of Type C products to serve the general office worker.

In this book we have illustrated the systems psychology approach to products in these various categories by reference to the following examples:

- — electronic meetings (as an example of Type A systems)
- — personal information systems and shared information systems (as examples of Type B systems)
- — decision systems (as an example of Type C systems).

Electronic meetings

The psychological research on electronic meetings had its heyday in the mid-1970s. Electronic meeting systems were installed in various countries on a small scale but never took off in the way the market forecasts of the time suggested they might. There is a resurgence of interest in the systems now with the advent of wideband local area networks, satellite communications, and other communications systems. The psychological research done suggests the following conclusions.

Very many different sorts of meetings take place in organizations but there are three major facets that are especially important in considering the potential for electronic meetings. These are: the aims of the meeting, what goes on at the meeting, and the atmosphere associated with the meeting. The research done has helped to identify the major dimensions which discriminate between meetings in regard to each of these facets.

Meetings which are relatively 'person-oriented' (e.g. where the emphasis is on persuasion, and other instances where interpersonal relations play an especially significant part such as when meeting people for the first time) are more affected by the medium of communication used than are meetings which are relatively 'task-oriented' (e.g. where the emphasis is on simple exchange of information). The most important distinction is between face to face communication and telecommunication. The differences between different telecommunication systems are relatively small, but where there are differences video meetings tend to be more like face to face than are audio meetings.

Users usually judge video systems more favourably than audio systems in general terms, and there are noticeable differences between different systems in each of these categories. However, their willingness to use electronic meeting systems depends much more on the type of meeting involved than the type of system. This is probably because the advantages and disadvantages of different systems tend to cancel out to some extent. For example, whilst audio systems tend to provide poorer 'social presence' than video systems (a disadvantage) they often result in a more 'business-like' approach to the meeting (which can be an advantage).

For many purposes, an optimal electronic meeting system would be one that provides high 'social presence' at low generalized cost (i.e. money spent, inconvenience involved, and other sorts of costs). In many circumstances this would probably be based on an audio system of some sort rather than video, but this situation is changing as satellite communications, local area networks, and other technical developments are making video a more affordable option. It is certainly the case that video is not as important as has sometimes been assumed and that audio systems can be effective for many purposes.

Personal information systems

By the close of the 1970s, interest in psychological research was shifting away from Type A (especially electronic meeting) systems to Type B systems. One of the key issues to emerge in this area has been how best to support the user's need to deal with items in his or her immediate work area (e.g. on the desk, in personal filing cabinets). The conclusions we have come to in this book are as follows. At the present stage of research they should be regarded as working hypotheses, to be tested and elaborated in the light of future research findings.

The USI should be image-rich, allowing the user to develop multiple associations and use other psychological mechanisms that are so useful in a paper-based environment. The imagery should incorporate a virtual space of at least two and preferably more dimensions. It should be as concrete as possible, giving an impression of 'reality'. It should be familiar, so the user can bring established psychological schemas to bear, and learning can be minimized. It should incorporate a hierarchy of 'levels of information', reflecting the psychological constraint that the user cannot give equal attention to everything all at the same time. It should be capable of being personalized to fit in with the user's own particular preferences. Finally, it should not normally be a literal interpretation of the conventional office environment, except perhaps for casual users with a low workload.

At the time of filing items, the USI should provide opportunities for deep and elaborate cognitive processing. When the user needs to find an item later, the USI should be designed to help the user make use of cognitive strategies that are as 'natural' as possible (e.g. providing opportunities for recognition, not just recall of items).

Shared information systems

Users do not only have to access information which they themselves have filed. They also need to access shared or public sources of information. There are important psychological differences between these two situations. For example, in accessing personal files the user normally knows more or less what is on file and what form it is in, but in accessing shared (e.g. corporate) or public sources the user normally does not know what is available, where, or in what form. The research reviewed suggests the following conclusions concerning systems for accessing shared information.

Two broad categories of information seeking can be defined on the basis of their psychological antecedents. These are 'diversive information seeking', which can be related to general arousal theory, and 'specific information seeking'. We have concentrated on the latter in our discussion of shared information systems.

People cannot be relied upon to be entirely 'rational' about when they seek information, or what information they are prepared to look for, so there is a case to be made for the electronic system to take a more active role in information seeking than has been the case conventionally.

People often need to brief themselves before engaging in information seeking proper, so as to formulate their needs more clearly. They use and in an electronic environment would continue to use multiple strategies in this. Electronic systems—including 'expert systems'—might possibly play a part, either by identifying 'special communicators' the user may wish to contact for advice, or by mimicking what goes on when a person talks to a special communicator.

Once the user has formulated his or her needs, (s)he must select an appropriate source of information. Ease of use, effectiveness, cost, and other factors are all important to a greater or lesser degree depending on the situation.

Unfortunately, an adequate taxonomy of situations does not exist for predicting which factors will be important, and when.

Conventional retrieval systems perform significantly below the ideal in terms of recall and precision. More important even than this, they are passive and unintelligent, putting the onus completely on the user to tell the system exactly what is needed.

Future systems will be more intelligent, have an understanding of psychologically different types of information and their relevance to user needs. They will also be more active, taking more of an initiative in finding and presenting the user with relevant information. This poses new kinds of problems for the systems designer. It will not be possible to program future systems to be entirely 'objective' or 'value free' in what information they decide to present to the user. It will be necessary for the designer to be very clear about the value judgements that are made in programming the system.

Decision systems

The development of artificial intelligence techniques has made a new kind of communication possible, different from either Type A (person–person) or Type B (person–'paper'/'paper'–person); This is Type C—person to 'intelligent' machine. The technology is only just now emerging and little is known about the detailed psychology of what is involved. Decision systems are likely to prove to be an important application area for artificial intelligence—an area where much work needs to be done and where we can expect interesting developments during the coming five to ten years. We have outlined, in Chapter 8, an approach to the use of artificial intelligence in decision systems based on a view of decision making as a knowledge rich activity combining the skills of the decision analyst, the domain specialist and the organization analyst—vested, notionally, in the decision maker.

The future

As we move into the mid-1980s and beyond we can discern several important trends that will interact to shape the office systems of tomorrow. These include especially

- an improvement in applications software to take better account of the psychology of the intended users
- greater possibilities for applying artificial intelligence
- better integration of services at the user–system interface.

In moving forward in the development of more user-oriented systems it will be important for systems designers to be in a position to make use of the most recent research findings on the psychology of user–system interaction. They will need to have the freedom to incorporate these findings in their designs and to avoid well-intentioned but premature standardization. Whatever human factors standards are adopted need to be based on adequate research, not simple consensus.

PRODUCT USABILITY

Whatever the particular product, it is of no use if it is not usable by its intended users—yet it is only now, after two decades of computer systems, that we are finally reaching a point where real efforts are being made to design systems that are usable by ordinary people with no specialist knowledge of computers and who simply want to be able to use them as tools in their work.

Assessing usability: a psychophysiological approach

The psychophysiological approach to the assessment of usability emphasizes the need to take account of the whole user, including physiology, behaviour and experience—and not just a single (e.g. cognitive) subsystem of the person. Psychophysiological measures have been used to assess product usability in terms of stress, overload, effects on job satisfaction, illness, and relationships within work groups. Accumulating evidence from such studies suggests that—whilst it is important to strive for the best possible working conditions no matter what technology is used—possible negative effects of electronic systems *per se* (as against organizational factors, for example) have been somewhat overstressed. Some authorities believe that governments and other agencies have rushed into the introduction of legislation and new regulations on the basis of incomplete research evidence. It is, of course, important to legislate in these areas, but it is equally important to get the legislation right, and that requires obtaining sufficient information from adequate research.

Physiological, behavioural and experiential data can all provide information about the suitability of a product for use by a defined type of user in defined situations. Among the physiological measures, the following are among the more useful:

- eye movements, e.g. for assessing users' use of visual displays, and emotional responses
- pupillary responses, e.g. for assessing the information load put on the user
- EEG, e.g. for assessing information load
- heart rate acceleration–deceleration, e.g. for assessing the level of external attention required of the user, and the level of information processing required
- event-related potentials, e.g. for assessing the mental loading associated with different dialogue styles in particular situations

— and other measures identified in Chapter 9.

The psychophysiological research has pointed up the complexity of what is involved in assessing usability meaningfully. It has also, partly as a corollory to this, emphasized the importance of complementing controlled laboratory tests of usability with the detailed study of users at work, within normal working contexts, and over extended periods of sampling. This approach complements the laboratory-based work, which is also important, in various ways. In particular, it allows the design team to work *with* the user in an effort to understand the meaning of events, rather than simply to observe the user as a 'black box'. Such work requires, however, a systematic approach within a controlled (but adaptive and realistic or 'ecologically valid' environment).

The 'Cafe of Eve' is advocated as a suitably Controlled, Adaptive, Flexible, Experimental Office of the Future in an Ecologically Valid Environment where product usability can be assessed adequately, and new USI concepts developed and existing products enhanced.

The capability to assess product usability adequately is essential for making meaningful comparisons between the products offered by different suppliers, and for deciding where improvements most need to be made.

The lack of adequate usability assessment tools to date has been one important factor in slowing progress in the development of USI design guidelines and standards. How can one meaningfully decide what a guideline or standard should be if one cannot adequately assess the results of applying or not applying it?

Despite this fundamental difficulty, some progress has been made, based on a mixture of 'practical experience' and controlled laboratory experiments. Progress on the ergonomics of systems hardware has been well documented elsewhere (e.g. Cakir, Hart and Stewart, 1980) and in this book we have concentrated on dialogue design guidelines (Chapter 10) and mental models (Chapter 11).

Dialogue design guidelines

The dialogue is the structure within which the user and the system exchange messages. The user and the system use the dialogue to interact with one another. It is the dynamic aspect of the user–system interface. The design of the dialogue is therefore of crucial importance in contributing to product usability—in converting the potential benefits of the underlying technology into actual benefits for the user.

A range of different dialogue types have been described in Chapter 10. Which is appropriate to use depends upon the particular application— especially the type of user the product is intended for, and the characteristics of the task. Guidelines have been presented for four of the more widely used types of dialogues, based on:

— form filling
— menus
— function keys
— command languages.

Mental models

Dialogue design guidelines can be very helpful to the systems designer and much research effort needs to be directed towards their further development. However, in Chapter 11 we have argued that we need to go beyond such guidelines—not discarding them, but going further to consider not just the catalogue of recommendations represented by the guidelines but an overall mental model of the USI within which the specific guidelines can be applied. Many different terms relating to mental models are often used rather vaguely in the literature. We have attempted to provide some degree of clarification by distinguishing between three major types of mental models: conceptual representation models; implied dialogue models; and cognitive models. We have also pointed to the importance of consistency as a key factor in facilitating usability.

INTRODUCING SYSTEMS INTO ORGANIZATIONS

Identifying user needs and designing usable products to serve them is only part of the problem addressed by systems psychology. Unless those products are accepted into organizations and become an integral part of the total organizational system, the effort put into designing them will come to nought. We considered this aspect of the problem in Chapters 12 and 13. In Chapter 12 we presented a general framework—overall guidelines for introducing new systems, and impacts they might be expected to have. In Chapter 13 we focused more directly on one particular part of the problem within that general framework—the need to identify the appropriate range of products to bring into the organizational system to create the 'electronic office'. Much of the discussion, whilst primarily aimed at user organizations, is also of relevance to equipment manufacturers who need to develop ever better understandings of the organizational needs for which they develop their products.

It is important for a user organization to consider carefully what office systems are actually all about—what the organization hopes to achieve by introducing them—and how to avoid the triple traps associated with

— the hard-sell
— the self-fulfilling prophecy
— the straightjacket.

Equipment manufacturers can also benefit by coming to a better understanding of what types of systems really need to be designed to serve organizations better.

This is not something that can be done simply. It requires an understanding of and commitment to the reality that different groups in an organization have different frames of reference; there is no single objectively true picture of the way any particular organization works or what it needs in order to work well. In coping with this actual complexity (rather than some simplistic theory of what the organization 'ought' to be like), it is necessary to take a complex approach. A threefold approach has been discussed which includes

- formulation and discussion of key organizational issues with the various groups involved
- consideration of the interrelationships between technology, tasks and people
- involvement of individuals in a consideration of the benefits of new systems for them.

Some practical guidelines are presented which reflect the general approach advocated and which supplement the more general guidelines presented in Chapter 12.

CONCLUSIONS

In this book we have only been able to scrape the surface of what must eventually turn out to be one of the largest areas of applied research ever undertaken. The Alvey and ESPRIT programmes recognize that the economies of Europe, the USA and Japan depend crucially on 'information technology'. The success of this technology—however sophisticated it may be in purely 'technical' terms, and whatever the size of the investment in optimizing the technical aspects—the success of the technology in practice depends fundamentally on how well it is matched to the needs of the humans it is designed to serve. It depends on solutions to the problems of which we have only been able to give a flavour in this book. Ultimately, however, it is not just economic success or failure that is at stake. When the missiles are launched they will be under computer control and will be a blazing symbol of the failure of humans to communicate. The very survival of our species and many others depends upon us developing the capability to use knowledge wisely.

References

(*See also the bibliographies at the end of Chapters 1 and 10*)

Adams, J. S. (1965). Inequity in social exchange. In: L. Berkoivitz (Ed.) *Advances in Experimental Social Psychology*. London: Academic Press, 276–299.

de Alberdi, M. J. I. (1982). *More Opinion Change Over Audio: Process or Pseudo-process?* Paper presented to the British Psychological Society, Social Psychology Section, Annual Conference, Edinburgh. 24–26th September.

Allen, R. B. (1982). Cognitive factors in human interaction with computers. *Behaviour and Information Technology*, **1**, (3), 257–278.

Allport, D. A. (1980). Patterns and actions: cognitive mechanisms are content-specific. In: A. Claxton (Ed.) *Cognitive Psychology*. London: Routledge and Kegan Paul, 26–64.

Allport, F. H. (1920). The influence of the group upon association and thought. *Journal of Experimental Psychology*, **3**, 159–182.

Anshen, M. (1962). Managerial decisions. In J. T. Dunlop (Ed.) *Automation and Echnological Change*. Englewood Cliffs, N.J.: Prentice Hall, 56–83.

Argyle, M. (1969). *Social Interaction*. London: Methuen.

Argyle, M., and Dean, J. (1965). Eye contact, distance and affiliation. *Sociometry*, **28**, 289–304.

Argyle, M., Furnham, A., and Graham, J. A. (1981). *Social Situations*. Cambridge: Cambridge University Press.

Argyle, M., and McHenry, R. (1971). Do spectacles really affect judgements of intelligence? *British Journal of Social and Clinical Psychology*, **10**, 27–29.

Baecker, R. (1979). Human–Computer interactive systems: a state-of-the-art review. *Proceedings of the 2nd International Conference on the Processing of Visible Language.*

Baird, J. E., Jr. (1977). *The Dynamics of Organizational Communication*. London: Harper and Row.

Bales, R. F. (1955). How people interact in conferences. *Scientific American*, March, 3–7.

Ballantine, M. (1980). Conversing with computers—the dream and the controversy. *Ergonomics*, **23**, (9), 935–945.

Barnard, P. J. and Hammond, N. V. (1982). Usability and its multiple determination for the occasional user of interactive systems. In: M. B. Williams (Ed.), *Pathways to the Information Society*. North-Holland, 543–548. Also available as IBM Human Factors Report HF059*.

Barnard, P. J., Hammond, N. V., MacLean, A. and Morton, J. (1982). Learning and remembering interactive commands in a text-editing task. *Behaviour and Information Technology*, **1**, 347–358. Also available as IBM Hursley Human Factors Report HF055*.

Barnard, P., Hammond, N. V., Morton, J., Long, J.B., and Clark, I. A. (1981). Consistency and compatibility in human–computer dialogue. *International Journal of Man–Machine Studies*, **15**, (1), 87–134.

Bateman, T. S. (1980). *A Longitudinal Investigation of Role Overload and its Relationship with Work Behaviours and Job Satisfaction*. DBA Dissertation, Indiana University, Graduate School of Business. *DAI*, **41**, (7), 3175-A.

Beatty, J. (1982). Task-evoked pupillary responses, processing load, and the structure of processing resources. *Psychological Bulletin*, **91**, 276–292.

Beech, D. (1982). Criteria for a standard command language based on data abstraction. *AFIPS Proceedings of the 1982 National Computer Conference*, **51**, 493–499.

Bennett, J. L. (1982). Managing to meet usability goals in the development of office systems. Office Automation Conference 1982.

Berlyne, D. E. (1960). *Conflict, Arousal and Curiosity*. London: McGraw-Hill.

Berlyne, D. E., and McDonnell, P. (1965). Effects of stimulus complexity and incongruity on duration of EEG desynchronization. *Electroencephalography and Clinical Neurophysiology*, **18**, 156–161.

Berman, H. J., Shulman, A. D., and Marwit, S. J. (1975). Comparison of multidimensional decoding of affect for audio, video and audio video recordings. *Sociometry*, 38.

Bertrand, U. S. (1981). *Personal and Organizational Correlates of Role Stress and Job Satisfaction in Female Managers*. PhD. Dissertation, The University of Wisconsin, Madison. *DAI*, **42**, (3), 1051-A.

Birchall, D. W., and Hammond, V. J. (1981). *Tomorrow's Office Today: Managing Technological Change*. London: Business Books.

Birdwhistell, R. L. (1970). *Kenesics and Context*. Philadelphia: University of Philadelphia Press.

Bjorn-Anderson, N. (1983). The changing roles of secretaries and clerks. In: H. J. Otway and M. Peltu (Eds) *New Office Technology: Human and Organizational Aspects*. London: Frances Pinter (Publishers) for the Commission of the European Communities, pp. 120–137.

Black, J. and Moran, T. (1982). *Learning and Remembering Command Names*. Paper presented at the Conference on Human Factors in Computer Systems, Gaithersburg, Maryland.

Black, J. B., and Sebrechts, M. M. (1981). Facilitating human–computer communication. *Applied Psycholinguistics*, **2**, (2), 149–177.

Blythe, P. (1973). *Stress Disease: The Growing Plague*. London: Arthur Barker.

Bo-Linn, C. (1980). *Implementation of a Computer-Based Management Information System: An Analysis of Organizational Change*. Ed.D. Dissertation, University of Houston. *DAI*, **41**, (5), 1854-A.

Bolt, (1979). *Spatial Data Management System*. Library of Congress No. 78-78256.

Borman, L., and Mittman, B. (1972). Interactive search of bibliographic data bases in an academic environment. *Journal of the American Society for Information Science*, May–June, 164–171.

Bouma, H. (1982). Visual reading processes and the quality of text displays. In: E. Grandjean and E. Vigliani (Eds.) *Ergonomic Aspects of Visual Display Terminals*. London: Taylor and Francis.

Bower, G. H. (1961). Application of a model to paired-associate learning. *Psychometrica*, **26**, 255–280.

Bower, G. H., Monteiro, K. P., and Gilligan, S. G. (1978). Emotional mood as a context for learning and recall. *Journal of Verbal Learning and Verbal Behaviour*, **17**, 573–587.

Bransford, J. D., and Johnson, M. K. (1972). Contextual prerequistion for understanding some investigations of comprehension and recall. *Journal of Verbal Learning and Verbal Behaviour*, **61**, 717–726.

Brauer, M. A. (1980). *Role Conflict, Role Ambiguity, and Job Satisfaction: A Study of Middle Management Positions in an Educational Organization*. PhD. Dissertation, The University of Nebraska, Lincoln. *DAI*, **41**, (7), 2845-A.

Brown, I. D. (1982). Measurement of mental effort: some theoretical and practical issues. In: G. A. Harrison (Ed.) *Energy and Effort*. London: Taylor and Francis.

Brown, I. D., and Poulton, E. C. (1961). Measuring the spare 'mental capacity' of car drivers by a subsidiary task analysis. *Ergonomics*, **4**, 35–40.

Buchanan, D. A., and Boddy, D. (1983). *Organizations in the Computer Age: Technological Imperatives and Strategic Choice*. Aldershot: Gower Publishing Company Ltd.

Burlingame, J. F. (1958). Thinking ahead: how near is the automatic office? *Harvard Business Review*, **36**, (2).

Bush, G. (1977). *Viewdata: Design of Structure of Post Office Routing Pages*. Interim Report. Unpublished British Telecom Report TSS6.2.

Bush, G., and Williams, E. (1978). *Viewdata: The Systematic Development and Testing of Post Office Routing Trees*. Unpublished manuscript, February 28, 1978: British Telecom.

Butler–Cox (1982). *The Market for Office Technology*. A Multiclient Study by Butler Cox, London.

Cadbury, Sir Adrian (1981). *The Times*, December 9, 1981.

Cakir, A., Hart, D. J., and Stewart, T. F. M. (1980). *Visual Display Terminals: A Manual Covering Ergonomics, Workplace Design, Health and Safety, and Task Organization*. Chichester/New York: John Wiley and Sons.

Campbell, J. P., and Pritchard, R. D. (1976). Motivation theory in industrial and organizational psychology. In: M. D. Dunnette (Ed.) *Handbook of Industrial and Organizational Psychology*. Chicago: Rand McNally, 1976, 63–130.

Caplan, R. D., Cobb, S., French, J. R. P., Harrison, R. V. and Pinneau, S. R. Jr. (1975). *Job Demands and Worker Health: Main effects and occupational differences*. Department of Health, Education and Welfare, NIOSH publication No. 75-160. Washington, D.C.: U.S. Government Printing Office.

Carbonnel, J. G., Michalski, R. S., and Mitchell, T. M. (1983). Machine learning, a Historical and methodological analysis. *AI Magazine*, **IV** (Fall), 69–79.

Card, S. K., Moran, T. P., and Newell, A. (1983). *The Psychology of Human–Computer Interaction*. London/New Jersey: Lawrence Erlbaum Associates.

Carroll, J. M. (1982). Learning, using and designing filenames and command paradigms. *BIT*, **1**, (4), 327–346.

Carroll, J. M., and Mack, R. L. (1982). *Metaphor, Computing Systems and Active Learning*. IBM Research Report, Yorktown Heights, N.Y.

Carroll, S. M., and Thomas, J. (1980). *Metaphor and the Cognitive Representation of Computing Systems*. IBM Research Report, RC-8302, Yorktown Heights, N.Y.

Champness, B. G. (1972a). *Attitudes Towards Person–Person Communication Media*. Unpublished Communications Studies Group paper no. E/72011/CH, available from British Telecom, 88 Hills Road, Cambridge.

Champness, B. G. (1972b). *Feelings Towards Media in Group Situations*. Unpublished Communications Studies Group paper no. E/72160/CH, available from British Telecom, 88 Hills Road, Cambridge.

Champness, B. G. (1972c). *The Perceived Adequacy of Four Communications Systems for a Variety of Tasks*. Unpublished Communications Studies Group paper no. E/72245/CH, available from British Telecom, 88 Hills Road, Cambridge.

Champness, B. G. (1973). *The Assessment of Users' Reactions to Confravision: II Analysis and Conclusions*. Unpublished Communications Studies Group paper no. E/73250/CH, available from British Telecom, 88 Hills Road, Cambridge.

Chandler, A. (1977). *The Visible Hand*. Harvard University Press.

Chapanis, A. (1971). The search for relevance in applied research. In: W. T.

Singleton, J. G. Fox and D. W. Whitfield (Eds.) *Measurement of Man at Work: An Appraisal of Physiological and Psychological Criteria in Man–Machine Systems.* London: Taylor and Francis.

Chapanis, A., Ochsman, R., Parrish, R., and Weeks, G. (1972). Studies in interactive communication: the effects of four communication modes on the behaviour of teams during cooperative problem solving. *Human Factors,* 14, 487–509.

Cherry, C. (1978). *World Communications: Threat or Promise? A Socio-technical Approach.* Chichester/New York: John Wiley and Sons Ltd.

Christie, B. (1973a). *The 1972/3 New Rural Society Project, Appendices M and N.* Unpublished report available on request from the US Department of Housing and Urban Development, Washington, D.C. Reference H-1694.

Christie, B. (1973b). *An Evaluation of the Audio-Video Conference System Installed in the Department of the Environment.* Unpublished Communications Studies Group paper no. W/73360/CR, available from British Telecom, 88 Hills Road, Cambridge.

Christie, B. (1974). Perceived usefulness of person–person telecommunications media as a function of the intended application. *European Journal of Social Psychology,* 4, (3), 366–368.

Christie, B. (1981). *Face To File Communication: A Psychological Approach. To Information Systems.* Chichester/New York: John Wiley and Sons Ltd.

Christie, B., Delafield, G., Lucas, B., Winwood, M., and Gale, A. (1972). Stimulus complexity and the EEG: differential effects of the number and variety of display elements. *Canadian Journal of Psychology,* 26, (2), 155–170.

Christie, B., and Holloway, S. (1975). Factors affecting the use of telecommunications by management. *Journal of Occupational Psychology,* 48, 3–9.

Christie, B., and Kingan, S. (1975). Electronic alternatives to the business meeting: managers' choices. *Journal of Occupational Psychology,* 50, 265–273.

Churchman, C., and West, (1968). *The Systems Approach.* New York: Delta Publishing Company.

Cole, I. (1982). Human aspects of office filing: implications for the electronic office. *Proceedings of the Human Factors Society,* 26th Annual Meeting, Seattle USA.

Coles, M. G. H. (1983). Situational determinants and psychological significance of heart rate change. In: A. Gale and J. Edwards (Eds.) *Physiological Correlates of Human Behaviour. Vol. 2 Attention and Performance.* London: Academic Press.

Collins, H. (1972). *The Telecommunications Impact Model: Stages I and II.* Unpublished Communication Studies Group paper no. P/72356/CL, available from British Telecom, 88 Hills Road, Cambridge.

Conroy, T. R., and Ewbank, R. V. K. (1980). *Perspectives of Interfacing People with Technology in the Development of Office Automation.* Paper presented at the International Communication Association Convention, Acapulco, Mexico, May 1980.

Cooper, C. L. (1980). Work stress in white- and blue-collar jobs. *Bulletin of the British Psychological Society,* 33, 49–51.

Cooper, C. L., and Marshall, J. (1978). Sources of managerial and white collar stress. In: C. L. Cooper and R. Payne (Eds.) *Stress at Work.* Chichester/New York: John Wiley and Sons Ltd., 81–105.

Coulouris, G. F. (1982). Designing interactive systems for the office of the future. *Behaviour and Information Technology,* 1, (1), 37–42.

Cox, T., Cox, S., and Thirlaway, M. (1983). The psychological and physiological response to stress. In: A. Gale and J. Edwards (Eds.) *Physiological Correlates of Human Behaviour: Vol. 1. Basic Issues.* London: Academic Press.

Craik, F. I. M., and Lockhart, R. S. (1972). Levels of processing: a framework for memory research. *Journal of Verbal Learning and Verbal Behaviour,* II, 671–684.

Craik, F. I. M., and Tulving, E. (1975). Depth of processing and the retention of words in episodic memory. *Journal of Experimental Psychology: General*, **104**, 268–294.

Cuff, R. N. (1980). On casual users. *International Journal of Man–Machine Studies*, **12**, 163–187.

Dainoff, M. J., Hurrell, J. J. Jr., and Happ, A. (1981). A taxonomic framework for the description and evaluation of paced work. In: G. Salvendy and M. J. Smith (Eds.) *Machine Pacing and Occupational Stress*. London: Taylor and Francis.

Dashiell, J. F. (1935). Experimental studies of the influence of social situations on the behaviour of individual human adults. In: C. Murchison (Ed.) *Handbook of Social Psychology*. Massachusetts: Clark University Press.

Deutsch, M., and Gerard, H. B. (1955). A study of normative and informational social influence upon individual judgement. *Journal of Abnormal Social Psychology*, **51**, 629–636.

Dorris, J. W., Gentry, G. C., and Kelley, H. H. (1972). *The Effects on Bargaining of Problem Difficulty, Mode of Interaction, and Initial Orientation*. Unpublished paper, University of Massachusetts.

Doswell, A. (1983). *Office Automation*. Chichester/New York: John Wiley and Sons.

Douglas, A. (1957). The peaceful settlement of industrial and intergroup disputes. *Journal of Conflict Resolution*, **1**, 69–81.

Douglas, S. (1982). *What Features Make a Text Editor Easy to Learn?* Office Automation Conference 1982.

Eason, K. D. (1976). Understanding the naive computer user. *The Computer Journal*, **19**, (1), 3–7.

Eason, K. D. (1981). *Manager–Computer Interaction: A study of a task–tool relationship*. PhD. Thesis; Loughborough University of Technology, UK.

Eberts, R. E. (1982). The development of an accurate internal model for high order systems. *Proceedings of the Human Factors Society 26th Annual Meeting*, Seattle, USA, 1982.

Eliot, T. S. (1934). *The Rock*. London: Faber and Faber.

Engel, G. M., Groppuso, J., Lowenstein, R. A., and Traub, W. G. (1979). An office communications system. *IBM Systems Journal*, **18**, (3), 402–431.

English, R. W., and Jelenevsky, S. (1971). Counsellor behaviour as judged under audio, visual, and audiovisual communication conditions. *Journal of Counselling Psychology*, **18**, 509–513.

Eosys (1982). *Office Structures and Design*. Eosys Ltd., Clove House, The Broadway, Farnham Common, Slough.

Evans, J. (1983). Negotiating technological change. In: H. J. Otway and M. Peltu (Eds.) *New Office Technology: Human and Organizational Aspects*. London: Frances Pinter (Publishers) for the Commission of the European Communities, 152–168.

Evans, L. (1981). An experiment: search strategy variations in SDI profiles. In: K. Sparck Jones (Ed.) *Information Retrieval Experiment*. London: Butterworths, 285–315.

Eysenck, H. J. (1970). *The Structure of Human Personality*. London: Methuen.

Eysenck, M. W. (1982). *Attention and Arousal: Cognition and Performance*. Berlin: Springer-Verlag.

Feigenbaum, E. A., and McCorduck, P. (1983). *The Fifth Generation: Artificial Intelligence And Japan's Computer Challenge To The World*. London/California: Addison-Wesley.

Fishbein, M., and Ajzen, I. (1975). *Belief, Attitude Intention and Behaviour: An Introduction to Theory and Research*. Reading, Mass: Addison-Wesley.

Fitter, M. (1979). Towards more 'natural' interactive systems. *International Journal of Man–Machine Systems*, **11**, 339–350.

Frankenhaeuser, M., Nordheren, B., Myrsten, A., and Post, B. (1971). Psychophysiological reactions to under-stimulation and over-stimulation. *Acta Psychologica*, **35**, 298–308.

Furnas, G., Gomez, L., Landauer, T., and Dumais, S. (1982). *Statistical Semantics: How can a Computer use what People name Things to Guess what Things People mean when they Name Things?* Paper presented at the Conference on Human Factors in Computer Systems, Gaithersburg, Maryland.

Gale, A. (1969). 'Stimulus hunger': individual differences in operant strategy in a button-pressing task. *Behaviour Research and Therapy*, **18**, 156–161.

Gale, A., and Chapman, A. J. (1984). The nature of applied psychology. In: A. Gale and A.J. Chapman (Eds.) *Psychology and Social Problems: An Introduction to Applied Psychology.* Chichester: John Wiley and Sons.

Gale, A., and Edwards, J. (1983a). A short critique of the psychophysiology of individual differences. *Personality and Individual Differences*, **4**, 429–435.

Gale, A., and Edwards, J. (1983b). *Physiological Correlates of Human Behaviour. Vol. 3. Individual Differences and Psychopathology.* London: Academic Press.

Gale, A., and Edwards, J. (1983c). The EEG and human behaviour. In: A. Gale and J. Edwards (Eds.) *Physiological Correlates of Human Behaviour. Vol. 2. Attention and Performance.* London: Academic Press.

Gale, A., and Edwards, J. (1983d). Psychophysiology and individual differences: theory, research procedures and the interpretation of data. *Australian Journal of Psychology*, **35**, 361–379.

Gale, A., and Edwards, J. (1984). Individual differences. In: M. G. H. Coles, E. Donchin and S. W. Porges (Eds.) *Psychophysiology: Systems, Processes and Applications.* New York: Guilford.

Geidt, F. H. (1955). Comparison of visual content and auditory cues in interviewing. *Journal of Consulting Psychology*, **19**, 407–419.

Gentner, D., and Gentner, D. R. (1983). Flowing waters or teeming crowds: mental models of electricity. In: D. Gentner and A. L. Stevens (Eds.) *Mental Models.* Hillsdale, New Jersey: Erlbaum, 99–130.

Gentner, D., and Grudin, J. (1983). *Ninety Years of Mental Metaphors.* Paper given at 5th Annual Conference of the Cognitive Science Society, Rochester, New York 1983.

Gilb, T. (1977). *Software Metrics.* Massachusetts: Winthrop Publishers.

Goffman, E. (1976). *Interaction Ritual.* London: Allan Lane Penguin Press.

Goldmark, P. C. (1972a). Tomorrow we will communicate to our jobs. *The Futurist*, April, 55–57.

Goldmark, P. C. (1972b). Communication and the community. *Scientific American*, September. Reprinted in: Communication—A Scientific American Book (1972). San Francisco: W.H. Freeman and Company, 101–108.

Goldmark, P. C. (1973). *The New Rural Society.* Report from the Department of Communication Arts, Cornell University, Ithaca, New York. (Reference: Papers on Communication, No. 5).

Grandjean, E., and Vigliani, E. (1982). *Ergonomic Aspects of Visual Display Terminals.* London: Taylor and Francis.

Gratz, R. D., and Salem, P. J. (1981). *Organisational Communication and Higher Education.* University of Illinois Press.

Hackman, J. R., and Lawler, E. E. (1971). Employee reactions to job characteristics. *Journal of Applied Psychology*, **55**.

Haider, M., Kundi, M., and Weissenbock, M. (1982). Worker strain related to VDU's with differently coloured characters. In: E. Grandjean and E. Vigliani (Eds.) *Ergonomic Aspects of Visual Display Terminals.* London: Taylor and Francis.

Halasz, F., and Moran, T. P. (1982). Analogy considered harmful. In: *Human Factors In Computer Systems*, Gaithersberg, USA.

Hale, C. R. (1982). Representing human cognition in complex man–machine environments. *Proceedings of the Human Factors Society 26th Annual Meeting*, Seattle, USA, 1982.

Hall, E. T. (1963). A system for the notation of proxemic behaviour, *American Anthropologist*, **65**, 1003–1026.

Hammond, N., and Barnard, P. (1982). *Knowledge Fragments and User's Models of Systems*. Hursley Human Factors Laboratory, Report No. HF071.

Hardman, Sir. H. (1973). *The Dispersal of Government Work from London*. London: Her Majesty's Stationery Office.

Heilbronn, M., and Libby, W. L. (1973). *Comparative Effects of Technological and Social Immediacy upon Performance and Perceptions during a Two-Person Game*. Paper read at the Annual Convention of The American Psychological Association, Montreal.

Helmer, O. (1966). *Social Technology*. New York: Basic Books.

Hendler, J., Kehler, T. P., Michaelis, P. R., Phillips, B., Ross, K. M., and Tennant, H. R. (1981). Issues in the development of natural language front-ends. *AFIPS Conference Proceedings*, **50**, 643–647.

Hepworth, A., and Osbaldeston, M. (1979). *The Way We Work: a European Study of Changing Practice in Office Job Design*. Saxon House, Farnborough, England.

Herzberg, F., Mausner, B., and Snyderman, B. (1959). *The Motivation to Work*. London: John Wiley.

Hess, E. H. (1965). Attitude and pupil size. *Scientific American*, **22**, 46–55.

Hitchen, M., Brodie, D. A., and Harness, J. B. (1980). Cardiac responses to demanding mental load. *Ergonomics*, **23**, 379–385.

Holland, W. F. (1972). Information potential: a concept of the importance of information sources in a research and development environment. *The Journal of Communication*, **22**, 159–173.

Hollnagel, E. (1983). What we do not know about man–machine systems. *International Journal of Man–Machine Studies*, **18**, 135–143.

Hoos, I. R. (1960). When the computer takes over the office. *Harvard Business Review*, **38**, (4).

Hopwood, A. (1974). *Accounting and Human Behaviour*. Englewood Cliffs, N.J.: Prentice-Hall.

Hopwood, A. (1983). Private correspondence.

House, W. C. (1971). *The Impact of Information Technology*. Pennsaukem, New Jersey: Auerbach.

Howarth, C. I. (1980). The structure of effective psychology: man as problem-solver. In: A. J. Chapman and D. M. Jones (Eds.) *Models of Man*. Leicester: The British Psychological Society.

Humphreys, P., Wooler, S., Phillips, L. D. (1980). *Structuring Decisions: The Role of Structuring Heuristics*. Technical Report 80-1, Decision Analysis Unit, Brunel University.

Hunting, W., Laubli, Th., and Grandjean, E. (1982). Constrained postures of VDU operators. In: E. Grandjean and E. Vigliani (Eds.) *Ergonomic Aspects of Visual Display Terminals*. London: Taylor and Francis.

Hyde, T. S., and Jenkins, J. J. (1973). Recall for words as a function of semantic, graphic, and syntactic orienting tasks. *Journal of Verbal Learning and Verbal Behaviour*, **12**, 471–480.

Hyland, M. (1981). *Introduction to Theoretical Psychology*. London: Macmillan Press.

IMR (1982). *Offices in Britain*. Industrial Market Research Ltd., 17 Buckingham Gate, London SW1. Klein, J. G. (1982).

Israel, J. B., Wickens, C. D., Chesney, G. L., and Donchin, E. (1980). The event-related brain potential as an index of display-monitoring workload. *Human Factors*, **22**, 211–224.

Jahoda, M. (1979). The impact of unemployment in the thirties and the seventies. *Bulletin of the British Psychological Society*, **32**, 309–314.

Janis, I. L., and Mann, L. (1977). *Decision Making: A Psychological Analysis of Conflict, Choice, and Commitment.* London: Collier Macmillan.

Jarrett, D. (1982). *The Electronic Office: A Management Guide To The Office Of The Future.* Aldershot: Gower Publishing Company Ltd.

Jenner, D. A., Reynolds, V., and Harrison, G. A. (1980). Catecholamine excretion rates and occupation. *Ergonomics*, **23**, 237–246.

Johansson, G., Aronsson, G., and Lindstrom, B. O. (1978). Social psychological and neuroendocrine stress reactions in highly mechanized work. *Ergonomics*, **21**, 583–599.

Kahneman, D., Slovic, P., and Tversky, A., (1982). *Judgement under Uncertainty: Heuristics and Biases.* Cambridge: Cambridge University Press.

Kak, A. V. (1981). Stress: an analysis of physiological assessment devices. In: G. Salvendy and M. J. Smith (Eds.) *Machine Pacing and Occupational Stress.* London: Taylor and Francis.

Kalimo, R., and Leppanen (1981). Mental strain in computerized and traditional text preparation. In: G. Salvendy and M. J. Smith (Eds.) *Machine Pacing and Occupational Stress.* London: Taylor and Francis.

Kalsbeek, J. W. H., and Sykes, R. N. (1967). Objective measurement of mental load. *Acta Psychologica*, **27**, 253–161.

Kano, S. (1977). A change of effectiveness of communication networks under different amounts of information. *Japanese Journal of Experimental Social Psychology*, **17**, (1), 50–59.

Kantor, J. R. (1978). Man and machines in psychology: cybernetics and artificial intelligence. *Psychological Record*, **28**, (4), 575–583.

Katz, D., and Kahn, R. L. (1978). *The Social Psychology of Organizations (2nd. edition).* Chichester/New York: John Wiley and Sons Ltd.

Keen, E. M. (1981). Laboratory tests of manual systems. In: K. Sparck Jones (Ed.) *Information Retrieval Experiment.* London: Butterworth, 136–155.

Keen, P. G. W. (1979). *Decision Support Systems and the Marginal Economics of Effort.* Center for Information Systems Research, MIT Report CISR/48.

Keen, P. G. W. (1980). *Decision Support Systems and the Marginal Economics of Effort.* CISR WP 48. Centre for Information Systems Research MIT.

Keen, P. G. W. (1983). Strategic planning for the new system. In: H. J. Otway and M. Peltu (Eds.) *New Office Technology: Human and Organizational Aspects.* London: Francis Pinter (Publishers) for the Commission of the European Communities, 51–67.

Keen, P. G. W., and Gambino, T. J. (1980). *Building a Decision Support System: The Mythical Man Month Revisited.* CISR WP57. Centre for Information Systems Research MITT.

Keen, P. G. W., and Scott Morton, M. S. (1978). *Decision Support Systems: An Organizational Perspective.* London: Addison-Wesley.

Kent, A. (1974). Unsolvable problems. In: A. Debons (Ed.) *Information Science: Search for Identity.* New York: Marcel Dekker, 299–311.

Klein, J. G. (1982). *The Office Book.* London: Frederick Muller.

Klemmer, E. T., and Stocker, L. P. (1971). *Picturephone Versus Speakerphone for Conversation Between Strangers.* Unpublished company data.

Kline, P. (1983). *Personality Measurement and Theory.* London: Hutchinson University Library.

Kling, R. (1983). Social goals in systems planning and development. In: H. J. Otway and M. Peltu (Eds.) *New Office Technology: Human and Organizational*

Aspects. London: Frances Pinter (Publishers) for the Commission of the European Communities, 221–237.

Krueger, H. (1982). Opthalmological aspects of work with display work-stations. In: E. Grandjean and E. Vigliani (Eds.) *Ergonomic Aspects of Visual Display Terminals.* London: Taylor and Francis.

Kruglanski, A. W., and Ajzen, I. (1983). Bias and Error in Human Judgement. *European Journal of Social Psychology,* **13**, 1–44

Kutas, M., and Donchin, E. (1980). Preparation to respond as manifested by movement-related brain potentials. *Brain Research,* **202**, 95–115.

Lacey, J. I. (1967). Somatic response patterning and stress: some revisions of activation theory. In: M. H. Appley and R. Trumbull (Eds.) *Psychological Stress: Issues in Research.* New York: Appleton Century Crofts.

Lamble, G. W. (1980). *Role Conflict and Ambiguity of Agricultural Extension Agents.* PhD. Dissertation, Texas A & M University. *DAI,* **41**, (4), 1346-A.

Lamming, M. (1979). *An Office Information System Design Study.* An unpublished paper from the Computer Systems Laboratory, Queen Mary College, Mile End, London E1 4NS.

Landauer, T. K., Dumais, S. T., Gomez, L. M., and Furnas, G. W. (1982). Human Factors in Data Access. *Bell System Technical Journal,* **61**, (9).

Lang, K., and Auld, R. (1982). The goals and methods of computer users. *International Journal of Man–Machine Studies,* **17**, 375–399.

Lansdale, M. W. (1983). Long term recency effects in recalling precision answers. *Proceedings of the 5th Annual Conference of The Cognitive Science Society,* Rochester, New York.

Laplante, D. (1971). *Communication, Friendliness, Trust and the Prisoner's Dilemma.* M.A. Thesis, University of Windsor.

Leavitt, H. J., and Whisler, T. L. (1958). Management in the 1980s. *Harvard Business Review,* **36**, (6), 41–48.

Levenson, R. W., and Gottman, J. M. (1983). Marital interaction: physiological linkage and affective exchange. *Journal of Personality and Social Psychology,* **45**, 587–597.

Long, J., Hammond, N., Barnard, P., Morton, J., and Clark, I. (1982). *Introducing the Interactive Computer at Work. The Users Views.* Hursley Human Factors laboratory Report HF060.

Loveless, N. (1983). Event-related brain potentials and human performance. In: A. Gale and J. Edwards (Eds.) *Physiological Correlates of Human Behaviour. Vol. 2. Attention and Performance.* London: Academic Press.

Lynes, T. (1981). Realism on pensions, *New Society,* Feruary 12, 285.

McDermott, J. (1980). R1: The formative years. *AI Magazine,* **2**, 21–29.

Mackay, C. J. (1980). The measurement of mood and psychophysiological activity using self-report techniques. In: I. Martin and P.H. Venables (Eds.) *Techniques in Psychophysiology.* Chichester: John Wiley and Sons.

Mackay, C. J., and Cox, T. (1979). *Response to Stress: Occupational Aspects.* London: International Publishing Corporation.

Maguire, M. (1982). An evaluation of published recommendations on the design of man–computer dialogues. *International Journal of Man–Machine Studies,* **16**, (3), 237–262.

Maier, N. R. F., and Thurber, J. A. (1968). Accuracy of judgements of deception when an interview is watched, heard and read. *Personnel Psychology,* **21**, 23–30.

Malone, T. W. (1983). How do people organize their desks? Implications for the design of office information systems. *ACM Transactions on Office Information Systems,* **1**, (1), 99–112.

Maslow, A. H. (1965). *Eupsychian Management.* Irwin.

Mayer, R. E. (1979). Can Advance Organisers Influence Meaningful Learning?

Review of Educational Research, **49**, 371–383.

Mayer, R. E. (1981). The psychology of how novices learn computer programming. *Computing Surveys,* **13**, 121–141.

Mayer, E., and Bayman, P. (1981). Psychology of calculator languages: a framework for describing differences in user's knowledge. *Communications of the ACM,* **24**, (8).

Mead, M. (1965). The city as a point of confrontation. In: *Transactions of the Bartlett Society, Vol. 3, 1964–65.* Bartlett School of Architecture, University College London, 9–22.

Meadow, C. T. (1970). *Man–Machine Communication.* Chichester: John Wiley and Sons.

Michalski, R. S., Carbonnel, J. G., and Mitchell, T. M. (Eds.)(1983). *Machine Learning, an Artificial Intelligence Approach.* Tioga Press.

Michie, D. *Japan Fifth Generation Plan: Themes for the UK.* Presented at Fifth Generation Conference, July 1982, London.

Milgram, S. (1965). Some conditions of obedience to authority. *Human Relations,* **18**, 57–75.

Minzberg, H. (1973). *The Nature of Managerial Work.* Harper and Row.

Moran, T. P. (1981). The command language grammar: a representation for the user interface of interactive computer systems. *International Journal of Man–Machine Studies,* **15**, 5–30.

Morley, I. E., and Stephenson, G. M. (1969). Interpersonal and interparty exchange: a laboratory simulation of an industrial negotiation at the plant level. *British Journal of Psychology,* **60**, 543–545.

Morse, N., and Reimer, E. (1956). The experimental change of a major organisational variable. *Journal of Abnormal and Social Psychology,* **52**, 120–129.

Mouton, J. S., Blake, R. R., and Olmstead, J. A. (1956). The relationship between frequency of yielding and the disclosure of personal identity. *Journal of Personality,* **24**, 339–347.

Murakami, K. (1983). Archery discipline and fifth generation computer research. *ICOT Journal,* No. 2 (September). Tokyo 108, Japan: Institute for New Generation Computer Technology.

Nickerson, R. S. (1981). Why interactive systems are sometimes not used by people who might benefit from them. *International Journal of Man–Machine Studies,* **15**, 469–483.

Nisbett, R. E., and Ross, L. (1980). *Human Inference: Strategies and Shortcomings of Social Judgement.* Englewood Cliffs, N.J.: Prentice-Hall.

Noll, A. M. (1977). *Teleconferencing Communications Activities.* IEE Communications Society, November.

Norman, (1981). The trouble with UMX. *Datamation,* November, 139.

Norman, D. A. (1983a). Some observations on mental models. In: D. Gentner and A. L. Stevens (Eds.) *Mental Models.* Hillsdale, New Jersey: Lawrence Erlbaum Associates, 7–14.

Norman, D. A. (1983b). *Learning and Memory.* San Francisco: Freeman & Co.

O'Connor, K. (1983). Individual differences in components of slow cortical potentials: implications for models of information processing. *Personality and Individual Differences,* **4**, 403–410.

O'Connor, R. O. O. (1976). *Corporate Guides to Long-range Planning.* Conference Board Report No. 687, New York.

O'Gorman, J. (1977). Individual differences in habituation of human physiological responses: a review of theory, method and findings in the study of personality correlates in non-clinical populations. *Biological Psychology,* **5**, 257–318.

O'Hanlon, J. F. (1981). Stress in short-cycle repetitive work: general theory and empirical test. In: G. Salvendy and M. J. Smith (Eds.) *Machine Pacing and*

Occupational Stress. London: Taylor and Francis.

O'Reilly, C. A., III (1980). Individuals and information overload in organisations: Is more necessarily better? *Academy of Management Journal*, **23**, (4), 684–696.

Ogburn, W. F. (1964). Cultural lag as theory. In: O. D. Duncan (Ed.) On *Culture and Social Change*. University of Chicago Press.

Orcott, J. D., and Anderson, R. E. (1974). Human–computers relationships, interactions and attitudes. *Behaviour Research Methods and Instrumentation*, **6**, (2), 219–222.

Osborn, A., and Ramsey, R. (1918). *The Optimism Book for Offices*. Art-Metre.

Östberg, O. (1982). Accommodation and visual fatigue in display work. In: E. Grandjean and E. Vigliani (Eds.) *Ergonomic Aspects of Visual Display Terminals*. London: Taylor and Francis.

Oster, P. J., and Stern, J. A. (1980). Measurement of eye movement electrooculography. In: I. Martin and P. H. Venables (Eds.) *Techniques in Psychophysiology*. Chichester: John Wiley and Sons.

PACTEL (1981). *A Strategy For Information Technology*. A published summary of a report prepared for the National Enterprise Board. PACTEL, 33 Greycoat Street, London SW1.

Panko, R. R. (1982). *Proceedings of the Third American Federation of Information Processing (AFIPS)*, San Francisco.

Parker, E. G. (1976). Social implications of computer/telecoms systems. *Telecommunications Policy*, **1**, (1), 3–20.

Payne, R. (1981). Organizational behaviour. In: C. L. Cooper (Ed.) *Psychology and Management: A Text for Managers and Trade Unionists*. London: Macmillan Press Ltd., 155–197.

Peace, D., and Easterby, R. (1973). The evaluation of user interaction with computer-based management systems. *Human Factors*, **15**, (2), 163–177.

Pessin, J. (1933). The comparative effects of social and mechanical stimulation on memorising. *American Journal of Psychology*, **45**, 263–270.

Phillips, D. A., and Treuniet, W. C. (1978). *Man–Machine Interactions in the Hermes Experiments, Vol. III.* Report No. CRC 1320-3E, Communications Research Center, Department of Communications, Ottawa.

Post, E. L. (1943). Formal Reductions of the general combinatorial decision problem. *American Journal of Mathematics*, 197–268.

Postman, L. (1964). Short-term memory and incidental learning. In: A. W. Melton (Ed.) *Categories of Human Learning*. New York: Academic Press.

Rabbitt, P. (1979). Current paradigms and models in human information processing. In: V. Hamilton and D. M. Warburton (Eds.) *Human Stress and Cognition: An information processing approach*. Chichester: John Wiley and Sons.

Reece, M., and Gable, R. K. (1982). The development and validation of a measure of general attitudes towards computers. *Educational and Psychological Measurement*, **42**, (3), 913-9-16.

Reid, A. A. L. (1970). *Electronic Person–Person Communications*. Communications Studies Group paper No. P/70244/RD, available from British Telecom, 88 Hills Road, Cambridge.

Reisner, P. (1982). Further developments toward using forward grammar as a design tool. In: *Human Factors In Computer Systems*, Faithersburg, U.S.A., March 1982.

Rice, R. E., and Case, D. (1982). *Electronic Messaging in the University Organization*. Paper presented at the 32nd Annual Conference of the International Communication Association, Boston, May 1982.

Rich, E. (1982). Programs as data for their help systems. *Proceedings AFIPS National Computer Conference*, 481–485.

Roberts, T. L. (1980). *Evaluation of Computer Text Editors*. Dissertion submitted to the Computer Science Department. Stanford University. DAI, 40, 5338-B.

Roberts, T. L. and Moran, T. P. (1983). The evaluation of text editors: methodology and empirical results. *Communications of the ACM*, **26**, (4), 265–283.

Robertson, C. K. (1981). Experimental evaluation of an interactive information processing aid for an emergency poison centre. *Behaviour Science*, **26**, 265–271.

Roediger, H. L. III (1980). Memory metaphors in cognitive psychology. *Memory and Cognition*, **8**, (3), 231–246.

Rolfe, J. M. (1971). The secondary task as a measure of mental load. In: W. T. Singleton, J. G. Fox and D. W. Whitfield (Eds.) *Measurement of Man at Work: an appraisal of physiological and psychological criteria in man–machine systems.* London: Taylor and Francis.

Root, R. T., and Sadacca, R. (1967). Man–computer communication techniques: two experiments. *Human Factors*, **9**, (6), 521–528.

Rosenberg, J. U. (1982). Evaluating the suggestiveness of command names. *BIT*, **1**, (4), 371–400.

Rosenberg, V. (1967). Factors affecting the preferences of industrial personnel for information gathering methods. *Information Storage and Retrieval*, **3**, (3), 119–127.

Rouse, W. B. (1982). This reference is to be found in the proceedings, *Analysis, Design, & Evaluation of Man–machine Systems.* Baden-Baden, FDR, Sept. 1982.

Rouse, S. H., Rouse, W. B., and Hammer, J. M. (1982). Design and evaluation of an on-board computer-based information system for aircraft. *IEEE Transactions on Systems, Man and Cybernetics*, **SMC-12**, 4, 451–463.

Rumelhart, D. E., and Norman, D. A. (1981). Analogical processes in learning. In: J. R. Anderson, (Ed.) *Cognitive Skills and their Acquisition.* Hillsdale, New Jersey: Lawrence Erlbaum and Associates.

Rutter, D. R., Stephenson, G. H., and Dewey, M. E. (1981). Visual communication and the content and style of conversation. *British Journal of Social Psychology*, **20**, 41–52.

Salton, G. (1981). The Smart environment for retrieval system evaluation advantages and problem areas. In: K. Sparck Jones (Ed.) *Information Retrieval Experiment.* London: Butterworths, 316–329.

Sauter, S. L., Harding, G. E., Gottleib, M. S., and Quakenboss, J. J. (1981). VDT-computer automation of work practices as a stressor in information-processing jobs: some methodological considerations. In: G. Salvendy and M. J. Smith (Eds.) *Machine Pacing and Occupational Stress.* London: Taylor and Francis.

Scaerdoti, E. Language access to distributed data with error recovery. *Proceedings of the 6th International Joint Conference on Artificial Intelligence.*

Schumacher, E. F. (1973). *Small is Beautiful.* London: Block and Biggs.

Schwab, D. P., and Cummings, L. L. (1970). Theories of performance and satisfaction: a review. *Industrial Relations*, **7**, 408-4-30.

Seybold, P. B. (1982). Comparing the usability of office systems. *Proceedings of the AFIPS Office Automation Conference.*

Shackel, B. (1981). *The Concept of Usability.* Paper presented to the ITT Human Factors Symposium, ITT Shelton, Connecticut.

Shahnavaz, (1982). Lighting conditions and workplace dimensions of VDU-operators. *Ergonomics*, 25, 165–173.

Shannon, C. E., and Weaver, W. (1949). *The Mathematical Theory of Communication.* Urbana: University of Illinois Press.

Sharit, J., and Salvendy, G. (1982). External and internal environments. II. Reconsideration of the relationship between sinus arrhythmia and information load. *Ergonomics*, **25**, 121–132.

Sharit, J., Salvendy, G., and Diesenroth, M. P. (1982). External and internal attentional environments. I. The utilization of cardiac deceleratory and acceleratory response data for evaluating differences in mental workload between machine-paced and self-paced work. *Ergonomics*, **25**, 107–120.

Shneiderman, B. (1980). *Software Psychology: Human Factors in Computer Information Systems*. Cambridge, Massachusetts: Winthrop Publishers.

Shneiderman, B. (1982). The future of interactive systems and the emergence of direct manipulation. *Behaviour and Information Technology*, **1**, (3), 237–256.

Short, J. A. (1972a). *Conflicts of Opinion and Medium of Communication*. Unpublished Communications Studies Group paper No. E/72001/SH, available from British Telecom, 88 Hills Road, Cambridge.

Short, J. A. (1972b). *Medium of Communication and Consensus*. Unpublished Communications Studies Group paper No. E/72210/SH, available from British Telecom, 88 Hills Road, Cambridge.

Short, J. A. (1972c). *Medium of Communication, Opinion Change and Solution of a Problem of Priorities*. Unpublished communications Studies Group paper No. E/72245/SH, available from British Telecom, 88 Hills Road, Cambridge.

Short, J. A. (1973d). *The Effects of Medium of Communication in Persuasion, Bargaining and Perceptions of the Other*. Unpublished Communications Studies Group paper No. E/73100/SH, available from British Telecom, 88 Hills Road, Cambridge.

Short, J. A. (1974). Effect of medium of communication on experimental negotiation. *Human Relations*, **27**, 225–234.

Short, J., Williams, E., and Christie, B. (1976). *The Social Psychology of Telecommunications*. Chichester/New York: John Wiley and Sons Ltd.

Shrivastava, P. K., and Parmar, K. H. (1977). Relationship of role conflict, role ambiguity, and role accuracy to job satisfaction. *Indian Journal of Behaviour*, **1**, (3), 37–39.

Sime, M. E., and Coombs, M. J. (1983). *Designing For Human–Computer Communication*. London/New York: Academic Press.

Simon, H. A. (1957). *Models of Man*. John Wiley and Sons.

Simon, H. A. (1960). *The New Science of Management Decision*. New York: Harper and Row.

Simon, H. A. (1960). The corporation: will it be managed by machines? In: M. L. Aushen and G. L. Back (Eds.) *Management and the Corporations*. New York: McGraw Hill, 17–55.

Simon, H. A. (1982a). Observation of a business decision. In: H. A. Simon (Ed.) *Models of Bounded Rationality: Behavioural Economics and Business Organisation, Volume 2*. Massachusetts: MIT Press.

Simon, H. A. (1982b). Richard T. Ely Lecture. In: H. A. Simon (Ed.) *Models of Bounded Rationality: Behavioural Economics and Business Organisation, Volume 2*. Massachusetts: MIT Press.

Singh, M. G. (1982). *Decentralised Control*. Amsterdam: North-Holland.

Slater, R. E. (1958). Thinking ahead: how near is the automatic office? *Harvard Business Review*, **36**, (2),

Smeltzer, L. R. (1982). *The Relationship of Communication to Work Stress*. Paper presented at the 32nd Annual Conference of the International Communication Association, Boston, May.

Smith, D. C., Irby, C., Kimball, R., Verplank, W., and Harslam, E. (1982). Designing the Star user interface. *BYTE*, April.

Smith, D. H. (1969). Communication and negotiation outcome. *Journal of Communication*, **19**, 248–256.

Smith, M. J. (1981). Occupational stress: an overview of psychosocial factors. In: G. Salvendy and M. J. Smith (Eds.) *Machine Pacing and Occupational Stress*. London: Taylor and Francis.

Smith, S. M., Glenberg, A., and Bjork, R. A. (1978). Environmental context and human memory. *Memory and Cognition*, **6**, 342–353.

Snowberry, K., Parkinson, S. R., and Sisson, N. (1983). Computer display menus. *Ergonomics*, **26**, (7), 699–712.

Sommer, S. (1965). Further studies in small group ecology. *Sociometry*, **28**, 337–348.

Sparck Jones, K. (1981a). Natural language access to databases. Some questions and a specific approach. *Journal of Information Science*, **4**, 41–48.

Sparck Jones, K. (1981b). The Cranfield tests. In: K. Sparck Jones (Ed.) *Information Retrieval Experiment*. London: Butterworths, 256–284.

Sparck Jones, K. (1981c). Retrieval system tests 1958–1978. In: K. Sparck Jones (Ed.) *Information Retrieval Experiment*. London: Butterworths, 213–255.

Spence, R., and Apperley, M. (1982). Data base navigation: an office environment for the professional. *Behaviour and Information Technology*, **1**, (1), 43–45.

Spiegler, I. (1983). Modelling man–machine interface in a database environment. *International Journal of Man–Machine Studies*, **18**, 55–70.

Sterling, T. (1979). Consumer difficulties with computerised transactions. *Communications of the Association for Computing Machinery*, **22**, (5), 283–289.

Strelau, J. (1983). *Temperament—Personality—Activity*. London: Academic Press.

Taylor, T. W. (1982). Should a software interface adapt its behaviour to the developing expertise of its users? *Dissertion Abstracts International*.

Thompson, J. D., and Tuder, A. (1959). Strategies, structures and processes of organisational decision. In: J. D. Thompson (Ed.) *Comparative Studies in Administration*. University of Pittsburgh Press.

Thorngren, B. (1972). *Studier: Lokalisering*. Stockholm: Ekonomiska Forskningsinstitutet.

Tomey, J. F. (1974). *The Field Trial of Audio Conferencing with the Union Trust Company*. Unpublished report from the New Rural Society Project, Fairfield University, Connecticut.

Tulving, E., and Thomson, D. H. (1973). Encoding specificity and retrieval processes in episodic memory. *Psychological Review*, **80**, 352–373.

Tversky, A., and Kahneman, D. (1974). Judgement under uncertainty: heuristics and biases. *Science*, **185**, 1124–1131.

Waksman, A. (1979). The interface problem in interactive systems. *Behaviour Research Methods and Instrumentation*, **6**, (2), 235–236.

Wapner, S., and Alper, T. G. (1952). The effect of an audience on behaviour in a choice situation. *Journal of Abnormal Social Psychology*, **47**, 222–229.

Wastell, D. G., Brown, I. D., and Copeman, A. K. (1981). Evoked potential amplitude as a measure of attention in working environments: a comparative study of telephone switchboard design. *Human Factors*, **23**, 117–121.

Waterman, D. A. (1977). *Exemplary Programming*. RITA, Rand Corporation, October, 1977, p. 5861.

Waterworth, J. A. (1982). Man–machine speech 'dialogue acts'. *Applied Ergonomics*, **13**, (3), 203–207.

Weston, J. R., and Kristen, C. (1973). *Teleconferencing: A Comparison of Attitudes, Uncertainty and Interpersonal Atmospheres in Mediated and Face-to-Face Group Interaction*. Report from the department of Communications, Canada.

Whiteside, J., Archer, N., Wiscon, D., and Good, M. (1982). How do people really use text editors? *Proceedings of the ACM*.

Wibbenmeyer, R., Stern, J. A., and Chen, S. C. (1983). Elevation of visual threshold associated with eyeblink onset. *International Journal of Neuroscience*, **18**, 279–286.

Wichman, H. (1970). Effects of isolation and communication on cooperation in a two person game. *Journal of Personality and Social Psychology*, **16**, 114–120.

Wickens, C., Gill, R., Kramer, A., Ross, W., and Donchin, E. (1981). The cognitive demands of second order manual control: applications of the event-related

potential. *Proceedings of the 17th Annual Conference on Manual Control,* 7–16.
Wiener, M., and Mehrabian, A. (1968). *Language within Language: Immediacy, a Channel in Verbal Communication.* New York: Appleton-Century-Crofts.
Wierwille, W. W. (1979). Physiological measures of aircrew mental workload. *Human Factors,* **21**, 575–593.
Wild, R. (1974). Job restructuring and work organisation. *Management Decisions,* **12**, (3), 117–126.
Williams, E. (1973). *Coalition Formation in Three-person Group Communication via Telecommunications Media.* Unpublished Communications Studies Group paper No. E/73037/WL, available from British Telecom, 88 Hills Road, Cambridge.
Williams, E., and Holloway, S. (1974) *The Evaluation of Teleconferencing: Report of a Questionnaire Study on Users' Attitudes to the Bell Canada Conference Television System.* Unpublished Communications Studies Group paper No. P/74247/WL, available from British Telecom, 88 Hills Road, Cambridge.
Wolek, F. W. (1972). Preparation for interpersonal communication. *Journal of the American Society for Information Science,* Jan–Feb., 3–10.
Wolford, G. (1971). Function of distinct associations for paired-associate performance. *Psychological Review,* **73**, 303–313.
Woodward, Joan (1965). *Industrial Organisation: Theory and Practice.* Oxford University Press.
Wynne, B. (1983). The changing roles of managers. In: H. J. Otway and M. Peltu (Eds.) *New Office Technology: Human and organizational Aspects.* London: Frances Pinter (Publishers) for the Commission of the European Communities, 138–151.
Young, I. (1974). *Understanding the Other Person in Mediated Interactions.* Unpublished Communication Studies Group paper No. E/74266/YN, available from British Telecom, 88 Hills Road, Cambridge.
Young, I. (1975). *A Three Party Mixed-Media Business Game: A Progress Report on Results to Date.* Unpublished Communications Studies Group paper No. E/75189/YN, available from British Telecom, 88 Hills Road, Cambridge.
Young, P. (1983). *Power of Speech: A History of Standard Telephones and Cables 1883–1983.* London: George Allen and Unwin.
Young, R. M. (1981). The machine inside the machine: user's models of pocket calculators. *International Journal of Man–Machine Studies,* **15**, 51–85.
Young, R. M. (1983). Surrogates and mappings: two kinds of conceptual models for interactive devices. In: D. Gentner and A. L. Stevens (Eds.) *Mental Models.* Hillsdale, New Jersey: Erlbaum, 35-52.
Yovits, M. C., and Abilock, J. G. (1974). A semiotic framework for information science leading to the development of a quantitative measure of information. *Proceedings of the 37th Annual Meeting of ASIS,* **11**, 163–168.

Author Index

Abilock, J. G., 165, 167, *340*
Adams, J. S., *326*
Ajzen, I., 166, 176, *330, 334*
Albderdi, M. J .I. de, 120, *326*
Allen, R. B., *326*
Allport, D. A., 246, 249, *326*
Allport, F. H., 110, *326*
Alper, T. G., 110, *339*
Anderson, J. R., *337*
Anderson, R. E., *336*
Anshen, M., *326*
Apperley, M., 13, 138, 139, *339*
Appley, M. H., *334*
Argyle, M., 105, 106, 109, 157, 209, *326*
Aronsson, G., 206, *333*
Aushen, M. L., *338*

Back, G. L., *338*
Baird, J. E., Jr., 38, *326*
Bales, R. F., 103, *326*
Barnard, P. J., 254, 260, 263, 265, 267, *326, 327, 332*
Bateman, T. S., 30, *327*
Bayman, P., 253, 267, *335*
Beatty, J., 201, *327*
Bergman, H., *240*
Berkoivitz, L., *326*
Berlyne, D. E., 117, 148, *327*
Berman, H. J., 116, *327*
Bertrand, U. S., 30, *327*
Birchall, D. W., 273, 277, 278, *327*
Birdwhistell, R. L., 105, 107, *327*
Bjork, R. A., 132, *338*
Bjorn-Anderson, N., 292, *327*
Black, J. B., 260, 265, *327*

Blake, R. R., 110, *335*
Blythe, P., 56, *327*
Bo-Linn, C., 276, *327*
Boddy, D., 273, *328*
Bolt, R., 138, *327*
Borman, L., 283, *326*
Bouma, H., 202, *327*
Bower, G. H., 133, 269, *327*
Bransford, J. D., 254, 267, *327*
Brauer, M. A., 30, *328*
Brinkman, A., *240*
Brodie, D. A., 200, *332*
Brown, I. D., 190, 196, 207, 208, *339*
Buchanan, D. A., 273, *328*
Bush, G., 162, *328*
Butler Cox, 299

Cakir, A., 321, *328*
Campbell, J. P., 54, *328*
Caplan, R. D., 203, *328*
Carbonnel, J. G., 183, *328*
Card, S. K., 23, *328*
Carroll, J. M., 259, 262, 263, 264, *328*
Carroll, S. M., 257, *328*
Case, D., 38, *337*
Champness, B. G., 121, 123, *328*
Chandler, A., 65, *328*
Chapanis, A., 77, 210, *328, 329*
Chapman, A. J., 211, *331, 332*
Chen, S. C., 202, *339*
Cherry, C., 2, *329*
Christie, B., 23, 34, 43, 44, 104, 105, 107–109, 115, 117, 118, 120–124, 148, 153, 154, 156, 167, 273, 282, 290, 291, 293, *329, 338*
Churchman, C., 178, *329*

Clark. I. A., *240*, 265, *327*
Cobb, S., 203, *328*
Cole, I., 212, 269, *329*
Coles. M. G. H., 199, *329, 331*
Collins, H., 115, *329*
Conroy. T. R., 276, *329*
Coombs, M. J., 23, *338*
Cooper, C. L., 55, 56, *329, 336*
Copeman, A. K., 190, 196, 208, *339*
Cox, S., 204, *329*
Cox, T., 204, *329, 334*
Craik, F. I. M., 130, 131, 259. *329, 330*

Dainoff, M. J., 203, *328*
D'Andrade, 255
Dashiell, J. F., 110, 113, *329*
Dean, J., 106, 109, *326*
Debons, A., *333*
Delafield, G., 148, *329*
Deutsch, M., 110, *330*
Dewey, M. E., 108, *336*
Diesenroth, M. P., 200, *338*
Donchin, E., 198, *331, 334*
Dorris, J. W., 112, *330*
Doswell, A., 23, 43, 50, 51, 52, 62, 298, *330*
Douglas, A., 111, *330*
Duncan, O. D., *336*
Dunlop, J. T., *326*
Dunnette, M. D., *328*

Eason, K. D., *240*, 269, 285
Eberts, R. E., 245, *330*
Edwards, J., 191, 198, 208, 209, *329, 331*
Eliot, T. S., *330*
Engel, G. M., 35, 43, 44, 86, 236, *240, 330*
English, R. W., 116, *330*
Eosys, 66, *330*
Evans, J., 280, 281, *330*
Evans, L., 159. *330*
Ewbank, R. V. K., 276, *329*
Eysenck, H. J., 53, 147, *330*
Eysenck, M. W., 207, *330*

Feigenbaum, E. A., 23, *330*
Fishbein, M., 166, *330*
Flohrer, W., *240*
Foley, J. D., *240*
Fox., J. G., *329, 337*
Frankenhaeuser, M., 204, *329*
French, J. R. P., 203, *328*

Furnas, G. W., 259, *331*
Furnham, A., 157, 209, *326*

Gale, A., 23, 148, 191, 198, 208, 209, 211, *329, 331*
Galitz, W. O., *242*
Gallaway, G. R., *242*
Gambino, T. J., 171, *333*
Geidt, F. H., 116, *331*
Gentner, D. R., 256, 267, *331, 340*
Gentry, G. C., 112, *330*
Gerard, H. B., 110, *330*
Gilligan, S. G., 133, *327*
Glenberg, A., 132, *338*
Goffman, E., 58, *331*
Goldmark, P. C., 98, 292, 293, *331*
Gottlieb, M. S., 203, *337*
Gottman, J. M., 209, *334*
Graham, J. A., 157, 209, *326*
Granda, R. E., 236, *242*
Grandjean, E.. 202, 206, *331, 332, 334, 336*
Gratz, R. D., 70, *331*
Groppuso, J., *330*
Grudin, J., 256, *330*

Hackman, J. R., 303, *331*
Haider, M., 206, *331*
Halasz, F., 256, 257, 258, *331*
Hale, C. R., 245, *332*
Hall, E. T., 106, *332*
Hammond, N. V., 254, 265, 267, 273, 277, 278, *332*
Happ, A., 203, *330*
Harding, G. E., 203, *337*
Hardman, Sir. H., 98, *332*
Harness, J. B., 200, *332*
Harrison, G. A., 203, *333*
Harrison, R. V., 203, *328*
Harslam, E., 140, 245, 246, 247, 248, *338*
Hart, D. J., 321, *328*
Heilbronn, M., 109, *332*
Helmer, O., 301, *332*
Hess, E. H., 106, *332*
Hitchen, M., 200, *332*
Holland, W. F., 152, *332*
Holloway, S., 121, 123, *329, 340*
House, W. C., 71, *332*
Hopwood, A., 172, 173, 174, *332*
Howarth, C. I., 211, *332*
Humphreys, P., 176, 180, *332*
Hunting, W., 206, *332*
Hurrell, J. J., Jr., 203, *330*

Hyde, T. S., 129, 130, *332*
Hyland, M., 120, *332*

IMR, 62, *332*
Irby, C., 140, 245, 246, 247, 248, *338*

Jahoda, M., 208, *333*
Janis, I. L., 149, *333*
Jarrett, D., 23, 35, 43, 44, 127, 288, *333*
Jelenevsky, S., 116, *330*
Jenkins, J. J., 129, 130, *332*
Jenner, D. A., 203, *333*
Johansson, G., 206, *333*
Johnson, M. K., 254, *327*
Jones, D. M., *332*

Kahn, R. L., 31, 32, 35, *333*
Kahneman, D., 176, *339*
Kak, A. V., 193, 194, *333*
Kalimo, R., 206, *333*
Kalsbeek, J. W. H., 200, *333*
Kano, S., 37, *333*
Katz, D., 31, 32, 35, *333*
Keen, P. G. W., 160, 171, 176, 177, 269, *333*
Kelley, H. H., 112, *330*
Kent, A., 164, *333*
Kimball, R., 140, 245, 246, 247, 248, *338*
Kingan, S., 118, *329*
Klemmer, E. T., 116, 117, *333*
Kline, P., 54, 134, *333*
Kling, R., 273, 280, 286, *333*
Koelega, H. A., *240*
Kristen, C., 119, 121, *339*
Krueger, H., 206, *334*
Kruglanski, A. W., 176, *334*
Kundi, M., 206, *331*
Kutas, M., 198, *334*

Lacey, J. I., 199, *334*
Lamble, G. W., 30, *334*
Lamming, M., 13, 139, *334*
Landauer, T. K., 259, *334*
Lansdale, M. W., 212
Laplante, D., 116, *334*
Laubli, Th., 206, *332*
Lawler, E. E., 305, *331*
Lappanen, A., 206, *333*
Levenson, R. W., 209, *333*
Libby, W. L., 109, *332*
Lindstrom, B. O., 206, *333*
Lockhart, R. S., 130, 259, *329*

Long, J. B., 259, 265, *334*
Loveless, N., 195, *334*
Lucas, B., 148, *329*
Lynes, T., 68, *334*

Mackay, C. J., 204, 207, *334*
Maguire, M., 249, *334*
Maier, N. R. F., 116, *334*
Malone, T. W., 133, 134, 135, 136, 269, *334*
Mann, L., 149, *333*
Marshall, J., 55, 56, *329*
Martin, I., *334, 336*
Martin, J., *240*
Marwit, S. J., 116, *327*
Mayer, R. E., 253, 257, 267, *334*
McDermott, J., 23, 182, *330*
McDonnell, P., 148, *327*
McHenry, R., 106, *326*
Mead, M., 294, *335*
Meadow, C. T., 145, *335*
Mehrabian, A., 109, *340*
Melton, A. W., *336*
Michalski, R. S., 183, *328*
Michie, D., 181, *335*
Milgram, S., 110, *335*
Miller, R. B., 235, *242*, *335*
Minzberg, H., *242*, *335*
Mitchell, T. M., 183, *328*
Mittman, B., 284, *327*
Monteiro, K. P., 133, *327*
Moran, T. P., 23, 246, 248, 256, 257, 258, 260, 266, *327, 328, 332*
Morley, I. E., 111, 113, *335*
Morton, J., 265, *327*
Mouton, J. S., 110, *335*
Murakami, K., 9, *335*
Murchison, C., *330*
Myrsten, A., 204, *331*

Newell, A., 23, *328*
Nisbett, R. E., 176, *335*
Noll, A. M., 123, *335*
Nordheren, B., 204, *331*
Norman, D. A., 252, 254, 255, 256, 258, 259, 267, 268, *335, 337*

O'Connor, K., 209, *335*
O'Connor, R. O. O., 27, *335*
O'Gorman, J., 208, *335*
O'Hanlon, J. F., 203, 204, *335*
O'Reilly, C. A., III, 31, *336*
Ogburn, W. F., 60, *336*
Olmstead, J. A., 110, *335*

Osborn, A., 63, *336*
Ostberg, O., 206, *336*
Otway, H. J., *327, 330, 334, 340*

PACTEL, 5, 6, 80, *336*
Panko, R. R., 298, *336*
Parker, E. G., 4, *336*
Parkinson, S. R., 47, *339*
Parmar, K. H., 30, *338*
Payne, R., 30, 54, *329, 336*
Peltu, M., *327, 330, 334, 340*
Pessin, J., 110, *336*
Phillips, D. A., 121, *336*
Phillips, L. D., 176, 180, *332*
Pinneau, S. R., Jr., 203, *328*
Porges, S. W., *331*
Post, B., 204, *331*
Post, E. L., 182, *336*
Postman, L., 129, *336*
Poulton, E. C., 207, *339*
Prentice, J., 6
Pritchard, R. D., 54, *328*

Quakenboss, J. J., 203, *337*

Rabbitt, P., 207, *336*
Ramsey, H. R., 63, *240*
Reid, A. A. L., 116, *336*
Reisner, P., 266, *336*
Reynolds, V., 203, *333*
Rice, R. E., 38, *336*
Roediger, H. L., III, 256, *337*
Rolfe, J. M., 207, *337*
Rosenberg, J. U., 259, 262, *337*
Rosenberg, V., 153, 156, *337*
Ross, K. M., 176, *335*
Rouse, S. H., *337*
Rouse, W. B., 269, 270, *337*
Rumelhart, D. E., 254, 255, 256, 257, 267, *337*
Rutter, D. R., 108, *337*

Salem, P. J., 70, *331*
Salton, G., 159, 160, *337*
Salvendy, G., 200, *330, 333, 336, 337, 338*
Sauter, S. L., 203, *337*
Scott Morton, M. S., 171, *333*
Sebrechts, M. M., 265, *327*
Shackel, B., *240*
Shahnavaz, H., 206, *337*
Shannon, C. E., 69, *337*
Sharit, J., 200, *337*
Shneiderman, B., 215, *338*

Short, J. A., 104, 105, 107, 108, 109, 113, 114, 115, 117, 120, 293, *338*
Shrivastava, P. K., 30, *338*
Shulman, A. D., 116, *327*
Sime, M. E., 23, *338*
Simon, H. A., 171, 175, 176, 269, *338*
Singh, M. G., 68, *338*
Singleton, W. T., *329, 337*
Sisson, N., 47, *339*
Smeltzer, L. R., 30, *338*
Smith, D. C., 140, 245, 246, 247, 248, *336, 338*
Smith, D. H., 113, *338*
Smith, M. J., 203, *330, 333, 337, 338*
Smith, S. L., *240*
Smith, S. M., 132, *338*
Snowberry, K., 47, *339*
Sommer, S., 106, *339*
Sparck Jones, K., 157, 158, 160, *330, 337, 339*
Spence, R., 13, 138, 139, *339*
Stephenson, G. H., 108, 111, 113, *335, 337*
Stephenson, G. M., 111, 113, *335*
Stern, J. A., 202, *339*
Stevens, A. L., *331, 340*
Stewart, T. F. M., *240*, 321, *328*
Stocker, L. P., 116, 117, *333*
Strelau, J., 191, 208, *339*
Sutherland, S., *241*
Sykes, R. N., 200, *333*

Thirlaway, M., 204, *329*
Thomas, J., 257, *328*
Thompson, J. D., 172, *339*
Thorngren, B., 103, *339*
Thurber, J. A., 116, *334*
Tomey, J. F., 121, *339*
Treurniet, W. C., 121, *336*
Trumbull, R., *334*
Tsichritzis, D. C., *240*
Tuder, A., 172, *339*
Tulving, E., 131, *330*
Tversky, A., 176, *339*

Venables, P. H., *334, 336*
Verplank, W., 140, 245, 246, 247, 248, *338*
Vigliani, E., 202, *331, 332, 334, 336*

Wapner, S., 110, *339*
Wastell, D. G., 190, 196, 208, *339*
Waterman, D. A., 183, *339*

Weaver, W., 69, *338*
Weissenbock, M., 206, *331*
West, 178, *329*
Weston, J. R., 119, 121, *339*
Whitfield, D. W., *329, 337*
Wibbenmeyer, R., 202, *339*
Wichman, H., 112, *339*
Wiener, M., 109, *340*
Wierwille, W. W., 201, *340*
Wild, R., 309, *340*
Williams, E., 104, 105, 107, 108, 109, 117, 118, 119, 120, 121, 123, 162, 293, *328, 338, 340*
Williams, M. B., *326*

Williges, H. B., *241*
Williges, R. C., *241*
Wilson, P., *241*
Wolek, F. W., 151, *340*
Wolford, G., 128, *340*
Woodward, J., 66, *340*
Wooler, S., 176, 180, *332*
Wynne, B., 287, *340*

Young, I., 116, 119, *340*
Young, P., 75, *340*
Young, R. M., 246, 247, 250, 251, 252, 267, *340*
Yovits, M. C., 165, 167, *340*

Subject Index

1880s, office of the, 61
1920s, office of the, 64
1983 Office System Scenario, 89
2001: A Space Odyssey, 1

Acceptability, 125
 of measures, 194
Additive factors logic, 206
Advance organizers, 257
Advantages of people over
  technology, 306
Advantages of technology over
  people, 306
Algorithms, 309
Alternatives, 309
Alvey Programme, 3, 10, 316, 325
America, 10
Analogy and metaphor, 254
Analysis of the problem, 277
Antiope, 161
Apple, 11, 140
Application packages, 94
Arcovision, 100
Arousal, 51, 117, 147, 207
Arousal potential, 148
Artefact, freedom from, 193
Artificial intelligence, 87, 163, 170
Assessment tools, 323
Associative experts, 216
Atlantic Richfield, 100
Attentional constraints, 243
Attitudes, 121, 125
Audio-only systems, 101
Audio plus 'papers', 102
Audio-video systems, 100
Automatic indexing, 159

Automation, 297
Awareness, 277

Balance, 310
BCU (Binary Choice Unit), 165
Behvaiour tree, 45
Behavioural intention, 167
Bell Canada, 121, 123
Bell Telephone Laboratories, 123
Benefits of new systems, 307
Biases attecting decision making, 176
Bibliography, 239
Bildschirmtext, 161
Bit map display, 95
Boundaries, 308
British Telecom, 100, 102

CAFE OF EVE, 210, 323
Calculators, models of, 250
Calendar, 94
Capability to communicate, 315
Captain, 161
Career opportunities, 279
Carlton University, 121
Casual experts, 215
Catecholamine excretion, 203
Ceremonial aspects of office
  information, 69, 317
Channel, 36
Characteristics of paper-based
  systems, 133, 135
Clerical staff, 202, 291
Coalitions in small groups, 119
Coding, guidelines, 230–1
Cognitive economy, 268
Cognitive effort, categories of, 177

Cognitive load, 269
Cognitive models, 247, 266
Cognitive strain, 207
Collative properties of stimulation,
    148
Colour, as a formatting aid, 232
    as a visual code, 232
    discrimination, 231
    displays, 95
    graphics, 233
    guidelines, 231
    precautions, 233
Command languages, 222, 227
Command names, 259, 260
Command syntax, 262
Commission of the European
    Communities, 5, 10, 281, 287
Commitment, 275
Communication, 308
Communication structures, 35
Communications, 2, 26
Communications media, 38
Communications Studies Group, 98
Compatibility, 308
Complementary books, 23
Complexity and economy of
    functions, 268
Compression Labs., Inc., 100
Computer conferencing, 85
Computer mediated systems, 102
Computer model of decision making,
    183
Conceptual design, 213
Conceptual representation model,
    247, 250
Concretization, 138
Confidence, 116
Conflict, 112
Confravision, 100, 121, 123
Consistency, need for, 268
Content of the display, 228
Context, 132, 133, 191, 255, 317
Cooperation, 112
Copying, 83
Cranfield experiments, 158
Criteria, 192
Cuelessness, 108

DACOM (Description and
    Classification of Meetings), 104
Data tablet, 95
Databases, 82, 83, 94
Dataland, 137
Decision analyst, 170, 179

Decision level, 308
Decision maker, 170, 174
Decision making, 174, 179, 288
Decision making biases, 176
Decision processes, 171, 173, 174
Decision systems, 170, 321
Decision types, 171
Defaults, 238
Delphi technical forecasting, 303
Desktop, 89, 90
Dialogue choice matrix, 223, 224
Dialogue design guidelines, 212, 323
    bibliography, 239
Dialogue selecting, an appropriate
    type, 217, 224
Dialogue user control, 236
Dictation, 92
Display content, 228
Diversive information seeking, 147
Domain knowledge base, 183
Domain specialist, 170, 179

Ease of analysis, 194
Ecological validity, 210, 323
Economic success, 325
Economy of effort, 269
EEG (electroencephalogram), 148,
    198, 322
Effective electronic environments,
    137
Elaborateness of processing, 130
Electronic mail, 84, 93
Electronic meetings, 97, 318
Emotional context, 133
Emperor of Brazil, 75
Energy model of organizations, 31
Enhanced voice telephony, 79
Equipment selection, 283
Ergonomics, 50, 51, 52
Ergonomists, 21
ERPs (event related brain potentials),
    194, 322
Error control, 238
ESPRIT (European Strategic
    Programme of Research in
    Information Technology), 2, 10,
    316, 325
Euronet Diane, 156
Europe, 10
Evaluating change, 285
Evaluative biases, 117
Exemplary progreamming, 183
Eye movements, 201, 322
Experienced pro's 216

Facsimile, 84
Familiarity, 136, 138
Feedback, 235
Fifth Generation Computer, 8
Files and piles, 134
Financial and administrative support,
    275
Finding and reminding, 134
Fishbein's model, 167
Flexibility, 309
    in time spent working, 294
    complexity and power, 218
Form filling, 219, 225
Format, guidelines, 229
Formulating the need, 149
FORUM, 102
Frame of reference, 264
Friendly consultation, 151
Full text search, 157
Function keys, 221, 226
Functional aspects of office
    information, 69, 317
Functional representation, 140
Functional structure, 216
Functions, 26, 28, 40
Future office systems, 297
Future systems, trends, 95, 321
Fuzziness of information, 243

Garden of Eden, 210
General Systems Theory, 208
Generality of results, 265
Global USI factors, 283
Goals, 276
Graphic interaction, 222
Graphics, 95
    packages, 94
    use of colour, 233
Group cohesion, 115
Guidelines, 212, 273
    for input to the system, 223
    for input to the user, 228

HAL, 1
Hard-sell, 297
Headquarters city, 294
Heart rate change and variability,
    199, 322
Hierarchy, 139
Historical perspective, 58, 317
Human factors specialists, 21
Human memory constraints, 242
Hybrid dialogues, 222
Hypnosis, 132

ICL, 171
Image-rich environment, 135
Immediacy, 109
Impacts, 4, 19, 29, 34, 68, 273, 286,
    287, 291, 292, 293, 294
Impacts, on organizational
    information input, 34
    on organizational information
        output, 35
    on organizational information
        processing, 35
Imperial College, 13, 138
Implementation of organizational
    change, 285
Implications, 38, 148, 149, 152, 165,
    166, 167, 178, 179, 180, 181
Implied dialogue model, 247, 259
Incidental learning, 129
Indexing exhaustivity, 160
Indexing specificity, 160
Indexing systems, 158, 159
Individual differences, 191
Induction of rules by machines, 182
Industrial revolution, 59, 317
Industrial society, 315
Information characteristics, 217
Information database, 183
Information load, 218
Information management, 269
Information model of organizations,
    32
Information occupations, 3
Information overload, 30
Information processsing, 315
    model limitations, 49
    psychophysiological concepts, 206
Information seeking, 145
Information society, 315
Information technology, 325
    market, 3
Information Technology Year, 2
Initiative, 218
Input media, 95
INSIS, 287
INSPEC, 159
Institute of New Computer
    Technolocy (ICOT), 9
Integration, 308
Integration at the user interface, 88
Intelligent machine, 170, 318
Interpersonal attraction, 115
Interpersonal communication, 105
Interrupt, guidelines, 237
Intimacy, 109

Introducing office systems into organizations, 273, 324
Intrusiveness, 193
Involvement of key parties, 276
ITT Europe, 10

Japan, 6
Job security, 279
Joystick, 95

Key issues, 17, 99, 127, 146
Keyboard ergonomics, 50
Knowledge, 315
  acquisition by machines, 182
  based systems, 181, 318
  -rich decision making, 181

Laws, 279
Learning from examples, 183
Levels, of processing, 130
  of user interaction, 214
Lexical design, 213
Life form, 315
Light pen, 95
Lisa, 11, 12, 14, 140
Lists, 94
Literal representation, 140
Local area networks (LANs), 85
Log on, guidelines, 236

Macintosh, 12, 14, 140
Managerial stress, 55
Managers, 20, 55, 285
Manufacturers, 18
Mathematics packages, 94
Measures, 192
Media, effects of, 105
Meeting of minds, 308
Memory-driven processing, 206
Menials, 309
Mental entity, 267
Mental load, 207
Mental models, 242, 245, 248, 324
  methodological warnings, 264
Menu selection, 221, 225
Messy offices, 134
Metaperceptions, 119
Metaphor and analogy, 254
Metaphors, complex, 257
  literary, 258
Methodology for identifying needs, 297, 300
  content, 302
  core-task dimensions, 305

guidelines, 307
issues, 304
multiple definitions, 301
Microform systems, 81
Missiles, 315, 325
MIT (Massachusetts Institute of Technology), 13, 137
Models of calculators, 250
Monitoring change, 285
Moses, 58
Motivation, 53, 290
Mouse, 95
MRC-TV, 100
Multiple prompts, 131

Naive users, 215, 256
NAN System, 139
Natural language, 222
Neat offices, 134
NEC, 100
Need, 272
  for information, 146
Needs of different user groups, 284
Neighbourhood work centres, 293
Networks, 37, 38
New Rural Society, 98
Number of office workers, 62

Objectives, 26, 27
Office dimension, 68
Office environment, 139
Office systems, 5, 16
Offices of the future, 67, 317
Online systems, 156
Opinion change, 120
Opportunities for computer support, 178, 179, 180, 181
Optical discs, 81
Orator, 102
Organization analyst, 170, 179
Organization as information processor, 31
Organization knowledge base, 183
Organization of work, 282
Organizational boundary, 86
Organizational context, 26, 316
Organizational databases, 83
Organizational functions, 26, 28
Organizational issues, 303
Organizational objectives, 26, 27
Organizational roles, 26, 28
Output media, 95

Panorama Project, 13, 138

Paper-based systems, 133, 135
Parallel dialogues, 223
Patterns of working, 292
Payment systems, 278
Person perception, 115, 120
Person to 'intelligent' machine
  communication, 87
Person to 'paper' to person
  communication, 80
Person to person communication, 76
Person-orientation, 111
Personal computers, 81, 95
Personal filing, 94
Personal information space, 162
Personal information systems, 127,
  319
Personalization, 135
Pharaohs, 57, 317
Physiological responses, 190
Picture telephone, 97
Picturephone, 78
Piggybacking, 151
Piles and files, 134
PLANET, 102
Policy-makers, 20
Post-industrial revolution, 59, 317
Precision and recall, 156
Preferences for information, 150
Preliminary briefing, 151
Prestel, 154
Primeval slime, 315
Principle of Bounded Rationality,
  175
Printing, 83
Prior experience, 254
Procedural characteristics, 216
Process technology, 68
Production dimension, 68
Production systems, 182
Production technologies, 66, 68, 317
Productivity, 289
Professional peripheration, 151
Professionals, 287
Project management, 94
Prosper, 171
Prototype technology, 68
Psychologists, 21
Psychophysiology, 189, 322
Public databases, 82
Pupillometry, 201, 322

Queen Mary College, London, 13,
  139
Query languages, 220

Question-and-answer, 219

R1, 182
Recall, 128
  and precision, 156
Recognition, 128
Redundancy of information, 243
References, 239, 326
Regulations, 279, 281
Relevance, 163
Reliability, 193
Reminding, 134
Remote Meeting Table, 101, 123
Resource-driven processing, 206
Response times, 235, 236
Revolutions in the office, 59
Role conflicts, 30
Role overload, 30
Role perceptions, 29
Roles, 26, 28, 40
  in decision making, 174
Rule-based programming, 181

Satisficing, 175
Scenario for the near term, 295
Schemas, schemata, 254
Screen ergonomics, 50
SDMS (Spatial Data Management
  System), 137
Secretaries, 291
Selecting a source, 153
Selective processing, 244
Self-fulfilling prophecy, 299
Semantic design, 213
Semantic knowledge, 214
Semantic memory, 245
Sequence control, guidelines, 237
Services, 91
Shared information systems, 145, 320
Sink, 36
Situated Action Theory, 208
Social presence, 107, 120
Social-political milieu of the office,
  297
Source, 36
Space, 308
Spatiality, 138
Special communicators, 152
Specialized databases, 82
Specific information seeking, 149
Specific USI features, 284
Speech recognition, 95
Speech store and forward, 84
Speech synthesis, 95

Spread of effect, 277
Spreadsheet, 94
Star, 11, 12, 14, 140
Stimulus hunger, 147
Storage and retrieval, 81
Straightjacket, 299
Stress, 54–6, 202, 290
Surrogate models, 250
Surveys of office activities, 43,
  297
Survival, 325
Symbiosis, 26
Syntactic design, 213
Syntactic knowledge, 215
System performance, 162
Systems Psychology, 1, 14, 15
  key issues, 17
  starting assumptions, 16

Tactile displays, 95
Task analysis, 207
Task characteristics, 216
Taxonomy of meetings, 104
Taylorism, 63
Technology agreements, 279
Teleconferencing, 79, 97, 123, 124,
  125
Telephone, 78, 79
  directory, 93
Teletex, 84
Televerket, 102
Telidon, 161
Temperament, 147, 290
Time, 45
Timescale of change, 279
Tomb of Meket Re, 58
Touch screen, 95
Trades unions, 20
Training programmes, 282
Trends, 75, 316
Type A communication, 75, 76,
  318
Type A unit, 36
Type B communication, 75, 80,
  318
Type B unit, 36
Type C communication, 75, 87,
  318
Type of information use, 217

Types of decisions, 171
Types of information, 166
Types of meetings, 102
Typing, 291

Usability, 189, 322
  empirical assessment 189, 265, 322
User, 39
  activities, 43
  as information processor, 47
  behaviour, 45
  control of the dialogue, 236
  functions, 40
  interface, 11, 41
  requirements, 214
  requirements definition, 213
  skill migration, 216
User-friendly, 310
User's attitudes, 120, 123
Users' roles, 40
User–system 'meshing', 234
USI (user–system interface), 11, 41
USI, approaches to, 12

Validating input, 239
Validity, 193
VDUs and psychophysiology,
  205
Video telephone, 78, 97
Videodiscs, 81
Videotex, 82, 154, 161
Viewdata, 161
Vignette, 310
Vision, 105
Voice, 95
  annotation, 85, 93
  editing, 93
  messages, 93
  telephone, 78

Wang, 11
Wisdom, 325
Word processors, 80, 92
Work, changes in the content, 289
Working from home, 293
Workstation ergonomics, 51

XCON, 182
Xerox, 10, 11, 12, 140